J. P. M. Finberg, M. B. H. Youdim, P. Riederer,
K. F. Tipton (eds.)

MAO – The Mother of all
Amine Oxidases

SpringerWienNewYork

Prof. Dr. J. P. M. Finberg
Prof. Dr. M. B. H. Youdim
Department of Pharmacology, The Bruce Rappaport Faculty of Medicine,
Technion – Israel Institute of Technology, Haifa, Israel

Prof. Dr. P. Riederer
Clinical Neurochemistry, Department of Psychiatry, University of Würzburg,
Federal Republic of Germany

Prof. Dr. K. F. Tipton
Biochemistry Department, Trinity College, Dublin, Ireland

Product Liability: The publisher can give no guarantee for information about drug dosage
and application thereof contained in this book. In every individual case the respective user
must check its accuracy by consulting other pharmaceutical literature. The use of registered names, trademarks, etc. in this publication does not imply, even in the absence of a
specific statement, that such names are exempt from the relevant protective laws and reg-
ulations and therefore free for general use.

Printing: A. Holzhausens Nfg., A-1070 Wien
Graphic design: Ecke Bonk
Printed on acid-free and chlorine-free bleached paper
SPIN: 10636439

With 76 (partly coloured) Figures

ISBN 3-211-83038-3 (hard cover) Springer-Verlag Wien New York
ISBN 3-211-83037-5 Journal of Neural Transmission [Suppl 52]
(soft cover) Springer-Verlag Wien New York

Preface

The 6th Rappaport Symposium and 7th Amine Oxidase Workshop was held in June 1996 at Shavei Zion, on the northern coast of Israel, some 25 km from the Lebanese border. In view of the tense security situation which preceded the symposium, the organisers are particularly grateful to all the participants for their support.

Just one year before this conference we all learned of the death of our dear colleague, Dr Moise Da Prada, to whose memory this meeting was dedicated. Included in these Proceedings are two appreciations of his life's work by close colleagues. We were honoured to be able to host his widow, Maria Strauss-Da Prada, at the meeting and hope that all the Da Prada family receive comfort from this record to the memory of Moise, whose passing is such a loss to our field. He will be remembered for his kindness and exuberance as well for his scientific distinction and his steadfastness in promoting the therapeutic value of MAO inhibitors through a period when they had, temporarily, fallen into disfavour.

The monoamine oxidases are enzymes which constantly produce surprises, and one of the aspects which we chose to highlight in the meeting was the recent discovery of patient groups bearing selective genetic deficiencies in MAO-A or MAO-B. The importance of such genetic alterations in the pathogenesis of mental and neurological disorders is currently under debate. In particular, the possibility that selective MAO-A deficiency could be responsible for violent behaviours in general was raised in the popular press immediately before the meeting, and we were happy to receive important contributions from Drs. D. Murphy and H. Brunner and their groups which help to clarify the situation.

Another surprise in the MAO field has been the discovery of the neuroprotective or neuro-rescuing actions of some propargyl-derivative MAO-B inhibitors, and the role of MAO in oxidative stress. Such aspects are covered in several places within this book together with many other advances in our understanding of the structure, behaviour and functions of the amine oxidases. We hope this volume will be of value to all those wishing to learn about current knowledge and opinions of the roles of the amine oxidases in health and disease.

We wish to acknowledge with thanks the sponsorship and financial support of the following in enabling our meeting to take place:
Rappaport Family Institute for Research in the Medical Sciences; Bruce

Rappaport Faculty of Medicine, Technion, Israel Institute of Technology; the International Union of Biochemistry and Molecular Biology (Interest Group IG 156); Hoffman-La Roche Ltd, Switzerland; Synthelabo Recherche (L.E.R.S), France; DeVries & Co Ltd, Roche, Israel; Teva Pharmaceutical Industries Ltd, Israel.

J. P. M. Finberg
M. B. H. Youdim
P. Riederer
K. F. Tipton

Prof. Mosé Da Prada
1934–1995

Mosé Da Prada: Development of an outstanding scientist

A. Pletscher

In 1963 Mosé Da Prada became a member of my personal research group at Hoffmann-La Roche Basel. At that time, he did not have much research experience. Nevertheless, I decided to accept his application for a position in my group, because I was intrigued by his high motivation to enter the field of biochemical pharmacology. Thus, I could closely follow his remarkable development as a scientist and the rising recognition of his human qualities by his environment. These aspects will be briefly dealt with in the following.

From Bologna to Basel

Three circumstances were at the origin of Da Prada's decision to leave his job as assistant at the Medical University Clinic of Bologna and to join our research group at Hoffmann-La Roche in Basel.

Firstly, he was attracted by Roche's activities in the then new field of biochemical neuropsychopharmacology. With its first, long acting, synthetic monoamine oxidase (MAO)-inhibitor, iproniazid, Roche had opened a new approach to the treatment of mental depression and had entered the field of biogenic monoamines. Da Prada, as a graduate in medicine and pharmacy, was much interested in these developments.

A second attractor in Roche Basel was Mosé's close friend Giuseppe Bartholini, with whom he had been working in Bologna and who had joined our group about a year before. Bartholini very much liked the scientific approaches and the working facilities at Roche. Since he was aware of the "scientific potential" of Mosé, he wanted him to work with us and motivated him to join our group.

A third reason for Da Prada's decision to leave Bologna was the unfavourable research situation in the medical clinics of its university. Due to insufficient salaries, many of the young MD's had to earn their living by performing part-time private practice in addition to their daily clinical work. Also, the infrastructure for research was poor. This made serious scientific work impossible. No wonder that a research-motivated young man like Da Prada tended to leave such a place.

Thus, the decision of Mosé Da Prada to change from Bologna to Basel was motivated by a "push and pull" situation.

Scientific development

During this stay in our group, two lines of research were in the center of Da Prada's activities, i.e. the biology of monoamines at the cellular and subcellular level and the action of neuropsychotropic drugs on the monoaminergic systems of the central nervous system. Since, as already mentioned, he had no experience in biochemical pharmacology, I asked him to start work with isolated blood platelets, which were used in our laboratory as relatively simple, partial models for certain aspects of cerebral monoamine dynamics. The methodological skills of Mosé soon became apparent. He developed chromatographic methods by which he clarified the metabolism of 5-hydroxytryptamine (5-HT) in platelets and the effect of drugs, like reserpine, benzoquinolizines and phenylalkylamines, on the formation of metabolites, e.g. 5-hydroxyindolacetic acid and 5-hydroxytryptophol. Later, Da Prada and Tranzer, using electron microscopy and biochemical techniques, played an essential role in the discovery of the intracellular storage sites of 5-HT in platelets of various species and in megakariocytes. Mosé succeeded for the first time in isolating the 5-HT-storage organelles (storage granules, dense bodies) in pure form by sophisticated fractionation methods. This allowed the direct elucidation of the action of drugs at the level of the storage organelles, showing, for instance, that reserpine exerts its effect at the granular membrane. The collaboration of Da Prada with physico-chemists also led to the clarification of the intragranular storage mechanism of biogenic amines in platelets and chromaffine granules of the adrenal medulla. Their experiments presented evidence for the existence of polymeric complexes of the monoamines with the intragranular nucleotides like ATP, and in the case of chromaffine granules, also with proteins.

These findings, in which Da Prada played an essential role, helped to establish platelets as partial models for other tissues, including brain, with regard to monoamine uptake, storage and metabolism. The platelet model is still being used, especially in clinical studies.

Later, Mosé also proceeded to the central nervous system. His first activities in this field were devoted to the elucidation of the mode of action of neuropsychotropic drugs, in order to find clues for the design of novel compounds with therapeutic action. He started with work on neuroleptics and phenylethylamines. These investigations, carried out together with his colleagues, helped to clarify the mechanism of action of chlorpromazine on cerebral dopamine metabolism. They also showed that phenylethylamines had different effects on cerebral monoamines, depending on their substituents at the aromatic ring and the nitrogen of the side chain. For instance, methamphetamine, with a chlorine substituent in 4-position of the ring, caused a marked, specific decrease of cerebral 5-HT and 5-hydroxyindolacetic acid (indicating an inhibition of 5-HT-biosynthesis). Thus, the 4-chloro derivative has an effect similar to that of methylene-3,4-dioxymethamphetamine (MDMA, "ecstasy"), a derivative, missed in our series, which was discovered later by another group.

Mosé was now "grown up" and promoted to become head of his own research group. Equipped with experience gained in his more basic work, he became engaged in the development of new drugs. His efforts were concentrated on MAO-inhibitors and anti-Parkinson-drugs, both classical areas of Roche research. With the members of his group, he created sophisticated screening tools, including novel, highly sensitive and specific radioenzymatic tests for monoamines, which were soon also used by the international "monoamine community". With the help of these tests, potent and specific inhibitors of MAO-A and -B and catechol-3-O-methyltransferase (COMT) were discovered and developed in Da Prada's group. Several of these compounds show therapeutic actions in neuropsychiatric disorders. Moclobemide (inhibitor of MAO-A) is on the market as an antidepressant, Tolcapone (inhibitor of COMT) is being prepared for marketing as adjuvant in Parkinson's disease and Lazabemide shows promising clinical effects in Alzheimer's disease. In all these achievements Mosé played an essential role as creative spirit and leader of the experimental work, which got wide international recognition. These research activities will be reported in the following presentations.

The human side

The start in Basel was not easy for Mosé. His move from the lovely south to the "barbaric" north brought changes in his personal life which were not all to his liking. He complained about the quality of food, the lack of sunshine, the unfriendliness of the people, etc. Thus, he did not appreciate the fines for traffic violations, such as those imposed on him for making U-turns over double security lines with his Alfa Romeo. However, as time passed, Mosé started to adapt to his new environment. His generous and friendly character and his readiness for collaboration helped him to establish friendly relations with many of his colleagues. Due to his rising recognition in the international scientific community, he also attracted foreign collaborators, from his home country, Italy, and from other parts of the world. For them he was not only an inspiring and charismatic scientific leader, but he also showed great sympathy for their personal problems. Mosé's hospitality was overwhelming, his fabulous laboratory parties, at which he offered the delicacies of his home country, will certainly remain in our memories.

The culmination in Mosé's personal life was his late matrimonial union with Maria. In her he found a companion who showed great understanding for his work and whose love supported him in difficult situations, from which he was not spared. With Maria his successful adjustment to life in the north was accomplished.

Nevertheless, Mosé remained attached to his family in Italy. Above all, his close connection with his brother Don Giovanni, a priest and well-known painter in Fusine, a village of the valley Valtellina, has to be mentioned. The brothers had many interests in common, e.g. their love for arts and literature. A sign of the high esteem which Mosé enjoyed in the Valtellina was the creation in the public library of Fusine, of a wing "Mosé Da Prada", which contains Mosé's considerable collection of non-scientific literature.

Mosé Da Prada belonged to an excellent species of industrial, pharmaceutical scientists. His intelligence, creativity, methodological skills and intuition, combined with his endeavours to promote the progress of medicine by practical contributions, played an essential role for his professional achievements. Equally important for his success were his enthusiasm for research and his outstanding human qualities, which motivated many of his young collaborators to give the best in their work.

Mosé Da Prada and the discovery of Tolcapone

E. Borroni, J. Borgulya, and **G. Zürcher**

CNS Pharmaceutical Research Department, Hoffmann-La Roche Ltd,
Basel, Switzerland

Mosé Da Prada dedicated his scientific life to the study of the monoaminergic neurone. His early work was concerned with the study of the mechanisms used by neurons and platelets to store neurotransmitters in synaptic vesicles or storage granules. His interest was then attracted by the enzymatic machinery responsible for the biosynthesis and catabolism of monoamines. His work focused particularly on the enzymes involved in the catabolic pathways of monoamines: monoamine oxidases (MAO) and catechol-O-methyl transferase (COMT).

Mosé Da Prada had the ability to translate basic research into therapeutic applications. His work resulted in the discovery of three new therapeutic agents: moclobemide, lazabemide and tolcapone. The pharmacological properties of moclobemide and lazabemide are described in other chapters of this book. Here, we would like to briefly describe the work that lead to the discovery of tolcapone, a new COMT inhibitor to be used as adjuvant with levodopa in the therapy of Parkinson's disease.

The involvement of Mosé Da Prada in Parkinson's disease research began in the sixties when, as a member of the group of Prof. Pletscher at Hoffmann-La Roche, he took part in the studies done to characterise inhibitors of the enzyme aromatic L-aminoacid decarboxylase (Pletscher et al., 1964). This enzyme converts levodopa into dopamine.

The work of Pletscher and collaborators resulted in the clarification of the role of peripheral aromatic L-aminoacid decarboxylase in the metabolism of levodopa (Bartholini et al., 1967; Pletscher and Bartholini, 1971). Co-administration of peripheral aromatic L-aminoacid decarboxylase inhibitors (DCI) and levodopa dramatically improved the bioavailability of levodopa. Peripheral (extracerebral) metabolism of levodopa was reduced thus increasing the amount of levodopa available to the brain for conversion into dopamine. This discovery opened the way to the introduction of a new preparation based on the combination of levodopa with DCI such as MADOPAR (levodopa and benserazide) and SINEMET (levodopa and carbidopa). From a therapeutic point of view this meant a 70% reduction of the dose of levodopa and an improved tolerability due to the reduction of the adverse effects produced by the formation of dopamine in peripheral organs (Cedarbaum, 1987). More than 20 years after its introduction levodopa/DCI is still the most effective single treatment for Parkinson's disease.

Although DCI improved bioavailability of levodopa, the plasma half-life of levodopa remained short (approximately 1 hour; Cedarbaum, 1987). Subsequent studies revealed that inhibition of peripheral aromatic L-aminoacid decarboxylase shunted the metabolism of levodopa toward another metabolic pathway, namely O-methylation (Cedarbaum, 1987; Da Prada et al., 1984). This reaction is catalysed by the enzyme catechol-O-methyltransferase (COMT), an enzyme able to transfer a methyl group from S-adenosylmethionine (SAM) to the hydroxyl group of levodopa. The product of this reaction is 3-O-methyldopa (3-OMD), a compound that cannot be readily converted into dopamine and that accumulates in the body due to its low rate of elimination (Kuruma et al., 1971). Besides its role in the metabolism of levodopa, COMT plays also a role in the clearance of dopamine since, together with MAO, it is responsible for the metabolic inactivation of dopamine (Kopin, 1985). Therefore, the enzyme COMT plays a dual role in the regulation of the brain dopamine levels. By regulating the bioavailability of levodopa it controls the synthesis of dopamine in the brain and by participating in the catabolism of dopamine it regulates its clearance.

Mosé Da Prada quickly realised that COMT inhibitors had the potential to further improve the therapeutic benefits of levodopa. In the early eighties, while he was deeply involved in the development of MAO-inhibitors, he began a small exploratory project aimed at identifying new and well tolerated COMT inhibitors. Mosé Da Prada was able to transmit his enthusiasm and motivation to colleagues and the number of those who collaborated on the project rapidly increased. Progress was also rapid. A vigorous chemical program lead to the synthesis of Tolcapone, a potent, selective and reversible inhibitor of COMT (Zurcher et al., 1988; Borgulya et al., 1989). In parallel, monoclonal antibodies were raised against COMT (Bertocci et al., 1990), the enzyme was isolated and its aminoacidic sequence determined (Bertocci et al., 1991), and experiments begun to determine the crystal structure of the enzyme.

Experiments in animals confirmed theoretical considerations behind the project. Co-administration of tolcapone with levodopa/DCI strongly reduced the formation of 3-OMD (Colzi et al., 1989) and increased the amount of levodopa that was able to reach the plasma and subsequently the brain (Zurcher et al., 1990b). This means that less levodopa was wasted in the formation of 3-OMD, a compound that produces no benefit to parkinsonian patients, and more levodopa was available to promote the primary therapeutic aim of levodopa, namely the biosynthesis of dopamine. The improved availability of levodopa to the brain resulted in a stronger and longer lasting increase of the striatal levels of dopamine (Zurcher et al., 1990a; Napolitano et al., 1995). This translated into an enhancement and prolongation of the therapeutic effect of levodopa in animal models of Parkinson's disease (Maj et al., 1990; Da Prada et al., 1993).

The favourable preclinical data opened the way to the clinical development of tolcapone. Once again, the scientific experience of Mosé, his open character and willingness to collaborate, facilitated the development of the project. The data obtained in patients clearly showed the ability of tolcapone to inhibit formation of 3-OMD and to improve (approximately two fold) the bioavailability of the oral preparation of levodopa. Co-administration of levodopa/DCI and tolcapone did not increase the maximal plasma concentration of levodopa but prolonged its elimination time (Dingemanse et al., 1995; Jorga et al., 1994; Jorga and Sedek, 1995). Due to its ability to reduce the rate of disappearance of plasma levodopa, tolcapone produced a more constant and longer lasting supply of levodopa to the brain. The results of clinical trials indicated that tolcapone was able to reduce the daily dose of levodopa/DCI required for full symptom control in parkinsonian patients. Despite the reduction of levodopa dose, tolcapone prolonged the therapeutic benefit of levodopa/DCI and improved quality of life of patients (LeWitt, 1996; Rajput, 1996; Ransmayr and Poewe, 1996). The therapeutic benefits of tolcapone were particularly evident in fluctuating patients, i.e., patients that experience deterioration of the response to levodopa (Rajput, 1996; Ransmayr and Poewe, 1996; Kurth et al., 1997). Tolcapone will be commercially available in the near future.

Mosé Da Prada died in April 1995. He had just the time to see the positive clinical results obtained by tolcapone. We had the pleasure and privilege to work with him in the tolcapone project and in various other projects. His enthusiasm, natural scientific curiosity and open minded approach to science remains an inspiring example for all of us.

References

Bartholini G, Burkard WP, Pletscher A, Bates HM (1967) Increase of cerebral catecholamines caused by 3,4-dihydroxyphenylalanine after inhibition of peripheral decarboxylase. Nature 215: 852–853

Bertocci B, Garotta G, Zürcher G, Miggiano V, Da Prada M (1990) Monoclonal antibodies recognizing soluble and membrane bound catechol-O-methyltransferase. J Neural Transm [Suppl] 32: 369–374

Bertocci B, Miggiano V, Da Prada M, Dembic Z, Lahm HW, Mahlerbe P (1991) Human catechol-O-methyltransferase: cloning and expression of the membrane-associated form. Proc Natl Acad Sci 88: 1416–1420

Borgulya J, Bruder H, Bernauer K, Da Prada, Zürcher G (1989) Catechol-O-methyltransferase-inhibiting pyrocatechol derivates: synthesis and structure-activity studies. Helv Chim Acta 72: 952–968

Cedarbaum JM (1987) Clinical pharmacokinetics of anti-parkinsonian drugs. Clin Pharmacokinet 13: 141–178

Colzi A, Zürcher G, Da Prada M (1989) Plasma concentrations of endogenous DOPA and 3-O-methyl-DOPA in rats administered benserazide and carbidopa alone or in combination with the reversible COMT inhibitor Ro 41-0960. In: Przuntek H, Riederer P (eds) Early diagnosis and preventive therapy in Parkinson's disease. Springer, Wien New York, pp 191–196 (Key Topics in Brain Research)

Da Prada M, Keller HH, Pieri L, Kettler R, Haefely WE (1984) The pharmacology of Parkinson's disease: basic aspects and recent advances. Experientia 40: 1165–1172

Da Prada M, Zürcher G, Kettler R, Dingemanse J, Jorga KM, Dubuis R (1993) Remodelling the kinetics and dynamics of levodopa therapy in Parkinson's disease by inhibiting MAO-B with lazabemide and COMT with tolcapone. In: Poewe W, Lees AJ (eds) Twenty years of Madopar, new avenues. Editiones Roche, Basel, pp 99–117

Dingemanse J, Jorga K, Zürcher G, Schmitt M, Sedek G, Da Prada M, Van Brummelen P (1995) Pharmacokinetic-pharmacodynamic interaction between the COMT inhibitor tolcapone and single-dose levodopa. Br J Pharmacol 40: 253–262

Jorga KM, Sedek G (1995) Effect of the novel COMT inhibitor tolcapone on L-DOPA pharmacokinetics when combined with different Sinemet formulations. Neurology 45: 291

Jorga KM, Dingemanse J, Fotteler B, Schmitt M, Zürcher G, Da Prada M (1994) Effect of a novel COMT inhibitor tolcapone on L-DOPA pharmacokinetics when combined with different Madopar formulations. New Trends Clin Neuropharm 8: 274

Kopin IJ (1985) Catecholamine metabolism: basic aspects and clinical significance. Pharmacol Rev 37: 333–364

Kurth MC, Adler CH, St Hilaire M, Singer C, Waters C, LeWitt P, Chernik DA, Dorflinger EE, Yoo K, Brewer M, Perry LM, Thomas C, Turpin DL, O'Brien CF, Seeberger LC, Duncan KL, Caviness JN, Douglas M, Wheeler K, Riley D, Rainey P, Tanner CM, Kelke R (1997) Tolcapone improves motor function and reduces levodopa requirement in patients with Parkinson's disease experiencing motor fluctuations: a multicenter, double-blind, randomized, placebo-controlled trial. Neurology 48: 81–87

Kuruma I, Bartholini G, Tissot R, Pletscher A (1971) The metabolism of L-3-O-methyldopa, a precursor of dopa in man. Clin Pharmacol Ther 12: 678–682

LeWitt P (1996) Tolcapone produces sustained improvement for Parkinson's disease patients with a stable response to levodopa. Mov Disord [Suppl] 11: 272

Maj J, Rogoz Z, Skuza G, Sowinska H, Superata J (1990) Behavioural and neurochemical effects of Ro 40-7592, a new COMT inhibitor with a potential therapeutic activity in Parkinson's disease. J Neural Transm 2: 101–112

Napolitano A, Zürcher G, Da Prada M (1995) Effects of tolcapone, a novel catechol-O-methyltransferase inhibitor, on striatal metabolism of L-DOPA and dopamine in rats. Eur J Pharmacol 273: 215–221

Pletscher A, Bartholini G (1971) Selective rise in brain dopamine by inhibition of extracerebral levodopa decarboxylation. Clin Pharmacol Ther 12: 344–352

Pletscher A, Da Prada M, Steiner FA (1964) Inhibition of cerebral decarboxylase and behaviour. Int J Neuropharmacol 3: 559–564

Rajput A (1996) Efficacy of tolcapone in the treatment of Parkinson's disease patients experiencing wearing-off phenomenon: the north American experience. Mov Disord [Suppl] 11: 271

Ransmayr G, Poewe W (1996) A multicenter double-blind placebo-controlled study of tolcapone added to levodopa in fluctuating Parkinson's disease. Mov Disord [Suppl] 11: 270

Zürcher G, Keller HH, Bruderer H, Borgulya J, Da Pada M (1988) Caratteristiche neurochimiche di una nuova classe di inibitori della COMT attivi per via orale: livelli plasmatici di DOPA e 3-OMD nel ratto trattato con DOPA e benserazide. In: Agnoli A, Battistin L (eds) Lega italiana per la lotta contro il morbo di Parkinson e le malattie extrapiramidali (LIMPE). Proceedings of the 14th meeting: "Morbo di Parkinson e demenze: metodologie diagnostiche", October 22–24, 1987. Pubbl D Guanella, Roma, pp 15–29

Zürcher G, Colzi A, Da Prada M (1990a) Ro 40-7592: inhibition of COMT in rat brain and extracerebral tissues. J Neural Transm [Suppl] 32: 375–380

Zürcher G, Keller HH, Kettler R, Borgulya J, Bonetti EP, Eigenmann R, Da Prada M (1990b) Ro 40-7592, a novel, very potent, and orally active inhibitor of catechol-O-methyltransferase: a pharmacological study in rats. In: Streifler MB, Korczyn AD, Melamed E, Youdim MBH (eds) Parkinson's disease: anatomy, pathology and therapy. Raven Press, New York, pp 497–503 (Adv Neurol 53)

Contents

Determination of regions important for Monoamine Oxidase (MAO) A and B substrate and inhibitor selectivities

J. C. Shih, K. Chen, and **R. M. Geha**

Department of Molecular Pharmacology and Toxicology, School of Pharmacy,
University of Southern California, Los Angeles, California, U.S.A.

Summary. MAO-A and -B are defined by their substrate and inhibitor preferences. To determine which regions of the isoenzymes confer these preferences, we have constructed six chimeric MAO enzymes by reciprocally exchanging corresponding N-terminal, C-terminal, and internal segments of MAO-A and -B then determined the catalytic properties of these chimeric enzymes. N-terminal chimerics $A_{45}B$ and $B_{36}A$ were made by exchanging amino acid segments 1–45 and 1–36 of MAO-A and -B respectively. C-terminal chimerics $A_{402}B$ and $B_{393}A$ were made by exchanging amino acid segments 403–527 and 394–520 of MAO-A and -B respectively, and internal chimerics $AB_{161-375}A$ and $BA_{152-366}B$ were made by exchanging amino acid segments 161–375 and 152–366 of MAO-A and -B respectively.

The enzymatic properties observed for the chimerics suggest that the exchanged internal regions but not the N- or C-terminal regions confer substrate and inhibitor preferences.

Introduction

Monoamine oxidase A and B (MAO-A and -B, EC 1.4.3.4) catalyze the oxidative deamination of several biogenic and xenobiotic amines and have distinct pharmacological profiles: MAO-A preferentially oxidizes serotonin (5-hydroxytryptamine, 5-HT) and is inhibited by clorgyline; MAO-B preferentially oxidizes phenylethylamine (PEA) and benzylamine and is inhibited by L-deprenyl; dopamine, epinephrine, tyramine and tryptamine are oxidized by both forms of MAO. MAO-A and -B consist of 527 and 520 amino acids respectively and have a 70% amino acid identity (Bach et al., 1988). They are closely linked on the X-chromosome, map at positions p11.23–11.4 and are deleted in some Norrie disease patients (Lan et al., 1989b).

To determine which regions of the isoenzymes are important for substrate and inhibitor preferences, we have constructed 6 chimeric MAO enzymes by exchanging corresponding N-terminal, C-terminal and internal segments of the two isoforms. We then determined the catalytic constants and IC_{50} values to check if the exchanged segments resulted in an alteration in the enzymatic

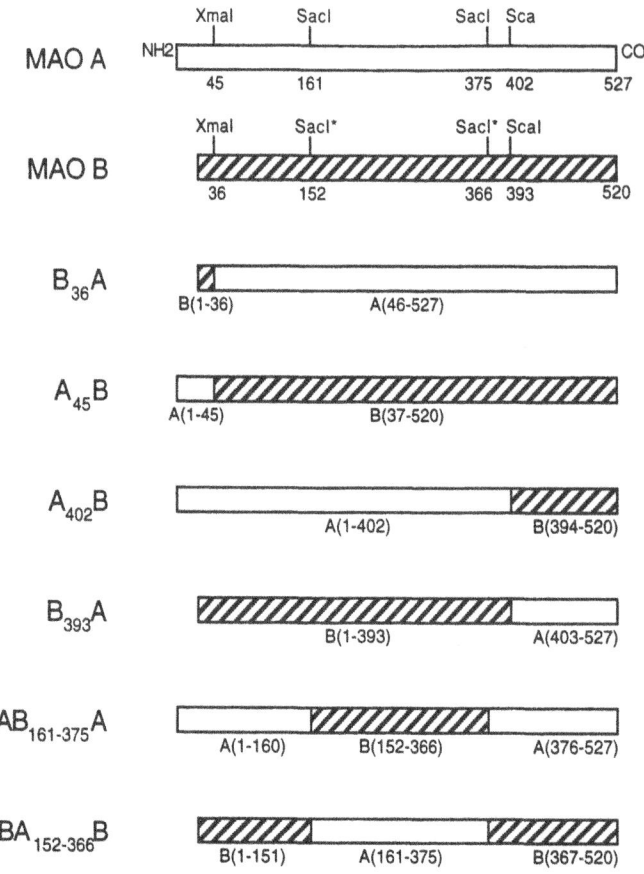

Fig. 1. Schematic diagram showing wild type and chimeric MAO forms. Open boxes represent MAO-A; hatched boxes represent MAO-B. Restriction endonuclease sites used for construction of chimeric forms are shown. SacI* sites were introduced by PCR-mutagenesis. Numbers represent amino acid positions

properties of the chimeric enzyme. A change in the catalytic properties would indicate an important function for the exchanged segment in conferring these preferences. As shown in Fig. 1, The N-terminal chimerics we constructed are $A_{45}B$ and $B_{36}A$, where amino acid segments 1–45 and 1–36 of MAO-A and -B respectively were exchanged. The C-terminal chimerics are $A_{402}B$ and $B_{393}A$ where amino acid segments 403–527 and 394–520 of MAO A and B respectively were exchanged; and the internal chimerics are $AB_{161–375}A$ and $BA_{152–366}B$ where amino acid segments 161–375 and 152–366 of MAO-A and -B respectively were exchanged.

Materials and methods

We constructed pECE plasmids containing chimerics $A_{45}B$, $B_{36}A$, $A_{402}B$ and $B_{393}A$ MAO cDNA, sequenced the inserts by the double stranded DNA method, transfected by the calcium phosphate technique into COS cells (Lan et al., 1989), determined the enzyme activity and performed Western analysis by the procedure of Chen et al. (1996).

Table 1. Kinetic parameters and IC_{50} values of wild-type and chimeric mutants expressed in COS cells

Enzyme	Substrate				IC_{50}(M)			
	[³H]5-HT		[¹⁴C]PEA		[³H]5-HT		[¹⁴C]PEA	
	K_m (10⁻⁶M)	V_{max} [a]	K_m (10⁻⁶M)	V_{max} [a]	Deprenyl (10⁻⁶M)	Clorgyline (10⁻¹⁰M)	Deprenyl (10⁻⁹M)	Clorgyline (10⁻⁷M)
MAOA	119.3 ± 11	36.8 ± 0.3	—	—	1.5 ± 0.1	3.7 ± 0.4	—	—
MAOB	—	—	4.6 ± 0.6	34.4 ± 0.4	—	—	2.7 ± 0.3	4.0 ± 0.4
B₃₆A	110.9 ± 1.0	34.1 ± 0.3	—	—	2.3 ± 0.2	3.0 ± 0.5	—	—
A₄₅B	—	—	3.6 ± 0.4	24.2 ± 0.2	—	—	3.6 ± 0.4	6.0 ± 0.5
A₄₀₂B	31.0 ± 0.4	11.9 ± 0.1	—	—	2.5 ± 0.3	3.0 ± 0.4	—	—
B₃₉₃A	—	—	—	—	—	—	—	—

Mean 1 S.D. of 3 determinations. —: not detectable. a, V_{max} is expressed as nanomoles per 20 min per milligram of protein

Chimerics $AB_{161-375}A$ and $BA_{152-366}B$ were constructed, expressed in yeast, and assayed by the procedure of Grimsby et al. (1996).

Results

The enzyme assays for the N- and C-terminal chimerics were performed on transfected mammalian COS cell homogenates and the enzymatic properties determined. The kinetic parameters exhibited by $B_{36}A$ were very similar to those of wild-type MAO-A. As shown in Table 1, their respective K_m values for [^{3}H]5-HT oxidation were 110.9×10^{-6}M and 119.3×10^{-6}M, and were both incapable of oxidizing [^{14}C]PEA. The IC_{50} values of clorgyline inhibition of $B_{36}A$ and MAO A were also similar (3.0×10^{-10}M and 3.7×10^{-10}M respectively). The IC_{50} value for deprenyl inhibition of 5-HT oxidation by $B_{36}A$ was 2.3×10^{-6}M, which is close to that of MAO A (1.5×10^{-6}M).

Chimeric enzyme $A_{45}B$ exhibited similar kinetic parameters to wild-type MAO-B. Table 1 shows their K_m values for PEA oxidation to be very close (3.6×10^{-6}M for $A_{45}B$ and 4.6×10^{-6}M for MAO-B). The IC_{50} values for the inhibition of PEA oxidation by $A_{45}B$ and MAO-B were 3.6×10^{-9}M and 2.7×10^{-9}M deprenyl, and 6.0×10^{-7}M and 4.0×10^{-7}M clorgyline, respectively. They were both incapable of oxidizing 5-HT. These results clearly suggest that amino acids 1–45 of MAO-A and 1–36 of MAO-B have no important function in conferring substrate and inhibitor preferences of the two isoenzymes.

The C-terminal chimeric $A_{402}B$ was, like MAO-A, able to oxidize 5-HT but not PEA and, interestingly, its K_m for 5-HT oxidation (31.0×10^{-6}M) was about one fourth that of MAO-A (119.3×10^{-6}M). These results suggest that replacing amino acids 403–527 of MAO A with amino acids 394–520 of MAO-B alters the conformation of the enzyme in a way that increases the affinity for 5-HT. The clorgyline and deprenyl IC_{50} values for the inhibition of 5-HT oxidation by $A_{402}B$ were similar to that of MAO A (clorgyline IC_{50}, 3.0×10^{-10}M for $A_{402}B$ and 3.7×10^{-10}M for MAO-A; deprenyl IC_{50}, 2.5×10^{-6}M for $A_{402}B$ and 1.5×10^{-6}M for MAO-A). Therefore $A_{402}B$ has a similar inhibition profile to MAO A but with a 4 fold increased affinity to 5-HT.

Chimeric $B_{393}A$ was found to be inactive. Western analysis demonstrated that chimeric $B_{393}A$ cross reacts with anti-MAO-B antibody and has a similar molecular mass as $A_{45}B$ and human brain homogenate MAO-B. The bands of $A_{45}B$ and $B_{393}A$ were of similar intensity which suggests that the lack of catalytic activity was not due to nonexpression (Chen et al., 1996).

The importance of the internal segments 161–375 and 152–366 of MAO-A and -B respectively in conferring substrate and inhibitor preferences, was determined by assaying the activity of chimerics $AB_{161-375}A$ and $BA_{152-366}B$, on crude mitochondrial fractions from yeast transformed with pYES 2.0 plasmid containing the chimeric cDNAs. As shown in Table 2, the k_{cat}/K_m values for PEA and benzylamine oxidation by $AB_{161-375}A$ have increased 8 and 11 fold respectively when compared to MAO-A ($AB_{161-375}B$ k_{cat}/K_m, $1.6 \times 10^5 s^{-1}M^{-1}$ and $2.4 \times 10^4 s^{-1}M^{-1}$ for PEA and benzylamine oxidation respectively; MAO-A k_{cat}/K_m, $0.21 \times 10^5 s^{-1}M^{-1}$ and $0.21 \times 10^4 s^{-1} M^{-1}$ for benzylamine and PEA

Table 2. Kinetic parameters of wild-type and internal chimeric mutants expressed in yeast

Enzyme	5-HT			PEA			Benzylamine		
	K_m $(10^{-6}M)$	k_{cat} (s^{-1})	k_{cat}/K_m $(10^5 s^{-1} M^{-1})$	K_m $(10^{-6}M)$	k_{cat} (s^{-1})	k_{cat}/K_m $(10^5 s^{-1} M^{-1})$	K_m $(10^{-6}M)$	k_{cat} (s^{-1})	k_{cat}/K_m $(10^4 s^{-1} M^{-1})$
MAOA	66.1 ± 2.3	9.89 ± 1.17	1.5	96.5 ± 8.4	2.07 ± 0.12	0.21	309 ± 36	0.65 ± 0.02	0.21
MAOB	—	—	—	2.2 ± 0.06	1.77 ± 0.24	8.0	305 ± 26	15.1 ± 1.5	4.9
AB$_{161-375}$A	—	—	—	19.7 ± 2.9	3.2 ± 1.0	1.6	1,398 ± 238	33.8 ± 3.3	2.4
BA$_{152-366}$B	—	—	—	—	—	—	—	—	—

Mean + S.E.M. of 3 determinations performed in duplicate. —: not detectable

Table 3. IC_{50} values of wild-type and internal chimeric mutants expressed in yeast

Enzyme	Substrate	Clorgyline IC_{50} (M)	Deprenyl IC_{50} (M)
MAOA	5-HT	$1.8 \pm 1.5 \times 10^{-10}$	$81 \pm 19 \times 10^{-8}$
MAOB	PEA	$6.4 \pm 1.5 \times 10^{-7}$	$0.47 \pm 0.06 \times 10^{-8}$
$AB_{161-375}A$	PEA	$5.4 \pm 0.8 \times 10^{-7}$	$2.7 \pm 0.4 \times 10^{-8}$
$BA_{152-366}B$	—	—	—

Values determined as in Table 2. —: not detectable

oxidation respectively). This indicates a clear shift in the catalytic properties of the chimeric enzyme towards those of MAO-B, becoming MAO-B like. Table 3 shows that the IC_{50} values of clorgyline and deprenyl inhibition of $AB_{161-375}A$ are much closer to those of MAO-B than MAO-A. These data show that amino acids 161–375 of MAO B have an important function in conferring substrate and inhibitor preferences. The other internal chimeric, $BA_{152-366}B$ was found to be inactive. As for $B_{393}A$, a Western analysis determined that the lack of catalytic activity was not due to nonexpression (Grimsby et al., 1996).

Discussion

MAO-A and -B share a 70% amino acid identity (Bach et al., 1988) yet they have distinct catalytic and inhibitory profiles. The two polypeptides have four conserved regions, two of which are located at the N- and C-terminals. We have reciprocally exchanged amino acids 1–45 and 1–36 of MAO-A and -B respectively resulting in chimerics $B_{36}A$ and $A_{45}B$. Their catalytic properties were almost identical to those of wild-type MAO-A and MAO-B respectively, clearly identifying the N-terminals as having no involvement in the substrate and inhibitor preferences of the MAO-A and -B isoenzymes. This is in agreement with the results of Gottowick et al. (1993) where other N-terminal chimerics also showed no alteration in enzymatic properties. However, since an AMP-binding site was identified at amino acids 15–29 and 6–20 in MAO-A and -B respectively (Chen et al., 1996), it must be stressed that the N-terminals still have an important function in catalysis. Therefore the N-terminal regions have an important function in the catalytic activity but not in conferring specificity to the two isoenzymes. Glu^{34} of MAO-B has recently been identified as essential for catalysis (Kwan et al., 1995).

When the 125 C-terminal amino acids of MAO-A were replaced with the corresponding 127 C-terminal amino acids of MAO-B, the resulting chimeric $A_{402}B$ had the same kinetic properties as MAO-A but with a 4 fold increased affinity for serotonin. Replacing the 127 C-terminal amino acids of MAO-B with the corresponding 125 C-terminal amino acids of MAO-A (chimeric $B_{393}A$) resulted in a complete loss of activity for the chimeric enzyme. These results suggest that C-terminals, like the N-terminals, are not important for specificity and that the MAO-B C-terminal when placed in MAO-A would alter the conformation of the enzyme to favor 5-HT oxidation. When compared to MAO-B, chimeric $B_{393}A$ had eight amino acid charge differences in its C-terminal which could have altered its conformation to obliterate its catalytic activity, but it is interesting that the introduction of the complementary charge differences into MAO-A (chimeric $A_{402}B$) did not affect its activity. This observation reinforces the view that the MAO-B active site may be more structurally restricted than that of MAO A (Kalir et al., 1981).

To study the function of the internal segments of MAO-A and -B, we have constructed internal chimerics by exchanging amino acids 161–375 of MAO-A with amino acids 152–366 of MAO-B. The resulting chimeric $AB_{161-375}A$ was more MAO-B-like in its catalytic properties than MAO-A-like despite a ~10 fold increase in the K_m for PEA oxidation, suggesting that at least part of a substrate/inhibitor conferring domain is present in this segment. The increased K_m is in agreement with a substrate binding site between amino acids 62–103 assigned by Gottowick et al. (1995) however since the k_{cat}/K_m values of $AB_{161-375}A$ are closer to those of MAO-B than MAO-A, amino acids 62–103 could constitute part of the binding site that is interacting with determinants located between residues 152–366. Chimeric $BA_{152-366}B$ displayed no catalytic activity despite a comparable expression level to MAO-B as was shown by Western analysis, and the fact that its activity was restored when it was recut and reverted back to MAO-B rules out cloning artifacts as the cause for the loss of activity.

Going from the active chimeric $A_{402}B$ to the inactive $BA_{152-366}B$, we notice that the only changes are the replacement of the N-terminus of $A_{402}B$ with the corresponding N-terminal amino acids 1–151 of MAO-B, and the replacement of amino acids 376–402 of the MAO-A portion of $A_{402}B$ with the corresponding amino acids 367–393 of MAO-B. The latter replacement produces only two amino acid changes for $A_{402}B$, Q384 (of MAO A)→L and H388→E. From this, we believe that the N-terminus of $BA_{152-366}B$, either alone or in cooperation with the two altered amino acids, changes the enzyme's conformation and renders it inactive (Grimsby et al., 1996).

In summary, we have shown that the N- and C-terminals of MAO-A and -B are devoid of a domain that confers substrate and inhibitor selectivities. We have also shown that segment 161–375 of MAO-A and segment 152–366 of MAO-B do contain such a domain.

Another set of internal chimerics with shorter exchanged segments would be able to better define the location of a substrate binding domain in segment 161–375 of MAO-B. Also, site-directed mutagenesis on amino acid residues that are conserved between species but differ between MAO-A and -B, especially non-conserved substitutions, may identify key functions for specific residues.

Acknowledgements

This study was supported by grants RO1 MH37020 and R37 MH39085 (MERIT Award) and Research Scientist Award KO5 MH00796 from the National Institute of Mental Health. Support from the Boyd and Elsie Welin Professorship is also appreciated.

References

Bach AWJ, Lan NC, Johnson DL, Abell CW, Bembenek ME, Kwan SW, Seeburg PH, Shih JC (1988) cDNA cloning of human liver monoamine oxidase A and B: molecular basis of differences in enzymatic properties. Proc Natl Acad Sci USA 85: 4934–4938

Chen K, Wu H-F, Shih JC (1996) Influence of C terminus on monoamine oxidase A and B catalytic activity. J Neurochem 66: 797–803

Gottowick J, Cesura AM, Malherbe P, Lang G, Prada MD (1993) Characterization of wild-type and mutant forms of human monoamine oxidase A and B expressed in a mammalian cell line. FEBS Lett 317: 152–156

Gottowick J, Malherbe P, Lang G, Da Prada M, Cesura AM (1995) Structure/function relationships of mitochondrial monoamine oxidase A and B chimeric forms. Eur J Biochem 230: 934–942

Grimsby J, Zentner M, Shih JC (1996) Identification of a region important for human monoamine oxidase B substrate and inhibitor selectivity. Life Sci 58: 777–787

Kalir A, Sabbagh A, Youdim MBH (1981) Selective acetylenic "suicide" and reversible inhibitors of monoamine oxidase type A and B. Br J Pharmacol 73: 55–64

Kwan SW, Lewis DA, Zhou BP, Abell CW (1995) Characterization of a dinucleotide-binding site in monoamine oxidase B by site-directed mutagenesis. Arch Biochem Biophys 316: 385–391

Lan NC, Chen C, Shih JC (1989a) Expression of functional human monoamine oxidase A and B cDNAs in mammalian cells. J Neurochem 52: 1652–1654

Lan NC, Heinzmann C, Gal A, Klisak I, Orth U, Lai E, Grimsby J, Sparkes RS, Mohandas T, Shih JC (1989b) Human monoamine oxidase A and B genes map to Xp11.23 and are deleted in a patient with Norrie disease. Genomics 4: 552–559

Authors' address: J. C. Shih, Ph.D., Department of Molecular Pharmacology and Toxicology, School of Pharmacy, University of Southern California, 1985 Zonal Avenue, Los Angeles, CA, U.S.A.

Monoamine oxidase A deficiency: biogenic amine metabolites in random urine samples

N. G. G. M. Abeling[1], **A. H. van Gennip**[1], **A. G. van Cruchten**[1], **H. Overmars**[1], and **H. G. Brunner**[2]

[1] Academic Medical Center, University of Amsterdam, Department of Clinical Chemistry and Department of Pediatrics, Amsterdam, and [2] Department of Human Genetics, University Hospital, Nijmegen, The Netherlands

Summary. We have recently described an association between abnormal behaviour and monoamine oxidase A (MAO-A) deficiency in several males from a single large Dutch kindred. A characteristically abnormal excretion pattern of biogenic amine metabolites was present in 24-hour urine of affected males. Because of this strikingly abnormal metabolite pattern observed in 24 hour urine samples of MAO-A deficient males we hypothesized that it should be possible to diagnose this condition by examining random urine samples. We therefore studied multiple urine samples obtained over a two-week study period from two males with selective MAO-A deficiency. The results demonstrate that the characteristic abnormalities in the excretion of biogenic amines and their metabolites were faithfully present in every one of 12 independent samples obtained from the MAO-A deficient males over the two-week study period. We conclude that MAO-A deficiency can be reliably diagnosed by measuring the ratio of normetanephrine (NMN) to VMA (or that of NMN to MHPG) in random urine samples.

Introduction

We have recently described an association between abnormal behaviour and MAO-A deficiency in several males from a single large Dutch kindred (Brunner et al., 1993a,b). Affected males differed from unaffected males by borderline mental retardation, and impaired regulation of impulsive behaviour (aggressive behaviour, abnormal sexual behaviour, and arson). The genetic defect for this condition was assigned to the short arm of the X chromosome, in the vicinity of the genes for MAO-A and -B. We evaluated these patients for MAO deficiency. Analysis of 24-hour urine samples indicated markedly disturbed monoamine metabolism. A characteristically abnormal excretion pattern of biogenic amine metabolites was present in 24-hour urine of affected males (Brunner et al., 1993a; Abeling et al., 1994; Lenders et al., 1996). Intermediate abnormalities were found in carrier females and normal patterns in normal males from the same family. All urine collections were done on a standardized diet, particularly devoid of bananas, walnuts and other biogenic amine-containing foodstuffs (Abeling et al., 1994). We found that this syndrome is associated with a complete deficiency of the enzyme activity of MAO-A. In each of 5 affected males, the same nonsense mutation was identified in the MAO-A gene. In 12 unaffected males from the same kindred the mutation was absent. Thus, isolated complete MAO-A deficiency in this family is associated with a recognizable behavioural phenotype that includes disturbed regulation of impulsive aggression.

The strikingly abnormal metabolite pattern observed in 24 hour urine samples of MAO-A deficient males suggests that it should be possible to diagnose this condition by examining random urine samples. However, this possibility has not yet been evaluated. We therefore decided to study urinary excretion patterns in multiple independent urine samples obtained from affected males over a two-week period. During the first week no dietary restrictions were imposed. During the second week, the patients were asked to avoid foodstuffs that are known to contain high amounts of biogenic amines.

Design of the study

Subjects

Of the index family with MAO-A deficiency six individuals, i.e. two affected males, two carrier females and two normal subjects (1 female, 1 male), participated in the study. Thirty unrelated healthy adults (15 females, 15 males) were also included as controls.

Protocol

The family members collected urine samples over a period of two weeks; the first week on a free diet, with allowance for consumption of catecholamine-rich foodstuffs, and the next week on a diet that was restricted in catecholamine-rich foodstuffs. The urine samples were collected at home, and local hospitals were instructed for sample handling and storage.

On Monday, Wednesday and Friday of each week two urine samples per day were collected, i.e. the first morning urine (around 8 a.m.), and the first urine produced after lunch (around 2 p.m.).

The samples were kept in a refrigerator at −4°C and delivered the next day at the laboratory. On receipt one half of each sample was frozen at −25°C and the other half acidified with concentrated HCl to pH2–3 and then also frozen. All samples were kept frozen until analysis.

The control subjects each produced one random urine sample which was kept frozen at −25°C until analysis, without further pretreatment.

Methods

Biogenic amines, their O-methylated, acidic and neutral metabolites were determined using the previously published HPLC-methods (Abeling et al., 1984; Stroomer

Table 1. Urinary excretion (nmol/mmol creatinine) of the biogenic amines and their metabolites in the urine samples of the MAO-A deficient subjects and carriers from the index family

| | Amines | | | | O-methylated metabolites | | | Deaminated metabolites | | | |
	NE	E	DA	5-HT	NMN	MN	3-MT	MHPG	VMA	HVA	5-HIAA
Case 1	211 ± 37	11 ± 4	1,411 ± 712	832 ± 162	919 ± 110	124 ± 13	278 ± 20	181 ± 30	401 ± 93	2,323 ± 453	722 ± 109
Case 2	87 ± 38	17 ± 6	823 ± 307	330 ± 114	541 ± 59	116 ± 28	257 ± 83	95 ± 25	198 ± 46	1,749 ± 616	708 ± 136
Carrier 1	123 ± 21	9 ± 6	769 ± 205	333 ± 89	306 ± 81	93 ± 20	149 ± 50		1,357 ± 487	2,288 ± 347	1,955 ± 499
Carrier 2	79 ± 18	11 ± 7	544 ± 181	142 ± 35	191 ± 26	88 ± 20	196 ± 88		1,813 ± 370	2,502 ± 492	2,631 ± 351
Controls	89 ± 34	14 ± 6	523 ± 170	51 ± 16	87 ± 28	40 ± 10	54 ± 24	916 ± 200	1,848 ± 382	2,656 ± 966	2,804 ± 1,074

NE Norepinephrine, E Epinephrine, DA Dopamine, 5-HT 5-hydroxytryptamine (= serotonine), NMN Normetanephrine, MN Metanephrine, 3-MT 3-methoxytyramine, MHPG 3-Methoxy-4-hydroxyphenylglycol, VMA Vanillylmandelic acid, HVA Homovanillic acid, 5-HIAA 5-hydroxyindoleacetic acid. Values are presented as mean ± SD. From each family member 12 values were averaged; 30 controls were measured

Table 2. MAO-A substrate/product ratios

	3-MT/HVA	NMN/VMA	NMN/MHPG	5-HT/5-HIAA
Case 1	0.125 ± 0.025	2.52 ± 0.86	5.17 ± 0.86	1.16 ± 0.21
Case 2	0.16 ± 0.06	2.98 ± 0.92	6.05 ± 1.60	0.53 ± 0.32
Controls	0.02 ± 0.01	0.05 ± 0.01	0.10 ± 0.03	0.02 ± 0.01

Abbreviations are the same as in Table 1. Values are presented as x ± S.D.

et al., 1990) as also applied in the original study of the index family (Brunner et al., 1993a).

Results and conclusions

Urinary concentrations of the catecholamines (dopamine, norepinephrine, epinephrine), serotonin (5-hydroxytryptamine, 5-HT) and their metabolites were measured in all samples. Results for affected males and for carrier females are presented in Table 1. Results for the normal males in the family were all within the range observed in unrelated controls. Selected substrate/product ratios for the affected males with MAO-A deficiency are presented in Table 2.

Based on the data presented in Table 2, the NMN/VMA* and NMN/MHPG* ratios appeared to be the most discriminative. These ratios were therefore plotted in a longitudinal graph for all 12 urine samples collected by the family members over the study period. Ratios measured in the 30 control persons are given as a reference range for comparison (Figs. 1, 2). Both ratios clearly discriminate the MAO-A deficient individuals from the normal controls irrespective of the time of day, or of dietary factors. Also, no differences were noted for samples that were acidified prior to freezing.

Although only two affected individuals participated in the present study, the conclusion seems justified that determination of the NMN/VMA or NMN/MHPG ratio in a random urine sample is a reliable and convenient screening tool to detect individuals with MAO-A deficiency.

Discussion

We studied multiple urine samples obtained over a two-week study period from two males with selective MAO-A deficiency. The results demonstrate that the characteristic abnormalities in the excretion of biogenic amines and

---

NMN normetanephrine; *VMA* vanillylmandelic acid; *MHPG* 3-methoxy-4-hydroxyphenylethyleneglycol

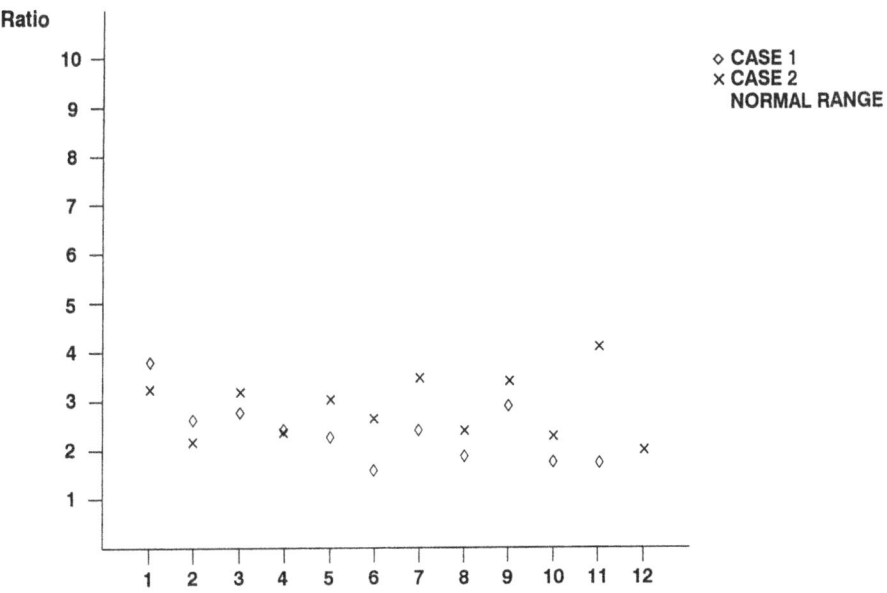

Fig. 1. Ratio of NMN to VMA in 12 consecutive urine samples from two males with MAO-A deficiency. On the X-axis, even numbers represent morning samples, and uneven samples represent afternoon samples. The normal range was determined from random samples obtained from 30 healthy unrelated controls. NMN represents total, i.e. free + conjugated normetanephrine

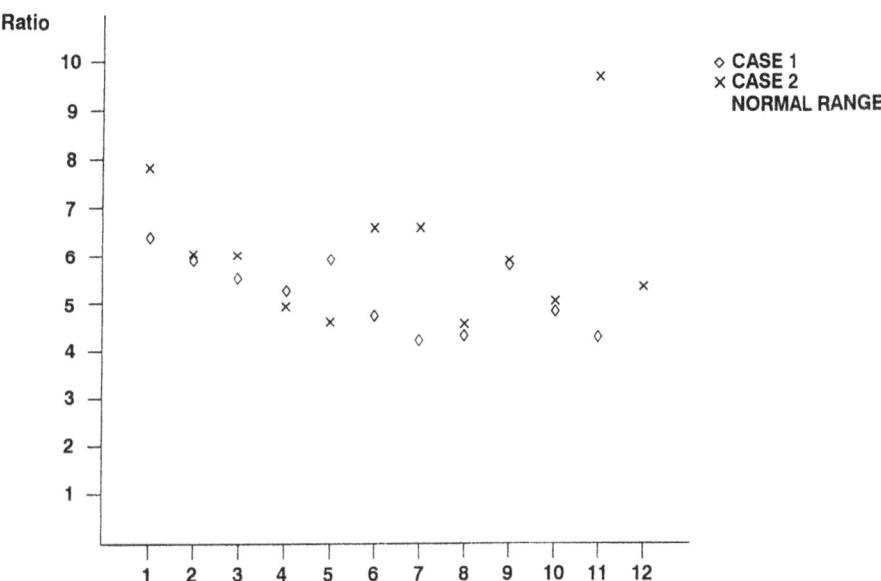

Fig. 2. Ratio of NMN to MHPG in 12 consecutive urine samples from two males with MAO-A deficiency. On the X-axis, even numbers represent morning samples, and uneven samples represent afternoon samples. The normal range was determined from random samples obtained from 30 healthy unrelated controls. NMN represents total, i.e. free + conjugated normetanephrine

their metabolites were faithfully present in every one of 12 independent samples obtained from the MAO-A deficient males over the two-week study period. Since MAO-A is involved in the degradation of norepinephrine, dopamine, and serotonin, several substrate to product ratios could be used for diagnostic purposes in these males. Consistent with our previous experience of 24-hour urine samples, and of plasma samples from affected males, we found that the most striking abnormalities are in the handling of norepinephrine, and more specifically, its O-methylated metabolite normetanephrine (NMN). Based on the results obtained in this study the ratio of NMN to VMA provides an accurate tool for the diagnosis of MAO-A deficiency. The NMN to VMA ratio was consistently higher than 1.5 in the affected subjects. This is in sharp contrast to the results obtained in normal controls (both within and outside this family) where the NMN to VMA ratio never exceeded 0.1. Similarly, for the NMN to MHPG ratio, patient values all exceeded 4.0 whereas they were below 0.2 in all controls. There was no appreciable effect on any of the substrate to product ratios measured in this study of either the time of day, or of dietary factors.

We conclude that the presence or absence of genetic MAO-A deficiency can be reliably diagnosed from the analysis of a single urine sample. For diagnostic purposes the NMN to VMA ratio as well as the NMN to MHPG ratio are highly robust. Furthermore, the effects of diurnal and dietary variations are relatively minor, and do not interfere with the diagnostic accuracy of this method.

Acknowledgements

We thank the staff of the Clinical Chemistry Departments of the Antonius Hospital (Dr. De Steenwinkel), Leidschendam, and of the Sint Joseph Hospital, Veldhoven (Dr. de Graaf) for their help in collecting urine samples. Thanks also to the family for their continued support.

References

Abeling NGGM, van Gennip AH, Overmars H, Voute P (1984) Simultaneous determination of catecholamines and metanephrines in urine by HPLC with fluorimetric detection. Clin Chim Acta 137: 211–226

Abeling NGGM, van Gennip AH, Overmars H, van Oost BA, Brunner HG (1994) Biogenic metabolite patterns in the urine of monoamine oxidase A-deficient patients. A possible tool for diagnosis. J Inher Metab Dis 17: 339–341

Brunner HG, Nelen MR, van Zandvoort P, Abeling NGGM, van Gennip AH, Wolters EC, Kuiper MA, Ropers HH, van Oost BA (1993a) X-linked borderline mental retardation with prominent behavioral disturbance: phenotype, genetic localization, and evidence for disturbed monoamine metabolism. Am J Hum Genet 52: 1032–1039

Brunner HG, Nelen M, Breakefield XO, Ropers HH, van Oost BA (1993b) Abnormal behavior associated with a point mutation in the structural gene for monoamine oxidase A. Science 262: 578–580

Lenders JWM, Eisenhofer G, Abeling NGGM, Berger W, Murphy DL, Konings CH, Bleeker Wagemakers LM, Kopin IJ, Karoum F, van Gennip AH, Brunner HG (1996)

Specific genetic deficiencies of the A and B isoenzymes of monoamine oxidase are characterized by distinct neurochemical and clinical phenotypes. J Clin Invest 97: 1010–1019

Stroomer AEM, Overmars H, Abeling NGGM, van Gennip AH (1990) Simultaneous detection of acidic 3,4,-dihydroxyphenylalanine metabolites and 5-hydroxy-indole-3-acetic acid in urine by high performance liquid chromatography. Clin Chim Acta 36: 1834–1837

Authors' address: N. G. G. M. Abeling, Academic Medical Center Laboratory for Genetic Metabolic Diseases, Fo-224, Meibergdreef 9, 1105 AZ Amsterdam, The Netherlands

Relationship between monoamine oxidase (MAO) A specific activity and proportion of human skin fibroblasts which express the enzyme in culture

R. M. Denney

Graduate School of Biomedical Sciences, Department of Human Biological Chemistry and Genetics, University of Texas Medical Branch, Galveston, TX, U.S.A.

Summary.   Total deficiency of monoamine oxidase A (MAO-A) in affected males of a single, human kindred appears to be associated with mild mental retardation and significant behavioral anomalies. Though total MAO-A deficiency appears to be rare, the extent and significance of individual variation in monoamine oxidase A activity in human populations is unclear. Since MAO-A activity is undetectable in blood cells, most systematic surveys of individual variation MAO-A activity have compared enzyme activity in human fibroblasts cultured from skin biopsies. Surprisingly, MAO-A activity in skin fibroblast cultures from unrelated donors ranges over 100-fold. It has been suggested that this extreme variation in fibroblast MAO-A activity between donors reflects individual, genetic variation in the regulation of MAO-A in fibroblasts. I have found from studies with immunofluorescence microscopy and flow cytometry that the proportion of MAO-A$^+$ cells in fibroblast cultures is (a) highly variable between cultures, (b) a reproducible characteristic of each culture and (c) the primary factor responsible for variation in MAO-A specific activity in whole cell, skin fibroblast homogenates. It has been shown previously that MAO-A activity of a skin fibroblast culture is relatively constant with continued passage prior to cellular senescence. Therefore, these new data raise the possibility that MAO-A expression is confined to a functionally distinct subset of human skin fibroblasts.

Abbreviations

MAO monoamine oxidase, *RFLP* restriction enzyme length polymorphism, *RITC* rhodamine isothiocyanate, *PBS* phosphate-buffered saline, *PBS-BSA* PBS containing 5 mg/ml bovine serum albumin

Introduction

Monoamine oxidases A and B (MAO; EC 1.4.3.4) oxidize important biogenic amines (see review by Weyler et al., 1990). Monoclonal antibodies discriminate human MAO-A and B (Denney et al., 1982a,b, 1983; Riley et al., 1989), and have been used in immunocytochemical studies to visualize the distribution of the enzymes in human (Westlund et al., 1988) and monkey brain (Westlund et al., 1985) and some peripheral tissues (Thorpe et al., 1987). The enzymes contain single, homologous subunit species encoded by genes which have identical exon/intron structure (Grimsby et al., 1991; Chen et al., 1991) and are arranged tail-to-tail near the centromere on the short arm of the X chromosome in humans (Levy et al., 1989; Lan et al., 1989; Chen et al., 1992).

A nonsense mutation in exon 8 of the MAO-A gene inactivates MAO-A in affected males from a unique Dutch kindred. Loss of MAO-A results in disturbed systemic amine metabolism, borderline mental retardation, and possible cardiovascular and behavioral abnormalities (Brunner et al., 1993a,b; reviewed by Chen et al., 1995). MAO-A-deficient, male transgenic mice generated by insertional mutagenesis in exon 2 of the MAO-A gene also exhibit developmental, and behavioral, abnormalities (Cases et al., 1995).

Current evidence indicates that MAO-A deficiency in human populations is probably rare. However, more subtle genetic differences which alter, but do not abolish, MAO-A activity might also affect human health. Unfortunately, data concerning individual variation in MAO-A expression are limited, in part because MAO-A activity is undetectable in most conveniently available human material (e.g., blood, excretions or secretions). However, fibroblasts cultured from skin biopsies express MAO-A activity (Breakefield et al., 1976; Groshong et al., 1977; Edelstein and Breakefield, 1978) which is relatively constant for each culture throughout its proliferative lifespan (30–50 doublings) prior to senescence (Edelstein et al., 1978; Edelstein and Breakefield, 1986). Since skin fibroblast MAO-A activity has been reported to be similar in human skin fibroblast cultures from three monozygotic twin pairs (Costa et al., 1980), it has been suggested that fibroblast MAO-A activity is a genetically determined, quantitative trait of the cell donor (Breakefield and Edelstein, 1980; Breakefield et al., 1981; Breakefield and Pintar, 1982). Based on this hypothesis, fibroblast MAO-A activity has been compared in cultures from normal individuals and individuals suffering from a variety of genetic disorders (Breakefield et al., 1976, 1980; Edelstein et al., 1978; Singh et al., 1979; Breakefield and Edelstein, 1980; Groshong et al., 1989; Giller et al., 1980; Giller et al., 1984). No reproducible disease associations were found. Surprisingly, however, MAO-A activity in different skin fibroblast cultures often differ by more than 100-fold (Hotamisligil and Breakefield, 1991; Hsu et al., 1995).

To better understand the nature of this extreme quantitative variation in MAO-A activity between skin fibroblast cultures, I have applied immunofluorescence microscopy and flow cytometry to test whether some of this between-culture variation in MAO-A activity might be a consequence of differences in the frequency of cells which express the enzyme.

Materials and methods

Source of cells and cell culture conditions

Diploid human fibroblasts were normal male human skin fibroblasts from the Human Cell Mutant Repository (Camden, N.J; see Table 1). Male cells were chosen to avoid potential heterogeneity in MAO-A expression which might result from expression of the enzyme from alternate alleles in different female cells within cultures. Cells were maintained antibiotic-free in Eagle's minimal essential medium (Gibco-BRL, Bethesda, MD) supplemented with 2X essential and non-essential amino acids and 10% fetal bovine serum (Irvine Scientific, Irvine, CA). Cultures were negative for mycoplasma infection by fluorescence microscopy after staining as described by Chen (1977). Cultures were fed with complete medium four days before harvest for immunofluorescence staining and MAO-A activity assays.

Immunofluorescence microscopy of MAO-A

The MAO-A-specific monoclonal antibody MAO A-4D3 has been described (Riley et al., 1989; 1991). Ascites fluids were prepared as described in Denney et al. (1982b). Cells for immunofluorescence staining were grown on 22×22 mm coverslips (#1) in 35 mm plastic Petri dishes. Cells were rinsed three times with phosphate-buffered saline (PBS; per liter, 0.2 g KCl, 0.2 g KH_2PO_4, 1.14 g Na_2PO_4, 8 g NaCl), and fixed for two hr on ice with 2% paraformaldehyde in PBS supplemented with 0.1 M potassium phosphate buffer, pH 7.5. Aldehyde groups were blocked with 0.1 M glycine in PBS (30 min, room temperature), and cells permeabilized by treatment with 0.5% Tween 20 (Sigma Chemical Co., St. Louis, MO; 15 min, room temperature). For antibody staining, cells were rinsed with PBS, and incubated for 30 min with dilutions of primary antibodies (1/100 for ascites fluids, 1/10 for conditioned media) diluted in PBS containing 5 mg/ml bovine serum albumin (PBS-BSA), rinsed with PBS, and incubated for 30 min with secondary antibody [1/10 diluted rhodamine isothiocyanate (RITC)-conjugated rabbit antimouse IgG (Cappell Laboratories, West Chester, PA)] in PBS-BSA. Stained specimens were mounted in PBS containing 50% (v/v) glycerol and 3% polyvinyl alcohol, and examined with epifluorescence optics in a Leitz Ortholux II microscope equipped with a 63X/oil Planapo fluorescence objective and filter cubes optimized for RITC. Photographs of RITC fluorescence were exposed for 7 sec on Kodak TMAX 3200 film (Kodak, Rochester, NY) and developed for ASA 3200.

Table 1. Skin fibroblasts used in the study[a]

Designation	Biopsy Site	Donor age	Passage no. acquired
GM288a	forearm	64 y	10
GM8333a	foreskin	5 mo	9
GM323a	upper arm	11 y	12

[a]Cultures were normal human male skin fibroblasts acquired at the passage numbers indicated from the NIGMS Human Genetic Mutant Cell Repository

Vital staining of fibroblast mitochondria with rhodamine 123

Living cells were incubated for 30 min with 10 μg/ml Rhodamine 123 (Sigma Chemical Co., St. Louis, MO) in complete growth medium (Johnson et al., 1980), rinsed three times with medium lacking dye, and observed in dye-free, complete medium by epifluorescence microscopy for RITC.

Determination of MAO-A specific activity

MAO-A activity was assayed in whole cell homogenates after sonication in 50 μl of 0.1 M potassium phosphate buffer (pH 7.5) as previously described (Denney et al., 1982a). Substrate was 100 μM ^{14}C-serotonin (54 mCi/mmol; New England Nuclear, Boston, MA). Protein concentrations were measured as described by Bradford (1976) with bovine serum albumin as standard.

Flow cytometry

Confluent cells from a T75 flask were dispersed by treatment with 0.25% Pancreatin 4X (Sigma Chemical Co., St. Louis, MO). A sample (30% of the washed suspension) was withdrawn from each cell suspension for assay of MAO-A. Subsequent centrifugations were 1 min at 100 xg. Cells for flow cytometry were prepared for antibody staining by a modification of the technique of Yang et al. (1994) for flow cytometry of intracellular neutrophil antigens (fixation with 2% paraformaldehyde, permeabilization with 0.5%, v/v, Tween 20). Cells treated with primary antibody were washed and exposed to fluorescein isothiocyanate (FITC)-conjugated rabbit antimouse IgG (whole molecule; Cappell Laboratories, West Chester, PA). Flow cytometry was performed on a Becton Dickinson FACS in the Flow Cytometry Laboratory of the Microbiology Department at UTMB by Mark Griffin (FACS technician), under the supervision of Laboratory Director, Louis Justement, Ph.D.

Results

MAO-A-4D3 stains mitochondria in human fibroblasts

GM288a fibroblasts express moderately high MAO-A activity (16–42 nmol/hr/mg). Mitochondria-specific staining could be readily detected when confluent GM288a fibroblasts grown on coverslips were stained with MAO-A 4D3 (Fig. 1A,B). Treatment of cells for four days prior to staining with 10^{-7} M dexamethasone yielded markedly enhanced staining, consistent with the known ability of this hormone to increase fibroblast MAO-A activity. The long structures stained by MAO-A-4D3 resemble mitochondria in living GM288a fibroblasts stained with the mitochondrion-specific, fluorescent vital dye, rhodamine 123 (Fig. 1C). Skin fibroblast mitochondria are known to be unusually long, compared to mitochondria in many other cell types (Lea et al., 1994).

30 μm

A   B

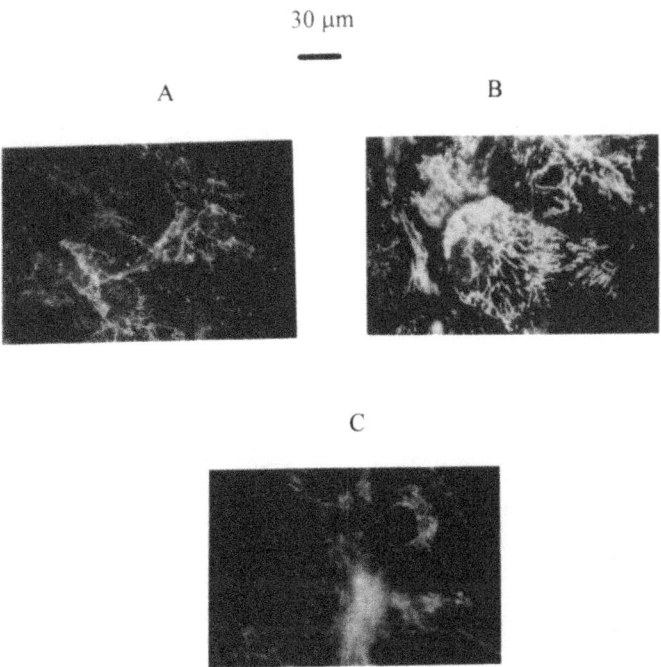

C

Fig. 1. GM288a fibroblasts stained for MAO A by immunofluorescence and for mitochondria by rhodamine 123. Cultures grown to confluence on cover slips were fed with fresh, serum-containing medium supplemented or not with 10^{-7} M dexamethasone. Cells in **A** (no dexamethasone) and **B** (with dexamethasone) were fixed with paraformaldehyde, permeabilized and stained with MAO A-4D3 and RITC-conjugated rabbit anti-mouse IgG. In **C**, living cells grown without dexamethasone treatment were stained with rhodamine 123

MAO-A is expressed by a subpopulation of cells in fibroblast cultures

Though mitochondria in most GM288a fibroblasts stained with MAO A-4D3 (Fig. 1), mitochondria in only occasional GM323a fibroblasts stained (Fig. 2A, no dexamethasone; Fig. 2B, after dexamethasone). Visual scanning of a third, MAO A-4D3 stained, skin fibroblast, GM8333a (Fig. 2C, no dexamethasone; Fig. 2D, after dexamethasone) suggested that the frequency of cells with detectable MAO A-specific staining was about 50%–60%. The rank order of frequencies of MAO-A$^+$ cells, as judged from immunofluorescence microscopy (GM288a > GM8333a > GM323a), was the same as the rank order of MAO-A specific activities of these cultures.

To better assess heterogeneity of MAO-A expression within skin fibroblast cultures, I adapted a flow cytometric method to quantitate the cellular distribution of MAO-A-specific immunofluorescence in these cells. Suspensions of paraformaldehyde-fixed, permeabilized cells were stained with MAO-A-4D3 or control antibody and analyzed for MAO-A-specific immunofluorescence by flow cytometry. Unfixed samples of each cell suspension were assayed in triplicate for MAO-A specific activity and protein. Table

30 μm

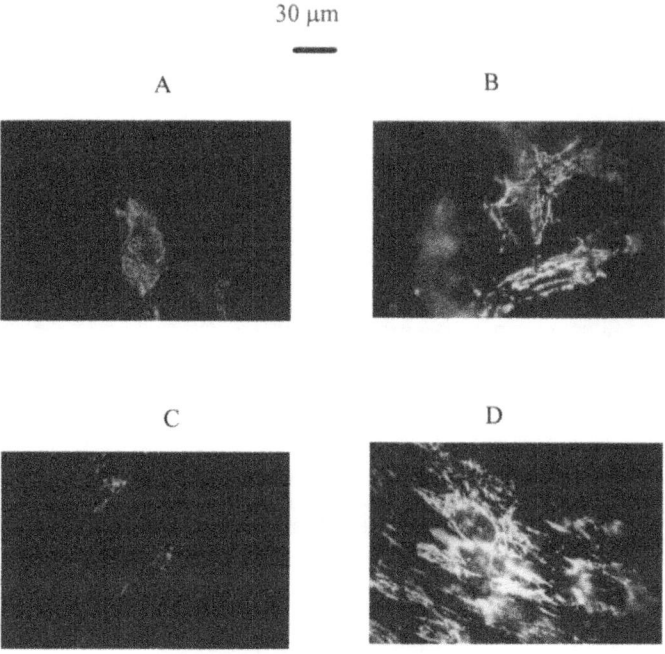

Fig. 2. Immunofluorescence microscopy of MAO A in mitochondria of GM323a and GM8333a fibroblasts. Coverslip cultures of confluent cells were incubated for an additional four days in serum-containing medium with or without 10^{-7} M dexamethasone and stained by indirect immunofluorescence with MAO A-4D3. **a** GM323a, no dexamethasone; **b** GM323a, with dexamethasone; **c** GM8333a, no dexamethasone; **d** GM8333a, with dexamethasone

Table 2. Specific activity of MAO A in total and MAO A$^+$ fibroblast populations[a]

Culture	Dex treatment	Number of determinations	Specific activity (nmol/hr/mg)[b]	MAO A$^+$ cells (%)[c]	Specific activity/ MAO A$^+$ cells[d]
GM288a	no	4	29 ± 10.8	84 ± 7	35 ± 14.4
	yes	2	75 ± 15 (2.6)	98 ± 1 (1.2)	80 ± 20 (2.3)
GM8333a	no	2	10 ± 2.7	68 ± 9	14 ± 2.0
	yes	2	34 ± 2.2 (3.4)	87 ± 5 (1.3)	38 ± 1 (2.7)
GM323a	no	2	0.065 ± 0.005	0.5 ± 0	14 ± 0.9
	yes	2	0.53 ± 0.03 (8.1)	1 ± 0 (2.0)	47 ± 9 (3.6)
Range of activity		Max/min	446-fold (−dex) 141-fold (+dex)		2.5-fold (−dex) 2.4-fold (+dex)

[a] Specific activities were measured in samples of cells harvested from T75 flasks after growth in medium lacking or containing 10^{-7} M dexamethasone. The remaining cells were fixed, stained for MAO A and analyzed for MAO A-specific staining by flow cytometry as described in Materials and methods. [b] Mean ± standard deviation of triplicate assays of cell homogenates (100 μM serotonin substrate). Protein was determined by the method of Bradford (1976). Figures in parentheses give the fold increase in MAO A activity by dexamethasone. [c] Assessed by flow cytometry. Figures in parentheses give the fold increase in MAO A activity by dexamethasone. [d] Specific activity in cell homogenates divided by fraction of MAO A$^+$ cells

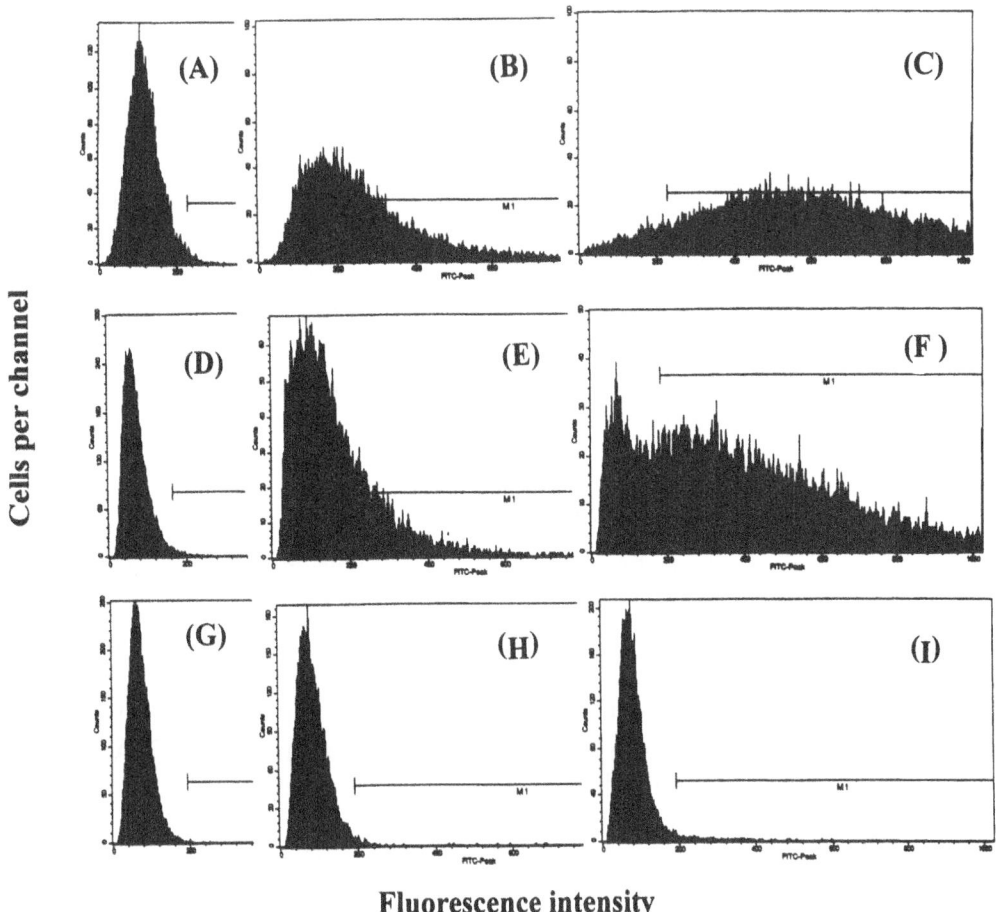

Cells per channel

Fluorescence intensity

Fig. 3. Flow cytometric assay of MAO A-specific immunofluorescence in GM288a, GM8333a and GM323a fibroblasts grown in culture. Confluent cells were incubated for four days in complete medium supplemented or not with 10^{-7} M dexamethasone. Cells were harvested, fixed, permeabilized, stained with MAO A-4D3 or irrelevant antibody, and analyzed for immunofluorescence by flow cytometry as described in Materials and methods. Histograms constitute plots of cell numhber per channel (Y-axis) versus peak fluorescence per cell (X-axis; linear scale.) **A** GM288a, no dexamethasone, control antibody; **B** GM288a, no dexamethasone, MAO A-4D3; **C** GM288a, puls dexamethasone, MAO A-4D3; **D-F** as A, B and C except cells were GM8333a fibroblasts. **G–I** as A, B and C, except cells were GM323a fibroblasts

2 summarizes flow cytometry and specific activity data from a minimum of two experiments for each line. Note that dexamethasone stimulated MAO-A activity 2.7–8.1 fold in the different cultures. The range of MAO-A specific activities in the three cultures was 446-fold (no dexamethasone), and 141-fold (dexamethasone-treated). Dexamethasone also appeared to increase slightly the observed proportion of MAO-A$^+$ cells, but to a much lesser extent than it increased MAO-A specific activities. Table 2, column 6, presents estimates of the specific activity of MAO-A in the MAO A$^+$ cells ($100\times$ specific activity/ percent of MAO-A$^+$ cells). These normalized specific activity values, while not identical between cultures, were much more similar than the unnormalized specific activities).

Discussion

The proportion of human skin fibroblast cells which expressed MAO-A detectable by immunofluorescence microscopy or flow cytometry differed widely among three normal human skin fibroblasts expressing widely different MAO-A activity. Normalization of homogenate specific activity by dividing by proportion of MAO-A$^+$ cells reduced the range of specific activities from 446-fold to only 2.5-fold (no dexamethasone), and 141-fold to 2.4-fold (dexamethasone-treated). These data, and more extensive studies of additional skin fibroblast cultures to be presented elsewhere, indicate that most of the variability in MAO-A specific activity results from variation in the frequency of MAO-A$^+$ cells. The MAO-A data presented here are consistent with other evidence that skin fibroblasts are metabolically diverse (see review by Fries et al., 1994).

Edelstein and Breakefield (1980) reported that the MAO-A specific activity in each of several confluent fibroblast cultures was stable over many passages prior to senescence. Stability of MAO-A specific activity in whole homogenates with passage would most easily be explained by stability with passage in both the frequency of MAO-A$^+$ cells, and the mean MAO-A content per MAO-A$^+$ cell. This hypothesis can be tested directly by monitoring the frequency of MAO A$^+$ cells and the mean MAO-A-specific fluorescence intensity in a culture at varying passage numbers by the flow cytometry technique described here.

The data presented here do not elucidate the mechanisms responsible for large differences in percent MAO-A$^+$ fibroblasts in different cultures. Such variation could result either from pre-existing phenotypic heterogeneity of fibroblasts in the skin (with maintenance of MAO-A phenotype in culture), or variable, culture-induced or spontaneous changes in MAO-A gene expression among progeny of fibroblasts generated after they are placed in culture. Consistent with possible in vivo heterogeneity is the report that fibroblasts cultivated from the papillary (superficial) and reticular (deeper) layers from a single individual differed 5- to 10-fold in MAO-A specific activity (Edelstein and Breakefield, 1986). However, in the light of my new data, this single example cannot be considered definitive since it has yet to be demonstrated that multiple, ostensibly identical cultures derived from the same individual have similar proportions of MAO-A$^+$ cells.

The expression of MAO-A by a variably-sized subpopulation of cells in cultures of skin fibroblasts may constitute a serious obstacle to the use of skin fibroblasts MAO-A activity as a meaningful genetic trait of fibroblast donors. However, these findings do not preclude the use of skin fibroblasts in comparisons of the structure of the MAO-A molecule between individuals. To data only fibroblasts from truly MAO-A-deficient individuals lack any detectable MAO-A activity. It appears that flow cytometric and/or flow sorting methodology may permit the measurement of fibroblast MAO-A specific activity in that subpopulation of cells which express the enzyme. Whether the relatively small, residual variation between cultures in MAO-A specific activity in the MAO-A$^+$ cell population reflects genetic differences between fibroblast donors, or proves to be due to technical matters, remains to be determined.

Acknowledgements

This work was supported by funds from the John Sealey Memorial Endowment and the Cancer Center of the University of Texas Medical Branch. I thank Miss A. Waguespack for growing the human fibroblast cultures used in this study.

References

Bradford MM (1976) A rapid and sensitive method for the quantitation of microgram quantities of protein utilizing the principal of protein-dye binding. Anal Biochem 72: 248–254

Breakefield XO, Edelstein SB (1980) Inherited levels of A and B types of monoamine oxidase activity. Schizophr Bull 6: 282–288

Breakefield XO, Pintar JE (1982) Monoamine oxidase (MAO) activity as a determinant in human neurophysiology. Behav Gen 12: 53–69

Breakefield XO, Castiglione C, Edelstein S (1976) Monoamine oxidase activity decreased in cells lacking hypoxanthine phosphoribosyltransferase activity. Science 192: 1018–1020

Breakefield XO, Giller EL Jr, Nurnberger JI Jr, Castiglione CM, Buchsbaum MS, Gershon ES (1980) Monoamine oxidase type A in fibroblasts from patients with bipolar depressive illness. Psychiat Res 2: 307–314

Breakefield XO, Edelstein SB, Grossman MH (1981) Variations in MAO and NGF in cultured human skin fibroblasts. In: Gershon ES, Matthysse S, Breakefield XO, Ciaranello RD (eds) Genetic research strategies in psychobiology and psychiatry. Boxwood Press, Pacific Grove CA, pp 129–142

Brunner HG, Nelen M, Breakefield XO, Ropers HH, van Oost BA (1993a) Abnormal behavior associated with a point mutation in the structural gene for monoamine oxidase A. Science 262: 578–580

Brunner HG, Nelen MR, van Zandvoort P, Abeling NGGM, van Gennip AH, Wolters EC, Kuiper MA, Ropers HH, van Oost BA (1993b) X-linked borderline mental retardation with prominent behavioral disturbance: phenotype, genetic localization, and evidence for disturbed monoamine metabolism. Am J Hum Genet 42: 1032–1039

Cases O, Seif I, Grimsby J, Gaspar P, Chen K, Pournin S, Muller Ü, Auet M, Babinet C, Shih JC, De Maeyer E (1995) Aggressive behavior and altered amounts of brain serotonin and norepinephrine in mice lacking MAOA. Science 268: 1763–1765

Chen TR (1977) In situ detection of mycoplasma contamination in cell cultures by fluorescent Hoechst 33258 stain. Exp Cell Res 104: 255–262

Chen Z-Y, Hotamisligil S, Huang J-K, Wen L, Ezzeddine D, Aydin-Muderrisoglu N, Powell JF, Huang RH, Breakefield XO, Craig IW, Hsu Y-PP (1991) Structure of the human gene for monoamine oxidase type A. Nucl Acid Res 19: 4537–4541

Chen Z-Y, Powell JF, Hsu Y-PP, Breakefield XO, Craig IW (1992) Organization of the human monoamine oxidase genes and long-range physical mapping around them. Genomics 14: 75–82

Chen Z-Y, Denney RM, Breakefield XO (1995) Norrie disease and MAO genes-nearest neighbors. Hum Mol Genet 4: 1729–1737

Costa MRC, Edelstein SB, Castiglione CM, Chao H, Breakefield XO (1980) Properties of monoamine oxidase in control and Lesch-Nyhan fibroblasts. Biochem Genet 18: 577–589

Denney RM (1995) The promoter of the human monoamine oxidase A gene. Prog Brain Res 106: 57–66

Denney RM, Fritz RR, Patel NT, Abell CW (1982a) Human liver MAO-A and MAO-B separated by immunoaffinity chromatography with MAO-B-specific monoclonal antibody. Science 215: 1400–1403

Denney RM, Patel NT, Fritz RR, Abell CW (1982b) A monoclonal antibody elicited to human platelet monoamine oxidase B. Isolation and specificity for human monoamine oxidase B but not A. Mol Pharmacol 22: 500–508

Denney RM, Fritz RR, Patel NT, Widen SS, Abell CW (1983) Use of a monoclonal antibody for comparative studies of monoamine oxidase B in mitochondrial extracts of human brain and peripheral tissues. Mol Pharmacol 24: 60–68

Denney RM, Sharma A, Dave SK, Waguespack MA (1994) A new look at the promoter of the human monoamine oxidase A gene. Mapping transcription initiation sites and capacity to drive luciferase expression. J Neurochem 63: 843–856

Edelstein SB, Breakefield XO (1981) Dexamethasone selectively increases monoamine oxidase type A in human skin fibroblasts. Biophys Biochem Res Commun 98: 836–843

Edelstein SB, Breakefield XO (1986) Monoamine oxidases A and B are differentially regulated by glucocorticoids and "aging" in human skin fibroblasts. Cell Mol Neurobiol 6: 121-150

Edelstein SB, Castiglione CM, Breakefield XO (1978) Monoamine oxidase activity in normal and Lesch-Nyhan fibroblasts. J Neurochem 31: 1247–1254

Fries KM, Blieden T, Looney RJ, Sempowski GD, Silvera MR, Willis RA, Phipps RP (1994) Evidence of fibroblast heterogeneity and the role of fibroblast subpopulations in fibrosis. Clin Immunol Immunopathol 72: 283–292

Giller EL Jr, Young JG, Breakefield XO, Carbonari C, Braverman M, Cohen DJ (1980) Monoamine oxidase and catechol-o-methyltransferase activities in cultured fibroblasts and blood cells from children with autism and the Gilles de la Tourette syndrome. Psychiat Res 2: 187–197

Giller EL Jr, Nocks J, Hall H, Stewart C, Schnitt J, Sherman B (1984) Platelet and fibroblast monoamine oxidase in alcoholism. Psychiat Res 12: 339–347

Grimsby J, Lan NC, Neve R, Chen K, Shih JC (1990) Tissue distribution of human monoamine oxidase A and B mRNA. J Neurochem 55: 1166–1169

Grimsby J, Chen K, Wang LJ, Lan NC, Shih JC (1991) Human monoamine oxidase A and B genes exhibit identical exon-intron organization. Proc Natl Acad Sci USA 88: 3637–3641

Groshong R, Gibson DA, Baldessarini RJ (1977) Monoamine oxidase activity in cultured human skin fibroblasts. Clin Chim Acta 80: 113–120

Groshong R, Baldessarini RJ, Gibson DA, Lipinski JF, Axelrod D, Pope A (1978) Activities of types A and B MAO and catechol-O-methyltransferase in blood cells and skin fibroblasts of normal and chronic schizophrenic subjects. Arch Gen Psychiatry 35: 1198–1205

Hotamisligil GS, Breakefield XO (1991) Human monoamine oxidase A gene determines levels of enzyme activity. Am J Hum Genet 49: 383–392

Hsu Y-PP, Weyler W, Chen S, Sims KB, Rinehart WB, Utterback MC, Powell JF, Breakefield XO (1988) Structural features of human monoamine oxidase A elucidated from cDNA and peptide sequences. J Neurochem 51: 1321–1324

Hsu Y-PP, Powell JF, Sims KB, Breakefield XO (1989) Molecular genetics of the monoamine oxidases. J Neurochem 53: 12–18

Hsu Y-PP, Schuback DE, Tivol EA, Shalish C, Murphy DL, Breakefield XO (1995) Analysis of MAOA mutations in humans. Prog Brain Res 106: 67–75

Johnson LV, Walsh ML, Chen LB (1980) Localization of mitochondria in living cells with rhodamine 123. Proc Natl Acad Sci USA 77: 990–994

Lan NC, Neinzmann C, Gal A, Klisak I, Orth U, Lai E, Grimbsby J, Sparkes RS, Mohandas T, Shih JC (1989) Human monoamine oxidase A and B genes map to Xp 11.23 and are deleted in a patient with Norrie disease. Genomics 4: 552–559

Lea PJ, Temkin RJ, Freeman KB, Mitchell GAm Robinson BH (1994) Variations in mitochondrial ultrastructure and dynamics observed by high resolution scanning electron microscopy (HRSEM). Microscop Res Tech 27: 269–277

Levy ER, Powell JF, Buckle VJ, Hsu Y-PP, Breakefield XO, Craig IW (1989) Localization of human monoamine oxidase-A gene to Xp11.23-11 by in situ hybridization: implications for Norrie disease. Genomics 5: 368–370

Riley LA, Denney RM (1991) Problems with the measurement of monoamine oxidase A protein concentrations in mitochondrial preparations. Revised molecular activities and implications for estimating ratios of MAO A:MAO B molecules from radiochemical assay data. Biochem Pharmacol 42: 1953–1959

Riley LA, Waguespack MA, Denney RM (1989) Characterization and quantitation of monoamine oxidases A and B in mitochondria from human placenta. Mol Pharmacol 36: 54–60

Singh S, Willers I, Kluss E-M, Goedde HW (1979) Monoamine oxidase and catechol-o-methyltransferase activity in cultured fibroblasts from patients with maple syrup disease, Lesch-Nyhan syndrome and healthy controls. Clin Genet 15: 153–159

Thorpe LW, Westlund KN, Kochersperger LM, Abell CW, Denney RM (1987) Immunocytochemical localization of monoamine oxidase A and B in human peripheral tissues and brain. J Histochem Cytochem 35: 23–32

Westlund KN, Denney RM, Kochersperger LM, Rose RM, Abell CW (1985) Distinct monoamine oxidase A and B populations in primate brain. Science 230: 181–183

Westlund KN, Denney RM, Rose RM, Abell CW (1988) Localization of distinct monoamine oxidase A and monoamine oxidase B cell populations in human brainstem. Neurosci 25: 439–456

Weyler W, Hsu Y-PP, Breakefield XO (1990) Biochemistry and genetics of monoamine oxidase. J Pharmacol Ther 47: 391–417

Yang YH, Hutchinson P, Littlejohn GO, Boyce N (1994) Flow cytometric detection of anti-neutrophil cytoplasmic autoantibodies. J Immun Meth 172: 77–84

Author's address: R. M. Denney, Ph.D., Graduate School of Biomedical Sciences, University of Texas Medical Branch, Galveston, TX 77555, U.S.A.

Are MAO-A deficiency states in the general population and in putative high-risk populations highly uncommon?*

D. L. Murphy[1], **K. Sims**[2], **G. Eisenhofer**[3], **B. D. Greenberg**[1], **T. George**[4], **F. Berlin**[5], **A. Zametkin**[6], **M. Ernst**[6], and **X. O. Breakefield**[2]

[1]Laboratory of Clinical Science, and [6]Child Psychiatry Branch, National Institute of Mental Health, NIH, Bethesda, MD, [2]Molecular Neurogenetics Laboratory, Massachusetts General Hospital, Boston, MA, [3]Clinical Neuroscience Branch, National Institute of Neurological Disorders and Stroke, NIH, Bethesda, MD, [4]Laboratory of Clinical Studies, National Institute on Alcohol Abuse and Alcoholism, NIH, Bethesda, MD, and [5]Department of Psychiatry, Johns Hopkins University School of Medicine, Baltimore, MD, U.S.A.

Summary. Lack of monoamine oxidase A (MAO-A) due to either Xp chromosomal deletions or alterations in the coding sequence of the gene for this enzyme are associated with marked changes in monoamine metabolism and appear to be associated with variable cognitive deficits and behavioral changes in humans and in transgenic mice. In mice, some of the most marked behavioral changes are ameliorated by pharmacologically-induced reductions in serotonin synthesis during early development, raising the question of possible therapeutic interventions in humans with MAO deficiency states. At the present time, only one multi-generational family and a few other individuals with marked MAO-A deficiency states have been identified and studied in detail. Although MAO deficiency states associated with Xp chromosomal deletions were identified by distinct symptoms (including blindness in infancy) produced by the contiguous Norrie disease gene, the primarily behavioral phenotype of individuals with the MAO mutation is less obvious. This paper reports a sequential research design and preliminary results from screening several hundred volunteers in the general population and from putative high-risk groups for possible MAO deficiency states. These preliminary results suggest that marked MAO deficiency states are very rare.

Considerable interest in the area of behavioral genetics followed the report of linkage in a large Dutch kindred between an x-chromosomal locus and a behavioral phenotype consisting of a characteristic syndrome including episodic impulsive, aggressive behavior and mild mental retardation (Brunner et al., 1993a). Written records and direct assessment revealed 14 male family

---
*Presented at the 6th Rappaport Symposium and 7th Amine Oxidase Workshop, Israel, June 1996

members spanning four generations with this behavioral phenotype strongly linked to Xp 11–21. Because three of these individuals had abnormal urinary monoamine metabolite excretion patterns, an abnormality in MAO-A or MAO-B was postulated (Brunner et al., 1993a). Shortly thereafter, a single base substitution resulting in a termination codon in the MAO-A gene and leading to the production of an incomplete MAO-A enzyme lacking functional activity was identified (Brunner et al., 1993b).

Somewhat earlier, several families with deficits in both MAO-A and MAO-B gene function had been described. These individuals all had microdeletions of Xp chromosomal areas encompassing both the MAO-A and MAO-B genes and the Norrie disease gene (Sims et al., 1989; Collins et al., 1992). In these individuals, the striking characteristics of the Norrie disease phenotype, which consists of blindness from infancy, variable mental retardation, plus some additional behavioral abnormalities, predominated over the more subtle behaviors and mild mental retardation observed in the individuals lacking MAO-A only. Marked neurochemical and physiological abnormalities similar to abnormalities found in individuals given MAO-inhibiting drugs were subsequently identified in these individuals, as well as in two of the individuals with the MAO-A point mutation (Murphy et al., 1990, 1991; Collins et al., 1992; Lenders et al., 1996). Subsequently, a family with an Xp-chromosomal deletion producing Norrie disease and MAO-B but not MAO-A deficiency was identified; in two affected males, only minor changes in excretion of phenylethylamine unassociated with any other neurochemical or apparent behavioral abnormalities were found (Lenders et al., 1996).

MAO deficit states have long been considered as potential contributors to abnormal behavior. MAO inhibitors, beginning with the anti-tuberculosis agent, iproniazid, and followed by MAO-inhibiting antidepressants such as phenelzine, tranylcypromine and isocarboxazid were recognized as capable of inducing increased psychomotor activity to the point of hypomania, physical aggressiveness, poor sleep and an increased response to external stimuli (Crane, 1956; Freymuth et al., 1959; Murphy, 1977; Rabkin et al., 1985). However, marked behavioral side-effects are uncommon and occur principally in individuals with neuropsychiatric disorders.

In early approaches to investigate whether endogenous differences in MAO activity might be associated with behavioral differences within the general population or in individuals with neuropsychiatric disorders, MAO activity in blood platelets were measured (Murphy et al., 1976; Fowler et al., 1982; Oreland and Hallman, 1995). However, platelets contain only MAO-B, and not MAO-A (Donnelly and Murphy, 1977), and hence possible associations between MAO-A and human behavior have remained largely unexplored.

The present study approached this question by measuring plasma concentrations of 3-methoxy, 4-hydroxyphenylglycol (MHPG), a metabolite of norepinephrine produced via deamination by MAO-A (Kopin et al., 1984). This measure seemed an appropriate marker for MAO-A functional activity as plasma and urinary MHPG was found to be decreased more than 90% or essentially absent in individuals lacking the MAO-A gene (Murphy et al.,

Table 1. How common are monoamine oxidase deficiency states in the general population and in selected possible "high-risk" groups? A three-stage study strategy

1. Primary strategy: Screening using plasma MHPG concentrations.
2. Second-level strategy: determination of normetanephrine/MHPG (or DHPG) ratios.
3. Third-level strategy: Analysis of MAO-A gene single strand conformational polymorphisms (SSCP) or coding region sequences.

1991; Lenders et al., 1996). It was also found to be reduced approximately 85% in human and non-human primate plasma, urine and cerebrospinal fluid following treatment with the selective inhibitor of MAO-A, clorgyline, but unaffected by doses of selegiline that selectively inhibited MAO-B (Murphy et al., 1981; Sunderland et al., 1985). As indicated in Table 1, initial screening using plasma MHPG was followed by two additional approaches as part of a three-stage strategy to identify and then verify possible MAO-A deficiency states in the general population and in putative "high-risk" groups. Our supposition, however, was that marked MAO-A deficiency states like those found in individuals with a lack of the MAO-A gene would be uncommon. Nonetheless, the success in reversing some of the aberrant behavioral manifestations of an absent MAO-A gene in knock-out transgenic mice by pharmacological means (Cases et al., 1995) suggested that since potential MAO-A deficiency states in humans might be treatable, it was important to identify them.

Subjects and methods

Subjects included the following groups: (A) 148 healthy male volunteers with an age range of 18 to 72 (mean 28 ± 5) who had blood collected for monoamine metabolites as part of other studies at the NIH Clinical Center. (B) 28 neuropsychiatric patients who were studied prior to and during treatment with the MAO-A-inhibiting antidepressant, clorgyline, 20–30 mg/day for 3–4 weeks, or treatment with the partially selective MAO-B-inhibitor, selegiline, 20 or 60 mg/day for 6 weeks (Murphy et al., 1979; Ernst et al., 1997). Of note, studies by our group had demonstrated that low selegiline doses (10 mg/day) had negligible effects on MHPG or other indices of MAO-A inhibition, while higher selegiline doses (30–60 mg/day) produced proportionately greater effects on these indices which approached those produced by clorgyline or by non-selective MAO-inhibitors (Murphy et al., 1981; Sunderland et al., 1985).

In addition, because the most prominent behaviors which drew attention to the individuals with the MAO-A deficit in the Dutch family included impulsive, aggressive acts, including inappropriate sexual attacks, volunteers attending three clinic treatment programs for such behaviors were sampled: (C) 46 male subjects (age range 21–56) with paraphilic disorders including pedophilia who were attending a sexual disorder clinic (Berlin et al., 1991); (D) 12 male subjects (age range 28–49) with histories of repeated episodes of domestic violence who were participants in group and individual therapy programs (George et al., 1997); 8 of these 12 individuals also had histories of alcohol abuse or dependency; and (E) 24 males with adult attention deficit hyperactivity disorder (ADHD) (age range 22–46) who were participants in a drug treatment study and had blood samples drawn for MHPG measurements after a drug-free period of 3–4 weeks

(Ernst et al., 1997). Most individuals in the sexual disorder and domestic violence programs were not taking medications, with the exception of depo-provera or lupron in the case of subjects in the sexual disorder clinic. These agents and a few other medications taken for medical conditions are not thought to affect catecholamine metabolites, and a direct comparison of the subjects receiving depo-provera, lupron or other miscellaneous medications versus the other clinic subjects revealed no significant difference in MHPG values between the groups.

Plasma MHPG, normetanephrine and other monoamine metabolites and monoamines were assayed as described previously (Eisenhofer et al., 1986; Murphy et al., 1991; Eisenhofer et al., 1995).

Results

As indicated in Fig. 1, plasma MHPG concentrations in the general population sample of 148 male subjects approximated a normal distribution, with most values ranging from 7 to 30 pmol/ml, and a few additional values at the upper end of the distribution. Not one value was as low as the MHPG values observed in the individuals treated with the selective inhibitor of MAO-A, clorgyline. Likewise, there were only a few individuals in the general population who had values as low as those observed in the selegiline-treated subjects.

As indicated in Table 2, 82 individuals from putative "high-risk" groups with behavioral disorders having some similarities to those described in individuals lacking the MAO-A gene had plasma MHPG values generally in the same range as those found in the general population sample. An exception was two individuals from the sexual offender clinic population who had no demonstrable peak at the retention time point for MHPG on the HPLC-ECD chromatogram.

When individuals in the lower 50% of the general population or the lower 20% of the "high-risk" group were re-examined using the second-stage of our three-stage strategy, namely the calculation of the plasma normetanephrine/ MHPG ratio (Table 3), individuals from these groups differed by approxi-

Table 2. Plasma MHPG concentrations (pmol/ml)

Subjects	Range of values
Gene-related reductions in MAO activity	
MAO-A/B Deficiency (N = 5)	0.1–0.4
MAO-A Deficiency (N = 2)	1.1–2.1
Other subjects	
Domestic Violence Clinic (N = 12)	5.3–18.2
(T. George)	
Sexual Offender Clinic (N = 46)	0.0**–21.7
(B. Greenberg, F. Berlin)	
Attention Deficit Hyperactivity Disorder (N = 24)	4.9–22.4
(A. Zametkin, M. Ernst)	

*Two individuals with no MHPG peak on HPLC-EC

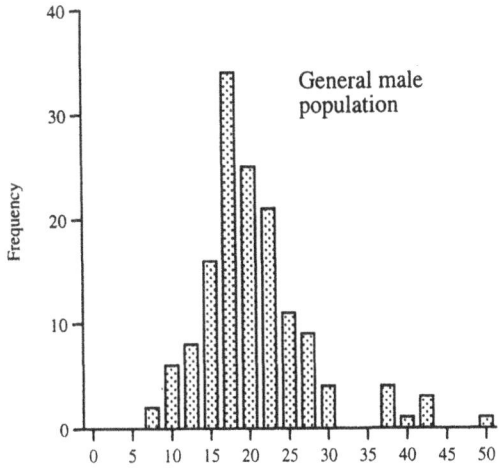

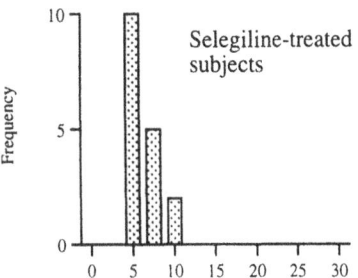

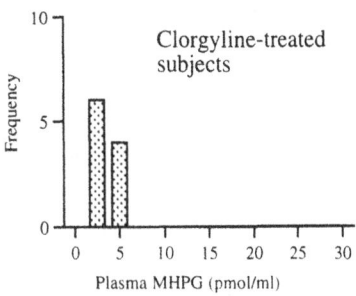

Fig. 1. Frequency distribution plots for plasma MHPG in 148 males (age range 18–72, mean 28 ± 5) from the general population (top panel), 18 neuropsychiatric patients treated with selegiline; 20 or 60 mg/day for six weeks (middle panel), and 10 neuropsychiatric patients treated with clorgyline, 20–30 mg/day for 3–4 weeks (lower panel)

mately 100-fold from values previously found in subjects with MAO-AB deficiencies related to Xp chromosomal deletions, thus indicating that marked MAO deficiency states were highly uncommon in these populations. It should be noted, however, that normetanephrine/MHPG ratios could not be calculated for the two individuals with no MHPG peak in the HPLC-ECD analysis. The "positive control" sample population — those treated with the MAO-inhibitor, selegiline — did exhibit, as expected, some normetanephrine/

Table 3. Normetanephrine/MHPG ratios in plasma from subjects with gene-related or drug-induced reductions in MAO activity

	Range of ratios
General population (N = 72)	0.46–1.24
Selected subjects from high-risk populations with lower plasma MHPG concentrations (N = 15)	0.39–3.12
Selegiline-treated subjects (N = 13)	1.26–94
MAO-AB deficiency related to Xp chromosomal deletions (N = 5)	215–530

MHPG ratios that approached those found in individuals lacking the MAO-AB genes.

Discussion

As anticipated, this screening study using plasma MHPG concentrations as a rapid, easily-obtained marker for identifying individuals with a possible functional deficiency in MAO-A activity revealed that MAO-A deficiency is highly uncommon in the general population and in putative "high-risk" groups with behavioral disorders. Direct comparisons of plasma MHPG and normetanephrine/MHPG ratios using identical analytical methodology in subjects from these populations and individuals lacking the MAO-AB genes or treated with selegiline in doses that inhibited both MAO-A and MAO-B provide firm evidence for this conclusion.

Only two individuals from the population of 230 subjects were identified with MHPG values or normetanephrine/MHPG ratios close to those of MAO-deficient or MAO-inhibited subjects. Third-stage evaluations employing MAO gene screening techniques (Hsu et al., 1995) have not yet been completed on samples from these two individuals. Also, evaluations of selected samples from subjects in the lower range of MHPG values in the present populations, nor in additional populations including individuals with mental retardation, have also not yet been completed (X. O. Breakefield, personal communication). However, limited additional analyses of plasma samples indicated that deaminated metabolites of serotonin and dopamine were within normal ranges in the two individuals with absent MHPG peaks in this study, rendering an MAO-A deficit state unlikely. In the last year, our laboratory also performed analyses of MHPG in cerebrospinal fluid samples from over 100 infants, and, as in the present study, found only one low MHPG value that differed substantially from the normal range of values in this population (Constantino and Murphy, 1996).

The present study was intended to demonstrate a screening approach and to provide preliminary data concerning the frequency of marked deficits in

MAO function similar to those reported in individuals lacking an intact MAOA gene (Brunner et al., 1993a,b; Lenders et al., 1996). Considerable evidence indicates that plasma MHPG reflects overall norepinephrine metabolism in humans (Kopin et al., 1983, 1984), although sympathetic nerves and peripheral organs other than the central nervous system provide the preponderance of plasma MHPG (Eisenhofer et al., 1996). In addition, twin and other genetic studies in humans and non-human primates have demonstrated substantial genetic contributions to MHPG levels (Oxenstierna et al., 1986; Higley et al., 1993).

The present study does not address the question of whether moderate reductions in MAO activity might contribute to dysfunctional behavior. It is of interest, however, that rather striking similarities in the type of abnormal behavior, particularly impulsive, aggressive and antisocial behaviors hypothesized to be related to serotonin system abnormalities have been reported in animals with genetic or drug-induced impairment of MAO-A activity on a genetic basis or produced by MAO-inhibition administration during early development (Whitaker-Azmitia et al., 1994; Cases et al., 1995) as well as in the Dutch kindred with MAO-A deficits on a genetic basis (Brunner et al., 1993a,b). Even studies examining platelet MAO-B activity reductions, (Schalling et al., 1987; Alm et al., 1996) including those dating back to some of the first studies in general population samples (Buchsbaum et al., 1976; Coursey et al., 1982) found associations with similar behaviors. It now seems unlikely that MAO-B deficits alone are involved in these behavioral abnormalities, as there is no evidence that recently identified individuals lacking the MAO-B gene have behavioral problems (Lenders et al., 1996). Nonetheless, a speculative basis for the commonality of behaviors observed might be found in the marked similarities in the structures of these two homologous genes, MAO-A and MAO-B, both in terms of an identical exon-intron organization, as well as the conservation of at least one polymorphic repeat (Grimsby et al., 1991; Shih et al., 1993). Alternatively, these two genes may share co-regulated gene expression factors or share other regulatory mechanisms, including common endogenous or exogenous inhibitory factors active during early development (Oreland and Hallman, 1995).

In summary, the question of whether and to what extent partial reductions in MAO activity might contribute to abnormal behavior or to the range of range of behaviors or personality traits found within the general population (Plomin et al., 1994; Lesch et al., 1996) remain controversial and unproven. The data presented in this paper, in contrast, demonstrate that marked reductions in MAO activity are highly uncommon and are unlikely to be involved in frequently-observed behavioral syndromes.

References

Alm PO, af Klinteberg B, Humble K, Leppert J, Sorensen S, Thorell L-H, Lidberg L, Oreland L (1996) Psychopathy, platelet MAO activity and criminality among former juvenile delinquents. Acta Psychiatr Scand 94: 105–111

Berlin FS, Hunt WP, Malin HM, Dyer A, Lahne GK, Dean S (1991) A five-year plus follow-up survey of criminal recidivism within a treated cohort of 406 pedophiles, 111 exhibitionists and 109 sexual aggressives: issues and outcome. Am J Forensic Psychiatry 12(3): 5–28

Brunner HG, Nelen M, Breakefield XO, Ropers HH, van Oost BA (1993a) Abnormal behavior associated with a point mutation in the structural gene for monoamine oxidase A. Science 262: 578–580

Brunner HG, Nelen MR, van Zandvoort P, Abeling NG, van Gennip AH, Wolters EC, Kuiper MA, Ropers HH, van Oost BA (1993b) X-linked borderline mental retardation with prominent behavioral disturbance: phenotype, genetic localization, and evidence for disturbed monoamine metabolism. Am J Hum Genet 52: 1032–1039

Buchsbaum MS, Coursey RD, Murphy DL (1976) The biochemical high-risk paradigm: behavioral and familial correlates of low platelet monoamine oxidase activity. Science 194: 339–341

Cases O, Self I, Grimsby J, Gaspar P, Chen K, Ournin S, Muller U, Aguet M, Babinet C, Shih JC, De Maeyer E (1995) Aggressive behavior and altered amounts of brain serotonin and norepinephrine in mice lacking MAOA. Science 268: 1763–1766

Collins FA, Murphy DL, Reiss AL, Sims KB, Lewis JG, Freund L, Karoum F, Zhu D, Maumenee IH, Antonarakis SE (1992) Clinical, biochemical, and neuropsychiatric evaluation of a patient with a contiguous gene syndrome due to a microdeletion Xp 11.3 including the Norrie disease locus and monoamine oxidase (MAOA and MAOB) genes. Am J Med Genet 42: 127–134

Constantino JN, Murphy DL (1996) Monoamine metabolites in "leftover" newborn human cerebrospinal fluid-a potential resource for biobehavioral research. Psychiatry Res 65: 129–142

Coursey RD, Buchsbaum MS, Murphy DL (1982) 2-year follow-up of subjects and their families defined as at risk for psychopathology on the basis of platelet MAO activities: 2-year follow-up of low platelet MAO. Neuropsychobiology 8: 51–56

Crane GE (1956) Further studies on iproniazid phosphate isonicotinil-isopropylhydrazine phosphate marsilid. J Nerv Ment Dis 124: 322

Donnelly CH, Murphy DL (1977) Substrate- and inhibitor-related characteristics of human platelet monoamine oxidase. Biochem Pharmacol 26: 853–858

Eisenhofer G, Goldstein DS, Stull R, Keiser HR, Sunderland T, Murphy DL, Kopin IJ (1986) Simultaneous liquid-chromatographic determination of 3,4-dihydroxyphenylglycol, catecholamines, and 3,4-dihydroxyphenylalanine in plasma, and their responses to inhibition of monoamine oxidase. Clin Chem 32: 2030–2033

Eisenhofer G, Friberg P, Pacak K, Goldstein DS, Murphy DL, Tsigos C, Quyyumi AA, Brunner HG, Lenders JW (1995) Plasma metadrenalines: do they provide useful information about sympatho-adrenal function and catecholamine metabolism? Clin Sci 88: 533–542

Eisenhofer G, Aneman A, Hooper D, Rundqvist B, Friberg P (1996) Mesenteric organ production, hepatic metabolism, and renal elimination of norepinephrine and its metabolites in humans. J Neurochem 66: 1565–1573

Ernst M, Liebenauer LL, Tebeka D, Jons PH, Eisenhofer G, Murphy DL, Zametkin AJ (1997) Selegiline in ADHD adults plasma monoamines and monoamine metabolites. Neuropsychopharmacology 16: 276–284

Fowler CJ, Tipton KF, MacKay AV, Youdim MB (1982) Human platelet monoamine oxidase-a useful enzyme in the study of psychiatric disorders? Neuroscience 7(7): 1577–1594

Freymuth HW, Walker H, Baumecker P, Stein H (1959) Effects of iproniazid on chronic and regressed schizophrenics. Dis Nerv Syst 20: 123

George DT, Benkelfat C, Rawlings RR, Eckardt MJ, Phillips MJ, Nutt DJ, Wynne D, Murphy DL, Linnoila M (1997) Behavioral and neuroendocrine responses to m-

Chlorophenylpiperazine in subtypes of alcoholics and in healthy comparison subjects. Am J Psychiatry 154(1): 81–87

Grimsby J, Chen K, Wang L-J, Lan NC, Shih JC (1991) Human monoamine oxidase A and B genes exhibit identical exon-intron organization. Proc Natl Acad Sci 88: 3637–3641

Higley JD, Thompson WW, Champoux M, Goldman D, Hasert MF, Kraemer GW, Scanlan JM, Suomi SJ, Linnoila M (1993) Paternal and maternal genetic and environmental contributions to cerebrospinal fluid monoamine metabolites in rhesus monkeys (macaca mulatta). Arch Gen Psychiatry 50: 615–623

Hsu YP, Schuback DE, Tivol EA, Shalish C, Murphy DL, Breakefield XO (1995) Analysis of MAOA mutations in humans. Prog Brain Res 106: 67–75

Kopin IJ, Gordon EK, Jimerson DC, Polinsky RJ (1983) Relationship between plasma and cerebrospinal fluid levels of 3-methoxy-4-hydroxy-phenyl-glycol. Science 219: 73–75

Kopin IJ, Blombery P, Ebert MH, Gordon EK, Jimerson DC, Markey SP, Polinsky RJ (1984) Disposition and metabolism of MHPG-CD3 in humans: plasma MHPG as the principal pathway of norepinephrine metabolism and as an important determinant of CSF levels of MHPG. Adv Biochem Psychopharmacol 39: 57–68

Lenders JWM, Eisenhofer G, Abeling NGGM, Berger W, Murphy DL, Konings CH, Wagemakers LMB, Kopin IJ, Karoum F, van Gennip AH, Brunner HG (1996) Specific genetic deficiencies of the A and B isoenzymes of monoamine oxidase are characterized by distinct neurochemical and clinical phenotypes. J Clin Invest 97: 1–10

Lesch K-P, Bengel D, Heils A, Sabol SZ, Greenberg BD, Petri S, Benjamin J, Muller CR, Hamer DH, Murphy DL (1996) Association of anxiety-related traits with a polymorphism in the serotonin transporter gene regulatory region. Science 274: 1527–1531

Murphy DL (1977) The behavioral toxicity of monoamine oxidase-inhibiting antidepressants. Adv Pharmacol Chemother 14: 71–105

Murphy DL, Wright C, Buchsbaum M, Costa J, Nichols A, Wyatt R (1976) Platelet and plasma amine activities in 650 normals: sex and age differences and stability over time. Biochem Med 16: 254–265

Murphy DL, Lipper S, Slater S, Shiling D (1979) Selectivity of clorgyline and pargyline as inhibitors of monoamine oxidases A and B in vivo in man. Psychopharmacology 62: 129–132

Murphy DL, Pickar D, Jimerson D, Cohen RM, Garrick NA, Karoum F, Wyatt RJ (1981) Biochemical indices of the effects of the selective MAO inhibitors clorgyline, pargyline and deprenyl in man. In: Usdin E, Dahl SG, Gram LF, Ligjaerde O (eds) Clinical pharmacology in psychiatry: neuroleptic and antidepressant research. Macmillan, London, pp 307–316

Murphy DL, Sims KB, Karoum F, de la Chapelle A, Norio R, Sankila E-M, Breakefield XO (1990) Marked amine and amine metabolite changes in Norrie disease patients with an X-chromosomal deletion affecting monoamine oxidase. J Neurochem 54: 242–247

Murphy DL, Sims KB, Karoum F, Garrick NA, de la Chapelle A, Sankila EM, Norio R, Breakefield XO (1991) Plasma amine oxidase activities in Norrie disease patients with an X-chromosomal deletion affecting monoamine oxidase. J Neural Transm [Gen Sect] 83: 1–12

Oreland L, Hallman J (1995) The correlation between platelet MAO activity and personality: short review of findings and a discussion on possible mechanisms. Prog Brain Res 106: 77–84

Oxenstierna G, Edman G, Iselius L, Oreland L, Ross SB, Sedvall G (1986) Concentrations of monoamine metabolites in the cerebrospinal fluid of twins and unrelated individuals: a genetic study. J Psychiatr Res 20: 19–29

Plomin R, Owen MJ, McGuffin P (1994) The genetic basis of complex human behaviors. Science 264: 1733–1739

Rabkin JG, Quitin FM, McGrath P, Harrison W, Tricamo E (1985) Adverse reactions to monoamine oxidase inhibitors, part II. Treatment correlates and clinical management. J Clin Psychopharmacol 5: 2–9

Schalling D, Asberg M, Edman G, Oreland L (1987) Markers for vulnerability to psychopathology: temperament traits associated with platelet MAO activity. Acta Psychiatr Scand 76(2): 172–182

Shih JC, Grimsby J, Chen K, Zhu Q (1993) Structure and promoter organization of the human monoamine oxidase A and B genes. J Psychiatry Neurosci 18: 25–32

Sims KB, de la Chapelle A, Norio R, Sankila E-M, Hsu Y-PP, Rinehart WB, Corey TJ, Ozelius L, Powell JF, Bruns G, Gusella JF, Murphy DL, Breakefield XO (1989) Monoamine oxidase deficiency in males with an X chromosome deletion. Neuron 2: 1069–1076

Sunderland T, Mueller EA, Cohen RM, Jimerson DC, Pickar D, Murphy DL (1985) Tyramine pressor sensitivity changes during deprenyl treatment. Psychopharmacology 86: 432–437

Whitaker-Azmitia PM, Zhang X, Clarke C (1994) Effects of gestational exposure to monoamine oxidase inhibitors in rats: preliminary behavioral and neurochemical studies. Neuropsychopharmacology 11: 125–132

Authors' address: Dr. D. L. Murphy, Laboratory of Clinical Science, NIMH, NIH Clinical Center, 10-3D41, 10 CENTER DR MSC 1264, Bethesda, MD 20892-1264, U.S.A.

Differential trace amine alterations in individuals receiving acetylenic inhibitors of MAO-A (clorgyline) or MAO-B (selegiline and pargyline)

D. L. Murphy[1], **F. Karoum**[2], **D. Pickar**[3], **R. M. Cohen**[4], **S. Lipper**[5], **A. M. Mellow**[6], **P. N. Tariot**[7], and **T. Sunderland**[1]

[1] Laboratory of Clinical Science, [3] Experimental Therapeutics Branch, and [4] Laboratory of Cerebral Metabolism, National Institute of Mental Health, Bethesda, MD, [2] Neuroscience Center, St. Elizabeth's Hospital, National Institute of Mental Health, Washington, DC, [5] Department of Psychiatry, Duke University School of Medicine, Durham, NC, [6] Department of Psychiatry, University of Michigan School of Medicine, Ann Arbor, MI, and [7] Department of Psychiatry, University of Rochester Medical Center, Rochester, NY, U.S.A.

Summary. Marked, dose-dependent elevations in the urinary excretion of phenylethylamine, para-tyramine, and meta-tyramine were observed in depressed patients treated for three or more weeks with 10, 30, or 60 mg/ day of the partially-selective inhibitor of MAO-B, selegiline (l-deprenyl). In comparative studies with other, structurally similar acetylenic inhibitors of MAO, pargyline, an MAO-B > MAO-A inhibitor used in doses of 90 mg/day for three or more weeks, produced elevations in these trace amines which were similar to those found with the highest dose of selegiline studied. Clorgyline, a selective inhibitor of MAO-A used in doses of 30 mg/ day for three or more weeks (a dose/time regimen previously reported to reduce urinary, plasma, and cerebrospinal fluid 3-methoxy-4-hydroxyphenylethyleneglycol (MHPG) > 80%, indicating a marked inhibitory effect on MAO-A in humans in vivo) produced negligible changes in trace amine excretion. In comparison to recent studies of individuals lacking the genes for MAO-A, MAO-B, or both MAO-A and MAO-B, the lack of change in trace amine excretion in individuals with a mutation affecting only MAO-A is in agreement with the observed lack of effect of clorgyline in the present study. Selegiline produced larger changes in trace amines — at least at the higher doses studied — than found in individuals lacking the gene for MAO-B, in agreement with other data suggesting a lesser selectivity for MAO-B inhibition when selegiline was given in doses higher than 10 mg/day. Overall, trace amine elevations in individuals receiving the highest dose of deprenyl or receiving pargyline were approximately three to five-fold lower than the elevations observed in individuals lacking the genes for both MAO-A and MAO-B, suggesting that these drug doses yield incomplete inhibition of MAO-A and MAO-B.

Introduction

Drugs which inhibit monoamine oxidase (MAO) affect the metabolism of many biogenic amines and their metabolites. A few MAO inhibitors (MAOIs) preferentially inhibit the deamination of some biogenic amines, with lesser effects on other amines. Clorgyline, for example, inhibits 50% of the deamination of serotonin (5-HT) by human brain cortex homogenates in vitro at concentrations of 0.2 pmol; in contrast, phenylethylamine metabolism remains unaffected by clorgyline until high concentrations (1 μmol) are used (Owen et al., 1979; Fowler et al., 1980). Other propargylamine derivatives such as selegiline and pargyline exhibit reverse selectivity, inhibiting phenylethylamine deamination in human brain in vitro at lower concentrations (selegiline, $IC_{50} = 30$ nmol; pargyline, 40 nmol) than those required to inhibit 5-HT deamination (selegiline, 500 nmol; pargyline, 160 nmol) (Fowler et al., 1982). Substrate selectivity with these drugs is not absolute, and at high MAO inhibitor concentrations in vivo, selectivity is lost (Felner and Waldmeier, 1979; Fowler and Tipton, 1984).

The basis for the preferential inhibition of the deamination of some substrates by these selective inhibitors results from the existence of two closely related isoenzymes, MAO-A and MAO-B, which are encoded by two adjacent genes on the X chromosome (Hsu et al., 1989; Weyler et al., 1990). Among the substrates for MAO-A and MAO-B, serotonin and, to a lesser extent, norepinephrine (NE), are preferentially deaminated by MAO-A, while phenylethylamine, phenylethanolamine, o-tyramine, telemethylhistamine, and benzylamine are preferential substrates for MAO-B (Murphy, 1978; Fowler and Tipton, 1984). Most studies examining the substrate-selective characteristics of different MAO inhibitors have examined changes in MAO-A versus MAO-B activity in vitro or following acute drug administration in vivo in rodents. A smaller number of studies have measured changes in tissue concentrations of monoamines following drug treatment in animals, principally rodents (Celada and Artigas, 1993). Only a few investigations of the chronic administration of these inhibitors have been accomplished, and some but not all of these have suggested a loss in selectivity with continued drug administration (Felner and Waldmeier, 1979; Garrick et al., 1984; Potter et al., 1985; Sunderland et al., 1994).

Only meager information is available concerning the changes in monoamine metabolism produced by selective MAOIs in humans. This paper focuses on some approaches our group has taken to evaluate to what extent the selectivity of several MAOIs is maintained in vivo in humans during longer term drug administration, particularly in reference to the effects on trace amine metabolism produced by the relatively selective inhibitor of MAO-B, selegiline. Several cohorts of depressed patients were treated for a minimum of three weeks with selegiline using three different doses (10, 30 and 60 mg/day) or with pargyline or clorgyline (Lipper et al., 1979; Murphy et al., 1979, 1987; Potter et al., 1982; Sunderland et al., 1985, 1994). Clinical data from these studies indicated that clorgyline had significant antidepressant and anti-anxiety effects (Lipper et al., 1979; Linnoila et al., 1982; Potter et al., 1982;

Murphy et al., 1983). Pargyline had minimal therapeutic effects, but, like clorgyline, produced quite marked orthostatic hypotension (Lipper et al., 1979; Murphy et al., 1979). Selegiline, at 10 mg/day, provided negligible antidepressant effects, while at higher doses it was an effective antidepressant (Sunderland et al., 1985, 1994; Mann et al., 1989).

Subjects and methods

Individuals hospitalized for depression from the studies reviewed (Lipper et al., 1979; Murphy et al., 1979, 1987; Potter et al., 1982; Sunderland et al., 1994) above were treated with the three acetylenic MAO inhibitors, selegiline, 10 mg/day (N = 12), 30 mg/day (N = 9), and 60 mg/day (N = 14); pargyline, 90 mg/day (N = 12); or clorgyline, 30 mg/day (N = 9), for three or more weeks. For most of the patients, the highest dose of selegiline and the doses of pargyline and clorgyline used were the maximally tolerated doses, with orthostatic hypotension representing the most common dose-limiting side effect.

Subjects with X-chromosomal deletions associated with a lack of the genes for MAO-A and MAO-B who also had Norrie disease were two brothers and two cousins from one family described in previous reports (de la Chapelle et al., 1985; Sims et al., 1989; Murphy et al., 1991). Urinary trace amine excretion for two of these individuals have been previously reported (Murphy et al., 1990).

Duplicate 24 hr urine samples were collected under careful supervision on two separate days. The subjects were receiving an institutional diet containing no caffeine. All samples were stored at $-70°C$ until assayed. Urine samples were assayed by mass fragmentography using methodology summarized previously (Karoum et al., 1979, 1985; Karoum and Neff, 1982) and used in earlier and ongoing studies of the effects of MAO inhibitors on urinary, cerebrospinal fluid, and plasma amines and their metabolites (Karoum et al., 1980; Murphy et al., 1981, 1984a,b, 1990, 1991; Karoum and Neff, 1982; Linnoila et al., 1982; Zametkin et al., 1985). For the data analysis, the mean results from the two sample days collected prior to treatment and again during treatment were compared using two-tailed t-tests.

Results

Pretreatment urinary excretion of the three trace amines investigated in the depressed patients, phenylethylamine, m-tyramine, and p-tyramine, were similar to those reported originally in healthy volunteers (Karoum et al., 1979) and in subsequent studies conducted by our group using mass fragmentographic methodology (Karoum et al., 1980; Murphy et al., 1981, 1984b; 1990; Karoum et al., 1985) (Tables 1–3).

As indicated in Table 1, even the lowest dose of selegiline, 10 mg/day, led to a marked, ten-fold increase in phenylethylamine excretion. The two higher doses of selegiline produced dose-proportionate increases in phenylethylamine excretion representing approximately 50-fold (at 30 mg/day) and 100-fold (at 60 mg/day) changes in urinary phenylethylamine. Pargyline also produced marked, 80-fold increases in phenylethylamine excretion. In contrast, clorgyline treatment was associated with essentially no change in urinary phenylethylamine excretion (Table 1).

D. L. Murphy et al.

Table 1. Phenylethylamine excretion

1A. Phenylethylamine excretion prior to and during MAO-
inhibitor administration

Drug	N	Dose (mg/d)	Phenylethylamine (µg/24h)
None	44	—	8 ± 1
Selegiline	12	10	126 ± 28***
	9	30	479 ± 91***
	14	60	812 ± 124***
Pargyline	12	90	680 ± 162***
Clorgyline	9	30	14 ± 4

***p < 0.001, two tailed paired t-test of pretreatment
versus MAO-inhibitor treatment periods

1B. Phenylethylamine excretion in MAO-A/B
deficient subjects with Norrie disease produced by an X-
chromosomal deletion. (data adapted from Murphy et al.,
1990; Lenders et al., 1996)

Subjects	N	Phenylethylamine (µg/24h)
Age-matched controls	9	2 ± 1
MAO-A/B deficient subjects	4	4,518 ± 4,068***· ++

***p < 0.001, Student's t-test of MAO-A/B deficient
subject compared to controls. ++p < 0.01, Student's t-test of
MAO-A/B deficient subjects compared to selegiline (60 mg/
d)-treated subjects

Dose-dependent increases in m-tyramine also followed selegiline administration, with two-fold (10 mg/day), three-fold (30 mg/day), and five-fold
(60 mg/day) changes in m-tyramine observed (Table 2). Pargyline also produced approximately three-fold increases in m-tyramine, while clorgyline had
no effect whatsoever on urinary m-tyramine excretion (Table 2).

A generally similar pattern of changes in urinary p-tyramine was observed, with two-fold increases produced by 10 mg/day selegiline administration. The higher selegiline doses produced three- to four-fold increases,
although doubling the dose from 30 to 60 mg/day produced only a modest
further mean elevation in p-tyramine excretion (Table 3). Pargyline (90 mg/
day) produced a two-fold increase in p-tyramine excretion. In contrast,
clorgyline (30 mg/day) did not affect p-tyramine excretion (Table 3). Overall,
the magnitude of increases produced by selegiline and pargyline on
phenylethylamine excretion were far greater (15 to 100-fold) than the two- to
five-fold increases in m-tyramine and p-tyramine excretion.

The data from the subjects treated with the three partially-selective
MAO-inhibitors were compared to trace amine changes found in individuals

Table 2. m-Tyramine excretion

1A. m-Tyramine excretion prior to and during MAO-inhibitor administration

Drug	N	Dose (mg/d)	M-Tyramine (μg/24h)
Pretreatment	None	42	64 ± 16
Selegiline	12	10	126 ± 33
	9	30	205 ± 49**
	14	60	328 ± 27***
Pargyline	12	90	214 ± 33**
Clorgyline	9	30	56 ± 19

**p < 0.01, ***p < 0.001, two tailed paired t-test of pretreatment versus MAO-inhibitor treatment periods

1B. m-Tyramine excretion in MAO-A/B deficient subjects with Norrie disease produced by an X-chromosomal deletion (data adapted from Murphy et al., 1990; Lenders et al., 1996)

Subjects	N	M-Tyramine (μ/24h)
Age-matched controls	9	30 ± 17
MAO-A/B deficient subjects	4	1,407 ± 1,610***,++

***p < 0.001, two tailed t-test of pretreatment versus MAO-inhibitor treatment periods. ++p < 0.01, Student's t-test of MAO-A/B deficient subjects compared to selegiline (60 mg/d)-treated subjects

with an Xp chromosomal deletion associated with complete lack of the genes for both MAO-A and MAO-B. As indicated in Tables 1–3, the MAO-deficient individuals manifested markedly higher urinary trace amine excretion than that found in patients treated for three or more weeks with the MAO-inhibitors. For phenylethylamine, the increase was more than 700-fold over baseline, and thus more than five-fold higher than that found with the highest dose of selegiline. For m-tyramine, the increase was more than 20-fold over baseline, and thus more than four-fold higher than that found with the highest dose of selegiline. For p-tyramine, the increase was more than ten-fold over baseline, and thus more than three-fold higher than that found with the highest dose of selegiline.

Discussion

The metabolism of the trace amine phenylethylamine, as well as m-tyramine and p-tyramine, are well-known to be affected by MAO inhibition in brain

Table 3. p-Tyramine excretion

1A. p-Tyramine excretion prior to and during MAO-inhibitor administration

Drug	N	Dose (mg/d)	P-Tyramine (μg/24h)
Pretreatment	41		905 ± 103**
Selegiline	12	10	790 ± 128*
	9	30	1,287 ± 194***
	14	60	1,426 ± 216***
Pargyline	12	90	436 ± 72
Clorgyline	9	30	453 ± 46

*p < 0.05, **p < 0.01, ***p < 0.001, two tailed paired t-test of pretreatment versus MAO-inhibitor treatment periods

1B. p-Tyramine in MAO-A/B deficient subjects with Norrie disease produced by an X-chromosomal deletion. (data adapted from Murphy et al., 1990; Lenders et al., 1996)

Subjects	N	P-Tyramine (μg/24h)
Age-matched controls	9	511 ± 187
MAO-A/B deficient subjects	4	4,777 ± 3,703***,++

***p < 0.001, two tailed t-test of pretreatment versus MAO-inhibitor treatment periods. ++p < 0.01, Student's t-test of MAO-A/B deficient subjects compared to selegiline (60 mg/d)-treated subjects

and other tissues and to lead to changes in the concentrations of these amines in cerebrospinal fluid, plasma, and urine (Boulton, 1978; Karoum et al., 1980; Murphy et al., 1984b,c; Davis and Boulton, 1994). Prior studies have reported increased phenylethylamine and tyramine after low doses of selegiline (Elsworth et al., 1978; Murphy et al., 1984b,c; Quitkin et al., 1984; Liebowitz et al., 1985).

The present study is the first to investigate the effects of a six-fold range of selegiline doses on trace amine excretion, and to compare the effects of selegiline with those of two structurally-similar acetylenic MAO-inhibitors, pargyline and clorgyline. In contrast to selegiline's selective effects on MAO-B, and pargyline's somewhat selective effects on MAO-B with some effects on MAO-A, clorgyline has been shown to be a highly selective inhibitor of MAO-A in studies in rodents, non-human primates, and, by indirect measures, in humans (Johnston, 1968; Murphy, 1978; Felner and Waldmeier, 1979; Major et al., 1979; Murphy et al., 1979, 1981; Garrick et al., 1984).

The lack of any effect of clorgyline on the urinary excretion of phenyl-ethylamine or tyramine provides confirmatory evidence that clorgyline maintains selective effects on MAO-A, with negligible effects on MAO-B, during chronic administration to humans at the dose used in this study. Our previous studies have demonstrated that this dose of clorgyline did not alter MAO-B activity, but did produce marked reductions in norepinephrine deamination, as reflected in decreases in MHPG concentration of >80% in CSF, plasma and urine (Major et al., 1979; Murphy et al., 1979). In these studies, reductions in the CSF concentrations of the metabolites of serotonin (5-HIAA, 50% decreased) and dopamine (HVA, 30% decreased) were less marked (Major et al., 1979). In another study which employed lower doses of clorgyline, a 70% reduction in urinary MHPG was accompanied by smaller reductions in HVA (40%) and 5-HIAA (30%), and again no change in urinary phenylethylamine excretion (Linnoila et al., 1982). The lack of effect of clorgyline on the urinary excretion of these three trace amines is also in agreement with the lack of change in their excretion in individuals with a mutation producing a complete lack of MAO-A activity without affecting MAO-B (Lenders et al., 1996).

The effects of pargyline on trace amine excretion were expected on the basis of its potent MAO-B inhibitory effects in vitro as well as in rodents and other species given pargyline in vivo (Fuentes and Neff, 1975; Lipper et al., 1979; Murphy et al., 1979). Another MAO inhibitor with approximately equal inhibitory effects on MAO-A and MAO-B in vitro, phenelzine, given in doses of 30–90 mg/day, also produced ten-fold increases in phenylethylamine excretion and three-fold increases in m-tyramine and p-tyramine excretion (McKenna et al., 1993).

The effects of selegiline were also expected on the basis of its MAO-B inhibitory potency and prior studies reporting increased phenylethylamine excretion following administration of low doses of selegiline (Elsworth et al., 1978; Eisler et al., 1981; Karoum and Wyatt, 1982; Liebowitz et al., 1985). Selegiline at a dose of 10 mg/day led to ten-fold, significant increase in urinary phenylethylamine and a two-fold increase in p-tyramine in a prior study using the same methodology (Karoum and Wyatt, 1982). It is of interest that the approximate ten-fold increase in phenylethylamine excretion found with the lowest, most MAO-B selective dose of selegiline in this study is essentially equivalent to the approximate ten-fold increases in phenylethylamine excretion found in our investigations of individuals with a selective genetic deficit in MAO-B and normal MAO-A function (Lenders et al., 1996).

The dose-related, greater increases in phenylethylamine and tyramine excretion found with higher doses of selegiline most likely represented the additive effects of increasing inhibition of MAO-A combined with essentially complete inhibition of MAO-B. Felner and Waldmeier (1979) reported that with higher doses and also with longer duration of systemic administration of selegiline in rodents, selegiline's selective effects as an MAO-B inhibitor were lost, and increasing inhibition of MAO-A occurred, confirming earlier in vitro findings (Fuentes and Neff, 1975). Similar evidence from studies in humans indicated that higher doses and a longer duration of treatment with selegiline led to a loss of MAO-B selective effects and a significant inhibition of MAO-A (Sunderland et al., 1985, 1994). The more marked increases in phenylethylamine and tyramine excretion found in individuals lacking both MAO-A and MAO-B on a genetic basis similarly suggests that MAO-A also contributes to the deamination of these trace amines when MAO-B activity is compromised (Murphy et al., 1990; Lenders et al., 1996).

References

Boulton AA (1978) The tyramines: functionally significant biogenic amines or metabolic accident. Life Sci 23: 659–672

Celada P, Artigas F (1993) Monoamine oxidase inhibitors increase preferentially extracellular 5-hydroxytryptamine in the midbrain raphe nuclei. A brain microdialysis study in the awake rat. Naunyn Schmiedebergs Arch Pharmacol 347: 583–590

Davis BA, Boulton AA (1994) The trace amines and their acidic metabolites in depression — an overview. Prog Neuropsychopharmacol Biol Psychiatry 18: 17–45

de la Chapelle A, Sankila E-M, Lindlof M, Aula P, Norio R (1985) Norrie disease caused by a gene deletion allowing carrier detection and prenatal diagnosis. Clin Genet 28: 317–320

Eisler T, Teravainen H, Nelson R, Krebs H, Weise V, Lake CR, Ebert MH, Whetzel N, Murphy DL, Kopin IJ, Calne DB (1981) Deprenyl in Parkinson disease. Neurology 31: 19–23

Elsworth JD, Glover V, Reynolds GP, Sandler M, Lees AJ, Puapradit P, Shaw KM, Stern KM, Kumar P (1978) Deprenyl administration in man: a selective monoamine oxidase B inhibitor without the "cheese effect". Psychopharmacology 57: 33–38

Felner AE, Waldmeier PC (1979) Cumulative effects of irreversible MAO inhibitors in vivo. Biochem Pharmacol 28: 995–1002

Fowler CJ, Tipton KF (1984) On the substrate specificities of the two forms of monoamine oxidase. J Pharm Pharmacol 36: 111–115

Fowler CJ, Oreland L, Marcusson J, Winblad B (1980) Titration of human brain monoamine oxidase -A and -B by clorgyline and L-deprenyl. Naunyn Schmiedebergs Arch Pharmacol 311: 263–272

Fowler CJ, Mantle TJ, Tipton KF (1982) The nature of the inhibition of rat liver monoamine oxidase types A and B by the acetylenic inhibitors clorgyline, 1-deprenyl and pargyline. Biochem Pharmacol 31: 3555–3561

Fuentes JA, Neff NH (1975) Selective monoamine oxidase inhibitor drugs as aids in evaluating the role of type A and B enzymes. Neuropharmacology 14: 819–825

Garrick NA, Scheinin M, Chang WH, Linnoila M, Murphy DL (1984) Differential effects of clorgyline on catecholamine and indoleamine metabolites in the cerebrospinal fluid of rhesus monkeys. Biochem Pharmacol 33: 1423–1427

Hsu YP, Powell JF, Sims KB, Breakefield XO (1989) Molecular genetics of the monoamine oxidases. J Neurochem 53: 12–18

Johnston JP (1968) Some observations upon a new inhibitor of monoamine oxidase in brain tissue. Biochem Pharmacol 17: 1285–1297

Karoum DF, Wyatt RJ (1982) Metabolism of (−)deprenyl's therapeutic benefit: a biochemical assessment. Neurology 32: 503–509

Karoum F, Neff NH (1982) Quantitative gas chromatography-mass spectrometry (GC-MS) of biogenic amines. In: Spector S, Back N (eds) Theory and practice. Alan R Liss, New York, pp 39–54 (Modern Methods in Pharmacology)

Karoum F, Nasrallah H, Potkin S, Chuang L, Moyer-Schwing J, Phillips I, Wyatt RJ (1979) Mass fragmentography of phenylethylamine m- and p-tyramine, and related amines in plasma cerebrospinal fluid, urine, and brain. J Neurochem 33: 201–212

Karoum F, Potkin SG, Murphy DL, Wyatt RJ (1980) Quantiation and metabolism of phenylethylamine and tyramine's three isomers in humnas. In: Mosnaim AD, Wolf ME (eds) Noncatecholic phenylethylamines. Marcel Dekker, New York, pp 177–191

Karoum F, Torrey EF, Murphy DL, Wyatt RJ (1985) The origin, drug interaction, urine, plasma and CSF concentrations of phenylacetic acid in normal and psychiatric subjects. In: Boulton AA, Baker GB, Dewhurst WG, Sandler M (eds) Neurobiology of the trace amine. Humana Press, Clifton NJ, pp 457–473

Lenders JWM, Eisenhofer G, Abeling NGGM, Berger W, Murphy DL, Konings CH, Wagemarkers LMB, Kopin IJ, Karoum F, van Gennip AH, Brunner HG (1996) Specific genetic deficiencies of the A and B isoenzymes of monoamine oxidase are characterized by distinct neurochemical and clinical phenotypes. J Clin Invest 97: 1–10

Liebowitz MR, Karoum Q FM, Davies SO, Schwartz D, Levitt M, Linnoila M (1985) Biochemical effects of L-deprenyl in atypical depressives. Biol Psychiatry 20: 558–565

Linnoila M, Karoum F, Potter WZ (1982) Effect of low-dose clorgyline on 24-hour urinary monoamine excretion in patients with rapidly cycling bipolar affective disorder. Arch Gen Psychiatry 39: 513–516

Lipper S, Murphy DL, Slater S, Buchsbaum MS (1979) Comparative behavioral effects of clorgyline and pargyline in man: a preliminary evaluation. Psychopharmacology 62: 123–128

Major LF, Murphy DL, Lipper S, Gordon E (1979) Effects of clorgyline and pargyline on deaminated metabolites of norepinephrine, dopamine and serotonin in human cerebrospinal fluid. J Neurochem 32: 229–231

Mann JJ, Aarons SF, Wilner PJ, Keilp JG, Sweeney JA, Pearlstein T, Frances AJ, Kocsis JH, Brown RP (1989) A controlled study of the antidepressant efficacy and side effects of (−)-deprenyl: a selective monoamine oxidase inhibitor. Arch Gen Psychiatry 46: 45–50

McKenna KF, Baker GB, Coutts RT (1993) Urinary excretion of bioactive amines and their metabolites in psychiatric patients receiving phenelzine. Neurochem Res 18: 1023–1027

Murphy DL (1978) Substrate-selective monoamine oxidases: inhibitor, tissue, species and functional differences. Biochem Pharmacol 27: 1889–1893

Murphy DL, Lipper S, Slater S, Shiling D (1979) Selectivity of clorgyline and pargyline as inhibitors of monoamine oxidases A and B in vivo in man. Psychopharmacology 62: 129–132

Murphy DL, Pickar D, Jimerson D, Cohen RM, Garrick NA, Karoum F, Wyatt RJ (1981) Biochemical indices of the effects of the selective MAO inhibitors clorgyline, pargyline and deprenyl in man. In: Usdin E, Dahl SG, Gram LF, Lingjaerde O (eds) Clinical pharmacology in psychiatry: neuroleptic and antidepressant research. Macmillan, London, pp 307–316

Murphy DL, Garrick NA, Cohen RM (1983) Monoamine oxidase inhibitors and monoamine oxidase: biochemical and physiological aspects relevant to human psychopharmacology. In: Burrows JD, Norman TR, Davies E (eds) Drugs in psychiatry. Antidepressants. Elsevier Press, Amsterdam, pp 209–227

Murphy DL, Garrick NA, Aulakh CS, Cohen RM (1984a) New contributions from basic science to understanding the effects of monoamine oxidase inhibiting antidepressants. J Clin Psychiatry 45: 37–43

Murphy DL, Karoum F, Alterman I, Lipper S, Wyatt RJ (1984b) Phenylethylamine, tyramine and other trace amines in patients with affective disorders: associations with clinical state and antidepressant drug treatment. In: Boulton AA, Baker GB, Dewhurst WG, Sandler M (eds) Neurobiology of the trace amines. Humana Press, Clifton NJ, pp 449–514

Murphy DT, Sunderland T, Cohen RM (1984c) Monoamine oxidase-inhibiting antidepressants: a clinical update. Psychiatr Clin North Am 7: 549–562

Murphy DL, Aulakh CS, Garrick NA, Sunderland T (1987) Monoamine oxidase inhibitors as antidepressants: implications for the mechanism of action of antidepressants and the psychobiology of the affective disorders and some related disorders. In: Meltzer HY (ed) Psychopharmacology: the third generation of progress. Raven Press, New York, pp 545–552

Murphy DL, Sims KB, Karoum F, de la Chapelle A, Norio R, Sankila E-M, Breakefield XO (1990) Marked amine and amine metabolite changes in Norrie disease patients with an X-chromosomal deletion affecting monoamine oxidase. J Neurochem 54: 242–247

Murphy DL, Sims KB, Karoum F, Garrick NA, de la Chapelle A, Sankila EM, Norio R, Breakefield XO (1991) Plasma amine oxidase activities in Norrie disease patients with an X-chromosomal deletion affecting monoamine oxidase. J Neural Transm [Gen Sect] 83: 1–12

Owen F, Cross AJ, Lofthouse R, Glover V (1979) Distribution and inhibition characteristics of human brain monoamine oxidase. Biochem Pharmacol 28: 1077–1080

Potter WZ, Murphy DL, Wehr TA, Linnoila M, Goodwin FK (1982) Clorgyline. A new treatment for patients with refractory rapid-cycling disorder. Arch Gen Psychiatry 39: 505–510

Potter WZ, Scheinin M, Golden RN, Rudorfer MV, Cowdry RW, Calil HM, Ross RJ, Linnoila M (1985) Selective antidepressants and cerebrospinal fluid. Arch Gen Psychiatry 42: 1171–1177

Quitkin FM, Liebowitz MR, Stewart JW, McGrath PJ, Harrison W, Rabkin JG, Markowitz J, Davies SO (1984) L-Deprenyl in atypical depressives. Arch Gen Psychiatry 41: 777–781

Sims KB, de la Chapelle A, Norio R, Sankila E-M, Hsu Y-PP, Rinehart WB, Corey TJ, Ozelius L, Powell JF, Bruns G, Gusella JF, Murphy DL, Breakefield XO (1989) Monoamine oxidase deficiency in males with an X chromosome deletion. Neuron 2: 1069–1076

Sunderland T, Mueller EA, Cohen RM, Jimerson DC, Pickar D, Murphy DL (1985) Tyramine pressor sensitivity changes during deprenyl treatment. Psychopharmacology 86: 432–437

Sunderland T, Cohen RM, Molchan S, Lawlor BA, Mellow AM, Newhouse PA, Tariot PN, Mueller EA, Murphy DL (1994) High dose selegiline in treatment-resistant older depressives. Arch Gen Psychiatry 51: 607–615

Weyler W, Hsu Y-P, Breakefield XO (1990) Biochemistry and genetics of monoamine oxidase. Pharmacol Ther 47: 391–417

Zametkin A, Rapoport JL, Murphy DL, Linnoila M, Karoum F, Potter WZ, Ismond D (1985) Treatment of hyperactive children with monoamine oxidase inhibitors. II. Plasma and urinary monoamine findings after treatment. Arch Gen Psychiatry 42: 969–973

Authors' address: Dr. D. L. Murphy, Laboratory of Clinical Science, NIMH, NIH Clinical Center, 10-3D41, 10 CENTER DR MSC 1264, Bethesda, MD 20892-1264, U.S.A.

Experience with tranylcypromine in early Parkinson's disease

S. Fahn[1] and **S. Chouinard**[2]

[1]Department of Neurology, Columbia University College of Physicians & Surgeons, and [2]The Neurological Institute of New York, Presbyterian Hospital, New York, NY, U.S.A.

Summary.   A leading hypothesis of the pathogenesis of neuronal degeneration of the substantia nigra dopamine-containing cells in Parkinson's disease (PD) is excessive oxidative stress. In part, this oxidative stress is the result of the oxidation of dopamine by the action of monoamine oxidases (MAO) A and B to generate hydrogen peroxide and subsequent oxygen free radicals. Because of this hypothesis we have treated patients with early PD, not yet requiring any symptomatic treatment, with tranylcypromine, a drug that inhibits both MAO's. These patients were required to observe a tyramine-restricted diet. Thirty-seven patients on tranylcypromine have been followed by us for up to 33 months. Four patients discontinued the drug because of pending surgery. Of the remaining 33, six had adverse effects that lead to discontinuation of the drug, mainly impotency in men. Another common adverse effect encountered was insomnia, but this problem was not a cause of stopping the drug. Depression lifted in all five patients who had this problem at the time tranylcypromine was initiated. Only two patients have so far required treatment with levodopa or a dopamine agonist, and this need occurred within the first 6 months of treatment. The evaluation of all 37 patients revealed that parkinsonian symptoms improved slightly on introduction of tranylcypromine as measured by the Unified Parkinson's Disease Rating Scale, the Hoehn & Yahr Staging Scale, and the Schwab & England Activities of Daily Living Scale. Follow-up evaluations for a minimum of 6 months between the first post-tranylcypromine visit and the most recent visit revealed only slight worsening of parkinsonian signs and symptoms, with a mean interval of almost 1.5 years. A longer period of follow-up is needed to determine how long the severity of PD will remain mild in this group of patients.

Introduction

Parkinson's disease (PD) is a progressively disabling neurodegenerative disorder manifested clinically by bradykinesia, tremor, rigidity, flexed posture, postural instability, and the freezing phenomenon. It is characterized pathologically by loss of pigmented neurons in the brainstem, particularly dopaminergic neurons in the substantia nigra pars compacta and noradrenergic neurons in the locus ceruleus. The neuronal degeneration is accompanied by the presence of intracytoplasmic eosinophilic inclusions known as Lewy bodies and by gliosis. The disease begins insidiously and steadily worsens. Slowness of thought (bradyphrenia), intellectual decline (dementia), affective illness (depression), loss of assertive drive, and autonomic instability are non-motor manifestations of the disease. Clinical features begin to emerge when approximately 80% of striatal dopamine content (or 60% of nigral dopamin-ergic neurons) are lost (Bernheimer et al., 1973). The course of clinical decline parallels the progressive degeneration of remaining neurons (Riederer and Wuketich, 1976). Neurochemical and pharmacologic studies implicate striatal dopamine deficiency as the basis of most of the motor features of PD (Fahn and Duffy, 1977; Hornykiewicz, 1982). This deficiency is the result of loss of striatal nerve terminals arising from the loss of mesencephalic dopaminergic neurons.

The cause of this neuronal loss in PD is unknown, but in recent years oxidant stress has come to the fore as a likely mechanism (Graham et al., 1978; Cohen, 1983, 1986; Fahn, 1989; Fornstedt et al., 1990; Olanow, 1990, 1992; Jenner, 1991; Jenner et al., 1992a,b; Fahn and Cohen, 1992; Zigmond et al., 1992; Spencer et al., 1995). The oxidant stress hypothesis as a mechanism in the cause of PD is based on the formation of oxyradicals produced in part by enzymatic oxidation of dopamine to form hydrogen peroxide via the action of monoamine oxidase (MAO). Hydrogen peroxide, in turn, can lead to the formation of oxyradicals unless hydrogen peroxide is quickly eliminated by the action of glutathione peroxidase.

It was this hypothesis that led to the DATATOP and similar studies, testing whether selegiline could protect against progression of PD. This study showed that treatment of patients with early PD with selegiline, a monoamine oxidase (MAO) type B inhibitor, could delay the introduction of levodopa by about 9 months (Parkinson Study Group, 1989a, 1993). Although the initial report of the DATATOP study (Parkinson Study Group, 1989a) suggested that selegiline could have both a protective and a symptomatic effect, newer data from the DATATOP study indicate that its symptomatic benefit can explain all the findings and, in fact, suggest a lack of protective effect from selegiline. First, selegiline has a long duration of mild symptomatic benefit, lasting at least 2 months after stopping the drug (Parkinson Study Group, 1996a). Second, when patients on selegiline or placebo were withdrawn from their assigned drugs for a 2 month duration and then both groups placed on selegiline, the group previously treated with placebo had a greater symptom-atic response (Parkinson Study Group, 1996a). If selegiline were protective, that group previously treated with selegiline should have continued to show less symptoms of PD than the group previously treated with placebo. This study also revealed that the symptomatic effect from selegiline seemed stron-gest when the patient is first exposed to the drug than after a long exposure.

Third, after levodopa is introduced, the presence of selegiline offered no benefit in the prevention of adverse effects, such as dyskinesias and the wearing-off phenomenon (Parkinson Study Group, 1996b). Because these adverse effects may be related to severity of disease, this finding suggests that selegiline did not retard disease severity. Finally, the analysis of the concentration of dopamine's metabolite, homovanillic acid (HVA), in cerebrospinal fluid (CSF) revealed no difference between selegiline and non-selegiline treated subjects after correcting for the prolonged MAO inhibition effect (Parkinson Study Group, 1995). If selegiline were protective, the CSF concentration of HVA in the selegiline-treated group should have been greater than that in the placebo-treated group. Thus, there is considerable evidence that selegiline has not had any measurable protective effect against PD.

The measurements of HVA concentration in CSF in the DATATOP study also revealed that selegiline, 10 mg/day, reduced the concentration of HVA by only 20% when the CSF was obtained the day selegiline treatment was stopped, i.e. interval = 0 days (Parkinson Study Group, 1995). This result indicates that HVA is being formed from dopamine in these patients by another mechanism other than through MAO type B. One possibility is that some of the HVA could have been formed non-enzymatically through auto-oxidation. Another explanation is that HVA was generated by MAO type A, which is not inhibited by selegiline 10 mg/day.

Whether PD progressed despite selegiline treatment because the oxidant stress hypothesis is not valid or because MAO type B inhibition alone is inadequate is uncertain. To provide possible benefit by inhibition of both MAO types A and B, we offered patients with mild PD not requiring symptomatic therapy the choice to be treated with tranylcypromine. By carefully following such patients we report on the tolerance and the symptomatic benefit of the drug. With longer follow-up, we could possibly determine if inhibition of both MAO A and B could be of value in slowing the progression of PD.

Methods

Beginning in July 1993, all patients seen by one of us (SF) with early PD not yet requiring levodopa or any other symptomatic drug therapy were explained the rationale of taking an inhibitor of both types A and B MAO instead of selegiline, which is a selective inhibitor of MAO B. Of 51 such patients, eight did not agree to take tranylcypromine because of the requirement of a tyramine-restricted diet. Of the 43 patients who agreed to take tranylcypromine, six decided to be followed by another neurologist after their initial consultation. No follow-up data are available on five of these patients, nor is it known if tranylcypromine was actually taken by these five individuals. The 37 patients who continued to be followed by us are the subject of this report (Table 1). Of these 37 patients, 19 had been taking selegiline, which was discontinued when tranylcypromine was started. Of the 37 patients, tranylcypromine was discontinued in four by their internist or surgeon because of the patients' need for a surgical procedure. Another patient had discontinued tranylcypromine just prior to open heart surgery but resumed the drug afterwards. This left a group of 33 patients which we also analyze for the development of the need for symptomatic therapy.

Tranylcypromine was initiated at a dose of 10 mg per day. Over the next 2 weeks the daily dose was increased by 10 mg weekly, until 10 mg t.i.d. was reached, which remained the recommended maintenance dose. As with previously treated patients (Fahn, 1992), they were also offered alpha-tocopherol 3,200 IU/day and ascorbate 3,000 mg/day. Patients were followed at 6-month intervals. At each evaluation, blood pressure, pulse rate, assessment for severity of PD using the Unified Parkinson's Disease Rating Scale (UPDRS) (Fahn and Elton, 1987), the Schwab & England Activities of Daily Living Scale

Table 1. Patients who were recommended to take tranylcypromine and their current status

	N
Total number of patients	51
Patients who agreed to take tranylcypromine	43
Patients who continued to be followed by SF	37
Patients not discontinued for medical reasons	33
Patients not terminated for adverse effects	27
Patients not terminated because of the need of levodopa therapy	26
Patients currently taking tranylcypromine for a period > 6 months	17
Patients currently taking tranylcypromine for a period > 6 months after the first post-tranylcypromine visit	17

(Schwab and England, 1969) and Hoehn & Yahr Staging (Hoehn and Yahr, 1967), and queries about adverse effects were made.

If a patient desired or needed to be placed on a dopamine agonist or on levodopa because of development of disability from PD, this was noted. Discontinuation of tranylcypromine because of adverse effects was also recorded. Patients were allowed to take anticholinergics and amantadine, but only 2 received either of those drugs.

Results

Table 2 lists the demographics of the 37 patients who were placed on tranylcypromine and followed by us. There were 29 men and 8 women. The mean age at onset ($\pm$S.D.) was 57.3 ± 9.7 years. The mean $\pm$ S.D. duration of PD before receiving tranylcypromine was 20.6 ± 37 months, with a range of 3 to 108 months. The mean $\pm$ S.D. Hoehn & Yahr stage was 1.4 ± 0.5, and the mean $\pm$ S.D. Schwab & England ADL score was 92.8 ± 6.3. Twenty-six patients desired to take alpha-tocopherol and ascorbate, at the dosages listed above. All patients adhered to the tyramine-restricted diet while on the medication. If patients experienced any adverse effects, the dose was reduced, which often helped. The maintenance dosage was 10 mg/day in one patient, 20 mg/day in two, 30 mg/day in 29, and 40 mg/day in one. Of the 26 patients who remained on tranylcypromine at the time of writing this report (Table 1), the mean $\pm$ SD duration on the drug was 17.8 ± 10.8 months, with a median of 19 months, and a range of 2 to 33 months.

Table 3 lists the adverse effects experienced by the group of 37 patients. Insomnia was the most common, and occurred in nine patients. Another two had previously existing insomnia which persisted on tranylcypromine. Eight men experienced some degree of decreased potency or decreased libido. One man noted that decreased potency was related to the timing of the medication. He was unable to sustain an erection within an hour after taking a dose, but by 4 hours after a dose, potency was restored. Other adverse effects were

Table 2. Demographics of the 37 patients followed on tranylcypromine

Dates tranylcypromine was started: July 1993 to April 1996

Men = 29; Women = 8

Previously taking selegiline: 19

Age at onset of Parkinson's disease: 57.3 ± 9.7 years
                                  range: 35–80 years

Age at start of tranylcypromine: 59.9 ± 9.7
                                  range: 38–80 years

Duration of Parkinson's disease before starting tranylcypromine: 20.6 ± 37 months
                        range: 3–108 months

Hoehn & Yahr Stage when tranylcypromine was started

	N
Stage 1	21
1.5	4
2	12
mean ± SD:	1.4 ± 0.5

Schwab & England Score when tranylcypromine was started

	N
100	10
95	11
90	10
85	2
80	4
mean ± SD:	92.8 ± 6.3

Also receiving alpha-tocopherol and ascorbate: 26

Table 3. Adverse effects experienced

	N
Insomnia	9
Continuation of existing insomnia	2
Decreased potency (in men)	8
Foggy-headed	2
Skin rash	2
Nausea	2
Outburst of anger	1
Disinhibited	1
HA/tight chest after Asian food	1
Blurry vision, temporary	1
Sedation	1
Fatigue	1

Table 4. Terminations because of adverse effects

	N	Duration (months)
Decreased libido	3	13, 6, 1
Fatigue, unsteady	1	0
Rash, nausea	1	3
Outburst of anger	1	8
	6	

Table 5. UPDRS, Schwab & England, and Hoehn & Yahr scores (N = 29)

Feature	At base line			At time of first visit after starting tranylcypromine		
	Mean	S.D.	Range	Mean	S.D.	Range
UPDRS Behavioral	1.1	1.5	0–5	0.6	1.1	0–3.5
UPDRS ADL	5.2	2.7	1.5–11	4.9	3.1	2–11
UPDRS Motor	12.3	6.8	4.5–34	12.4	9.0	2–44
UPDRS Total	18.5	8.7	8–43.5	17.2	11.1	6–53.5
Schwab & England	94.5	4.9	80–100	94.0	4.9	85–100
Hoehn & Yahr	1.3	0.4	1–2	1.4	0.5	1–2

uncommon, but one patient had a burst of disinhibited behavior which led to his stopping the drug. None of the patients experienced any meaningful change in blood pressure measured during office visits. But one patient had an experience of feeling tightness in the chest accompanied by a headache approximately an hour after eating East Asian food. The symptoms cleared within a hour, and the patient never saw a physician for it. Of note is that all five patients who had complained of depression before being placed on tranylcypromine reported relief of depression after taking the drug.

There were six patients who stopped taking tranylcypromine because of adverse effects (Table 4). All were men, and decreased libido was the most common problem (N = 3); these three men were able to remain on tranylcypromine for variable periods of time before being unable to tolerate this adverse effect and stopped the drug (Table 4). The other three patients who stopped the drug because of adverse effects were for unrelated reasons. One patient complained of fatigue and unsteadiness shortly after starting tranylcypromine. One developed a rash and nausea 3 months after taking the drug. One had an outburst of disinhibited behavior 8 months after being on the drug.

We evaluated the wash-in symptomatic effect of tranylcypromine. We have data on the 29 patients (out of the group of 37) who have had an

Table 6. Summary of changes in PD ratings, comparing pre and post tranylcypromine (first visit) scores (N = 29)

Rating	Worse	Better	No change
UPDRS Behavioral	3	9	17
UPDRS ADL	10	14	5
UPDRS Motor	10	17	2
UPDRS Total	10	19	0
Schwab & England	5	3	21
Hoehn & Yahr	3	0	26

The numbers reflect the number of patients who showed improvement, worsening, or no change in the various clinical ratings between the baseline and first post-tranylcypromine visits

Table 7. Changes in UPDRS, Schwab & England, and Hoehn & Yahr scores between first visit on tranylcypromine and most recent (last) visit, with an least a 6 months follow-up period (N = 17, interval = 16.1 ± 5.6 mo, interval range: 6–24 months)

Exams	First	Last
UPDRS Behavioral	0.4 ± 0.9	0.9 ± 0.9
UPDRS ADL	4.9 ± 2.5	6.1 ± 2.5
UPDRS Motor	11.0 ± 6.4	13.2 ± 6.8
UPDRS Total	16.2 ± 8.0	20.0 ± 7.7
Schwab & England	92.9 ± 4.9	91.1 ± 4.9
Hoehn & Yahr	1.5 ± 0.5	1.6 ± 0.5

examination after starting the drug (Table 5). The other 8 patients have either just started on the drug and have not yet returned (N = 5) or discontinued the drug due to adverse effects before they could be re-examined (N = 3). Overall, there was little change except for improvement in depression scores (UPDRS Behavioral) and concomitant feeling of being able to handle daily chores a little easier (UPDRS ADL). By comparing individual patients before and after starting tranylcypromine (Table 6), we see that the behavioral scores were predominantly not changed or improved; the other UPDRS subscores as well as UPDRS total were split between improvement or worsening, with the majority showing the former. The Schwab & England and Hoehn & Yahr scores were mostly unchanged. None had an improvement in Hoehn & Yahr, and only three had some improvement in Schwab & England. Worsening outperformed improvement on these two scales.

Although it is still early in the follow-up of these patients, we have analyzed the status of PD at the latest follow-up time point in order to determine any trend in the rate of progression of the severity of PD. We limited the analysis to the 17 patients who had a follow-up interval of at least 6 months after the first post-tranylcypromine visit (Table 7). The interval between the first visit after starting tranylcypromine and the last visit of record has a mean ± S.D. of 16.1 ± 5.6 months, with a range of 6–24 months. There is gradual worsening of severity of PD, with the mean UPDRS total score increasing from 16.2 to 20.0 UPDRS points. The Schwab & England scores and the Hoehn & Yahr scores worsened very little over this time period.

Two patients out of the group of 33 had worsening severity of PD that required introduction of levodopa or dopamine agonist therapy. One was a 57 year old man who had PD for 39 months before starting tranylcypromine. He had pronounced tremor. His total UPDRS score at baseline was the 2nd highest of all patients (total = 36.5) which increased to 53.5 when examined 5 months after the introduction of tranylcypromine. Because of the worsening of PD, we discontinued tranylcypromine and placed him on carbidopa/ levodopa. The second patient was a 53 year old lawyer who had PD for 26 months prior to starting tranylcypromine. He was unable to tolerate his parkinsonian symptoms. When he had a slight increase of bradykinesia when examined 6 months after starting on tranylcypromine, he was treated with pergolide without discontinuing the tranylcypromine.

Discussion

Prior to the levodopa-treatment era, Hornykiewicz (1966) used a nonspecific irreversible monoamine oxidase inhibitor (MAOI) in patients with advanced PD and found essentially no clinical improvement. The biochemical findings at autopsy on those patients were compared to patients with PD who did not receive the MAOI and to normal controls. There was an increase in brain concentrations of serotonin and norepinephrine, reaching levels even greater than that seen in normal individuals. But the very low concentrations of dopamine typical of PD were barely increased by the MAOI. In the presence of levodopa (and the absence of a peripheral decarboxylase inhibitor), the addition of MAOI's to patients with PD resulted in markedly fluctuating blood pressures, from very high to very low levels (Fahn, unpublished observations). Whether this blood pressure response would also occur if patients were receiving levodopa in the presence of a peripheral dopa decarboxylase inhibitor (PDI), thereby blocking formation of peripheral dopamine, is unknown. The next type of MAOI to be used in patients with PD being treated with levodopa in the presence of a PDI was the MAO-B inhibitor, selegiline (Birkmayer et al., 1983), because it does not induce the "cheese effect" (intense hypertension after the ingestion of a tyramine-rich meal).

An effort to utilize neuroprotection therapy for PD led to the testing of selegiline in patients with early PD, who were not receiving other anti-PD medications (Tetrud and Langston, 1989; Parkinson Study Group, 1989b; Myllala et al., 1989). This approach was based on the concept that reducing the enzymatic oxidation of dopamine, and thereby reducing the formation of hydrogen peroxide, would result in a reduction of reactive free radicals and thus slow the rate of worsening of PD. But, as explained in the Introduction, selegiline has now been shown to have mild symptomatic benefit and not slow the rate of PD progression. But selegiline appears to inhibit oxidation of CNS dopamine by only 20% (Parkinson Study Group, 1965). Thus, we advised patients with early PD to consider taking tranylcypromine, an inhibitor of both MAO-A and -B, in an effort to slow the rate of progression of their disease.

We have found so far in 39 patients, with a follow-up period as long as 33 months, no serious adverse effects (i.e., hypertensive crisis) with the use of tranylcypromine in the presence of a diet free of tyramine-rich foods and in the absence of levodopa. But tranylcypromine was found frequently to cause some degree of insomnia and male impotence or decreased libido in a considerable number of patients (Table 3). Fourteen out of 39 patients had at least one of these complaints. Most could adjust to these symptoms, and impotency may be dependent on the timing of the last dose. Nevertheless, this complaint has resulted in discontinuing the tranylcypromine in some of these patients (Table 4). On the other hand, depression, a fairly common feature in patients with PD, was relieved by tranylcypromine in all five patients who had this symptom. One patient complained of an incident of having a headache and

Table 8. Comparison of tranylcypromine and selegiline treatment in delaying the need for levodopa or a dopamine agonist. End point is the time when treatment with levodopa or a dopamine agonist is required

	Patients needing levodopa or a dopamine agonist		
	Tranylcypromine-treated		Selegiline-treated
Duration of treatment	Number on drug	Number reaching end point	Percent reaching end point (N = 400)
6 mo	21	2	5%
9 mo	18	0	11%
12 mo	16	0	18%
15 mo	14	0	26%
18 mo	12	0	33%
21 mo	12	0	39%
24 mo	10	0	47%
27 mo	6	0	54%
30 mo	3	0	59%
33 mo	2	0	65%
36 mo	0	0	70%
39 mo	0	0	77%
42 mo	0	0	82%

Table 9. The duration of treatment with tranylcypromine plus ascorbate and tocopherol (this group of patients) and those treated earlier with just ascorbate and tocopherol (Fahn, 1992) before reaching end point. End point is the time when treatment with levodopa or a dopamine agonist is required

Duration of Treatment	Tranylcypromine			No tranylcypromine		
	Number on Rx	Number reaching end point	Cumulative	Number on Rx	Number reaching end point	Cumulative
6 mo	21	2	2	21	0	0
9 mo	18	0	2	21	0	0
12 mo	16	0	2	19	2	2
15 mo	14	0	2	18	1	3
18 mo	12	0	2	18	0	3
21 mo	12	0	2	17	1	4
24 mo	10	0	2	16	1	5
27 mo	6	0	2	16	0	5
30 mo	3	0	2	15	1	6
33 mo	2	0	2	14	1	7
36 mo	0	—	—	13	1	8
39 mo	0	—	—	11	2	10
48 mo	0	—	—	9	2	12
51 mo	0	—	—	5	4	16
57 mo	0	—	—	4	1	17
>66 mo	0	—	—	3	1	18

chest tightness after eating Asian food. No other suggestive tyramine-like symptom was encountered. Patients are constantly reminded to stay on the diet. Blood pressure did not change after starting tranylcypromine.

The motoric clinical features of PD showed little symptomatic improvement after starting tranylcypromine (Tables 5 and 6). But over time, while remaining on tranylcypromine, these patients showed only slight worsening of their PD. The Schwab & England and Hoehn & Yahr scores have remained constant over an average of almost 1.5 years on tranylcypromine. The difference of approximately 4 UPDRS points over 16 months is less than the worsening of scores observed in the DATATOP study (Parkinson Study Group, 1993, Table 2). From that study which followed early PD patients, the annualized rates of worsening in the selegiline-treated and placebo-treated groups were, respectively, approximately 8 and 15 UPDRS points per year during the early stages of PD. The 15 units/year rate of decline in the placebo-treated subjects was calculated from the patients' baseline visit to their last evaluation before starting levodopa and may have been inflated by rapidly declining subjects who reached endpoint early in that study. Thus, we can estimate the decline would be approximately 12 UPDRS units per, which is 1 unit per month. In 16 months we would anticipate that untreated patients would have progressed by 16 UPDRS points and that selegiline-treated patients by about 10 points. With tranylcypromine we found a worsening of only

Table 10. The duration of Parkinson's disease before reaching end point in patients treated with tranylcypromine plus ascorbate and tocopherol (this group of patients) and those treated earlier with just ascorbate and tocopherol (Fahn, 1992). End point is the time when treatment with levodopa or a dopamine agonist is required

Duration of PD (months)	Tranylcypromine			No tranylcypromine		
	On Rx	Number reached end point	Cumulative	On Rx	reaching end point	Cumulative
<13	27	0	0	21	0	0
13–24	22	0	0	20	1	1
25–36	18	1	1	18	2	3
37–48	10	1	2	16	2	5
37–60	3	0	2	10	6	11
61–72	2	0	2	6	4	15
73–84	1	0	2	5	1	16
85–97	1	0	2	1	4	20
98–110	0	—	—	1	0	20
111–122	0	—	—	0	1	21

4 UPDRS points over the 16 months of mean follow-up (Table 7). Without withdrawing the tranylcypromine we cannot determine whether the slower worsening of clinical measures of PD is the result of any neuroprotection or from subtle, long-duration symptomatic benefit.

Although this tranylcypromine-treatment was open-label and cannot be accurately compared to the double-blind DATATOP study evaluating selegiline, our results suggest that inhibiting both MAO-A and-B may have merit. Another indication of beneficial effect is the fact that only two patients on tranylcypromine to date had to initiate levodopa or a dopamine agonist. We can compare this to the rate of reaching end point in DATATOP (the time when symptomatic treatment is needed) (Table 8). Except for two patients on tranylcypromine requiring symptomatic treatment within 6 months, no other patient required it among the 21 who have been followed for at least 6 months and who did not discontinue tranylcypromine because of adverse effects. Table 8 shows that selegiline-treated subjects in DATATOP had a worse profile.

Another comparison is with patients treated with ascorbate and tocopherol but without tranylcypromine. This group of patients was treated at an earlier date by one of us (Fahn, 1992), and is very similar in the state of their PD severity to the tranylcypromine-treated group reported here. Because our tranylcypromine-treated patients also received ascorbate and tocopherol, the major difference between these populations is that one is receiving tranylcypromine and the other did not. Another difference is that the tranylcypromine-treated group has not been followed as long, so it is premature to make a final comparison of the two groups. But the tranylcypromine group is faring well except for the two patients who required levodopa or a dopamine agonist within 6 months of treatment. Table 9 shows the duration of treatment with medications (ascorbate and tocopherol and with or without tranylcypromine) before end point was reached. End point is the time when carbidopa/levodopa or a dopamine agonist is needed. Because the durations of treatment are so different, it is difficult to make any statement about the relative effectiveness of each group to delay the need for carbidopa/levodopa.

Table 10 also compares the same two groups as in Table 9, but now compares them according to the duration of PD before end point is reached. We need to follow the tranylcypromine-treated group longer before we can determine if this treatment adds any advantage.

Acknowledgement

Supported in part by the Parkinson's Disease Foundation.

References

Bernheimer H, Birkmayer W, Hornykiewicz O, Jellinger K, Seitelberger F (1973) Brain dopamine and the syndromes of Parkinson and Huntington. J Neurol Sci 20: 415–455

Birkmayer W, Knoll J, Riederer P, Youdim MBH (1983) (−)-Deprenyl leads to prolongation of L-dopa efficacy in Parkinson's disease. Mod Probl Pharmacopsychiatry 19: 170–176

Cohen G (1983) The pathobiology of Parkinson's disease: biochemical aspects of dopamine neuron senescence. J Neural Transm [Suppl 19]: 89–103

Cohen G (1986) Monoamine oxidase, hydrogen peroxide, and Parkinson's disease. Adv Neurol 45: 119–125

Fahn S (1989) The endogenous toxin hypothesis of the etiology of Parkinson's disease and a pilot trial of high dosage antioxidants in an attempt to slow the progression of the illness. Ann NY Acad Sci 570: 186–196

Fahn S (1992) A pilot trial of high-dose alpha-tocopherol and ascorbate in early Parkinson's disease. Ann Neurol 32: S128–S132

Fahn S, Cohen G (1992) The oxidant stress hypothesis in Parkinson's disease: evidence supporting it. Ann Neurol 32: 804–812

Fahn S, Duffy P (1977) Parkinson's disease. In: Goldensohn ES, Appel SH (eds) Scientific approaches to clinical neurology. Lea & Febiger, Philadelphia, pp 1119–1158

Fahn S, Elton RL, Members of the UPDRS Development Committee (1987) The Unified Parkinson's Disease Rating Scale. In: Fahn S, Marsden CD, Calne DB, Goldstein M (eds) Recent developments in Parkinson's disease, vol 2. Macmillan Healthcare Information, Florham Park NJ, pp 153–163, 293–304

Fornstedt B, Pileblad E, Carlsson A (1990) In vivo autoxidation of dopamine in guinea pig striatum increases with age. J Neurochem 55: 655–659

Graham DG, Tiffamy SM, Bell WR, Gutknecht WF (1978) Autooxidation versus covalent binding of quinone as the mechanism of toxicity of dopamine, 6-hydroxydopamine and related compounds toward C1300 neuroblastoma cells in vitro. Mol Pharmacol 14: 644–653

Hoehn MM, Yahr MD (1967) Parkinsonism: onset, progression and mortality. Neurology 17: 427–442

Hornykiewicz O (1966) Metabolism of brain dopamine in human parkinsonism: neurochemical and clinical aspects. In: Costa E, Côté LJ, Yahr MD (eds) Biochemistry and pharmacology of the basal ganglia. Raven Press, Hewlett NY, pp 171–185

Hornykiewicz O (1982) Brain neurotransmitter changes in Parkinson's disease. In: Marsden CD, Fahn S (eds) Movement disorders. Butterworth Scientific, London, pp 41–58

Jenner P (1991) Oxidative stress as a cause of Parkinson's disease. Acta Neurol Scand 84: 6–15

Jenner P, Dexter DT, Sian J, Schapira AHV, Marsden CD (1992a) Oxidative stress as a cause of nigral cell death in Parkinson's disease and incidental Lewy body disease. Ann Neurol 32: S82–S87

Jenner P, Schapira AHV, Marsden CD (1992b) New insights into the cause of Parkinson's disease. Neurology 42: 2241–2250

Myllyla VV, Sotaniemi KA, Tuominen J, Heinonen EH (1989) Selegiline as primary treatment in early phase Parkinson's disease — an interim report. Acta Neurol Scand 126: 177–182

Olanow CW (1990) Oxidation reactions in Parkinson's disease. Neurology 40 [Suppl 3]: 32–37

Olanow CW (1992) An introduction to the free radical hypothesis in Parkinson's disease. Ann Neurol 32: S2–S9

Parkinson Study Group (1989a) Effect of deprenyl on the progression of disability in early Parkinson's disease. N Engl J Med 321: 1364–1371

Parkinson Study Group (1989b) DATATOP: a multicenter controlled clinical trial in early Parkinson's disease. Arch Neurol 46: 1052–1060

Parkinson Study Group (1993) Effects of tocopherol and deprenyl on the progression of disability in early parkinson's disease. N Engl J Med 328: 176–183

Parkinson Study Group (1995) Cerebrospinal fluid homovanillic acid in the DATATOP study on Parkinson's disease. Arch Neurol 52: 237–245

Parkinson Study Group (1996a) Impact of deprenyl and tocopherol treatment on Parkinson's disease in DATATOP subjects not requiring levodopa. Ann Neurol 39: 29–36

Parkinson Study Group (1996b) Impact of deprenyl and tocopherol treatment on Parkinson's disease in DATATOP patients requiring levodopa. Ann Neurol 39: 37–45

Riederer P, Wuketich S (1976) Time course of nigrostriatal degeneration in Parkinson's disease. J Neural Transm 38: 277–301

Schwab RS, England AC Jr (1969) Projection technique for evaluating surgery in Parkinson's disease. In: Gillingham FJ, Donaldson MC (eds) Third symposium on Parkinson's disease. E & S Livingstone, Edinburgh, pp 152–157

Spencer JPE, Jenner P, Halliwell B (1995) Superoxide-dependent depletion of reduced glutathione by L-DOPA and dopamine. Relevance to Parkinson's disease. Neuroreport 6: 1480–1484

Tetrud JW, Langston JW (1989) The effect of deprenyl (selegiline) on the natural history of Parkinson's disease. Science 245: 519–522

Zigmond MJ, Hastings TG, Abercrombie ED (1992) Neurochemical responses to 6-hydroxydopamine and L-dopa therapy: implications for Parkinson's disease. Ann NY Acad Sci 648: 71–86

Authors' address: Dr. S. Fahn, Neurological Institute, 710 West 168th Street, New York, NY 10032-3784, U.S.A.

Deprenyl monotherapy improves visuo-motor control in early parkinsonism

S. Hocherman[1], G. Levin[1], N. Giladi[2], and M. B. H. Youdim[3]

[1]Department of Physiology and [2]Department of Pharmacology, Faculty of Medicine, Technion, and [3]Department of Neurology, Carmel Hospital, Haifa, Israel

Summary.   Deprenyl is a potent MAO-B inhibitor which is commonly prescribed for treatment of parkinsonism. Despite prevalent use its effects on the symptoms and course of Parkinson's disease (PD) are still debated. The present study was therefore undertaken in order to measure quantitatively changes in visuo-motor control (VMC), consequent to deprenyl monotherapy in early PD. Previous work from our laboratory has shown typical VMC deterioration in PD patients, that correlates with disease severity. Thus, measurements of such changes provides a sensitive tool with which the symptomatic effects of drug treatment can be assessed quantitatively. Fourteen newly diagnosed, PD patients with light symptoms were studied. The VMC of all patients was tested after the first neurological examination, before drug treatment commenced. A second test was done after 30 days of treatment with deprenyl at a dose of 2.5 mg/day. Following this test, dosage was increased to 10 mg/day and a third VMC test was given after 30 more days of treatment. Our results show significant improvement in VMC functions following 30 days of 2.5 mg/day treatment and a continuing improvement after the next 30 days of 10 mg/day treatment. This improvement pertains mainly to directional control of self initiated movements and is smaller for movements that are guided externally. We conclude that deprenyl monotherapy has a clear symptomatic beneficial effect for patients with early PD.

Introduction

(−)Deprenyl is a selective and irreversible inhibitor of type B monoamine oxidase (Knoll, 1992; Okuda et al., 1992) and as such is believed to prolong the life of dopamine released at striatal dopaminergic nerve terminals. Apart from that it is metabolized into (−)amphetamine and (−)methamphetamine, which lead to the possibility of its action as a dopamine releasing agent (Okuda et al., 1992; Jenkins et al., 1992; Yasar et al., 1993). In addition, (−)deprenyl increases the levels of superoxide dismutase and catalase in the basal ganglia, thus attaining a possible neuroprotective role (Knoll, 1992;

Carrillo et al., 1992; Chia et al., 1992; Knoll et al., 1992; Tardos and Kanyo, 1992; Tatton and Greenwood, 1991).

(−)Deprenyl therapy has been shown to delay the need for initiation of L-DOPA treatment in new patients (Knoll, 1992; Shoulson, 1992; Shults, 1993; Vezina, 1992) but has not been shown to slow the overall rate of PD disease progression (Parkinson Study Group, 1996). Thus, the beneficial effect of (−)deprenyl in slowing the advance of PD is still debated.

A careful mathematical analysis of the DATATOP findings by Schulzer et al. (1992) revealed that the slower advance of parkinsonism in (−)deprenyl treated patients is more likely to represent a transient symptomatic effect than a long lasting neuroprotective effect. This agrees with the finding that the motor benefits of (−)deprenyl last for about 3 months (The Parkinson Study Group, 1993), and may be explained by the fact that under chronic (−)deprenyl administration the initial suppression of striatal tyrosine hydroxylase activity is replaced by an increase to above normal level (Vrana et al., 1992).

Several reports on improvement of the UPDRS score after initiation of (−)deprenyl treatment support the notion that (−)deprenyl may have beneficial effects which relate to a symptomatic amelioration of PD (Shults, 1993; Hubble et al., 1993; Schulzer et al., 1992). Contrary to the above findings, no significant symptomatic effects could be detected in a controlled quantitative study which involved this drug (Ziv et al., 1993). While a net beneficial effect of (−)deprenyl is evident, its influence on high-level motor functions is still not clear. The present study relates to this issue by evaluating the effects of (−)deprenyl treatment on visuo-motor control (VMC) in PD patients. In the past we have shown that the PD related deficit in VMC can be measured quantitatively (Hocherman and Aharon-Peretz, 1994). Recently, we have improved the quantification of this measurement (Hocherman and Giladi, 1995). Making it a sensitive tool with which even small changes in VMC can be detected. This improvement of our technique makes it especially suitable for detection of quantitative changes in symptoms, consequent to different regimens of drug treatment.

Methods

Subjects

Fourteen newly diagnosed PD patients were studied. All patients had light, mostly unilateral symptoms. Six patients had neurological symptoms mainly on the left side (left hemi-PD) and eight had neurological symptoms mainly on the right side (right hemi-PD) of the body.

Instrumentation

All visuo-motor tests are done by use of a computerized system that was described elsewhere (Hocherman and Aharon-Peretz, 1994; Hocherrman and Giladi, 1995). In brief, this system consists of a digitizing tablet which is laid at chest level and is hidden from the subject's view by an overlying board, upon which a computer monitor is placed. Paths for tracing and for tracking are displayed on the computer monitor. A screen cursor represents the location of an unseen hand-held handle which is situated on top of a digitizing tablet. The location of the handle is read by the computer at 100 Hz, with a resolution of 0.05 mm. A one to one correspondence between movement of the handle and movement of the cursor is maintained.

Tests

Tracing: A path (either sinewave, square or circle) is displayed on screen. The subject needs to bring the cursor to a designated starting point from which he/she moves the cursor along the entire path, as accurately as possible, by use of the unseen handle. No demands on speed are made. All three paths are traced and both hands are tested.

Tracking: The same paths are used, but a 1 cm target is moved by the computer program along the path, at a predetermined speed. The subject's task consists of maintaining the cursor within the target by moving the handle correspondingly. If the cursor comes out of the target circle the latter stops moving (tracking interruption) until the cursor is brought back into the target.

Scores

Performance is evaluated off-line. Six performance measures which belong to 3 classes are produced.

General criteria: These consist of 1: the total time (TT) of test performance, and 2: the mean distance (MnEr) between the model path and the path traversed by the hand.

Directional control: This is assessed by two measures. 1: The vectorial error (VcEr), which consists of the proportion of movement vector which is made in a direction perpendicular to that of the model path, at every sampling point. 2: The cumulative movement time during which the VcEr exceeded a level of 50% ($T_{50\%}$).

Velocity control: The two measures used to estimate this function are 1: velocity of hand movement (V) and 2: the number of tracking interruptions (N_{Ints}). Note that the second measure pertains only to tracking tests.

Statistics

Descriptive statistics of performance scores and repeated measures analyses of variance are done by use of the SAS package.

Results

The results of the present study are shown in 3 categories.

General criteria

A significant dose-related reduction in Total-Time (TT) consequent to deprenyl treatment is seen in tracing ($F(2,11) = 6.07$ $p = 0.017$). A 12.5%

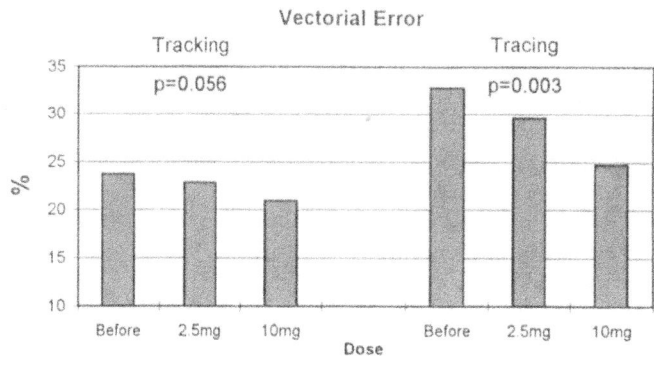

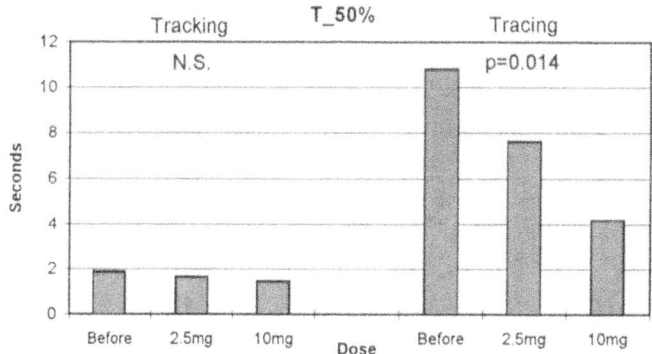

Fig. 1. Effects of treatment with deprenyl on the ability of early PD patients to control the direction of hand movement in visuo-motor testing. *Upper part*: Mean vectorial error of all patients before drug treatment was started (Before), 30 days after treatment with 2.5 mg deprenyl/day was started (2.5 mg) and 30 days after dosage was increased to 10 mg deprenyl/day (10 mg). Vectoial error scores are shown separately for tracking tests (left) and for tracing tests (right). *Lower part*: Mean $T_{50\%}$ scores (see methods) in tracking (left) and tracing (right), before and during drug treatment. P represents the indicated probability values of the effect of dose of drug treatment on the parameters shown

reduction in mean TT is seen after 30 days of treatment with 2.5 mg deprenyl/day. Another 14.3% reduction in mean TT is seen following additional 30 days of treatment with 10 mg deprenyl/day. No significant change in TT is found in tracking ($p > 0.05$).

No significant improvement is found in the mean error in either tracing or tracking.

Directional control

A significant ($F(2,11) = 10.37$ $p = 0.003$) dose-related reduction in the vectorial error (VcEr), consequent to deprenyl treatment, can be seen in tracing (Fig. 1, top). A 10% reduction is seen following 30 days of treatment with 2.5 mg deprenyl/day. A further 17% reduction in VcEr is seen after additional 30 days of treatment with 10 mg deprenyl/day.

In tracking the change in VcEr is small and only marginally significant $(F(2,11) = 3.76 \, p = 0.057)$.

The improvement in directional control which is seen in tracing is further accentuated by considering the change in $T_{50\%}$ (lower part of Fig. 1). The mean dose-related change in this variable is significant $(F(2,11) = 6.39 \, p = 0.014)$ and amounts to 30% after 30 days of treatment with 2.5 mg deprenyl/ day, and to 45% after the additional 30 days of treatment with 10 mg deprenyl/ day.

In tracking the $T_{50\%}$ is low to begin with and does not change significantly with treatment (Fig. 1, bottom).

Control of velocity

A small but significant increase in movement velocity during tracing is found $(F(2,11) = 4.28 \, p = 0.04)$. This change amounts to 3% and 9% after 2.5 mg deprenyl/day and 10 mg deprenyl/day respectively. No significant change occurs in movement velocity or number of interruptions, in tracking.

Discussion

The present work was undertaken in order to determine whether treatment with deprenyl improves the disease related symptoms in early parkinsonism. Previous studies have suggested that this is indeed the case. However, this suggestion relies mainly on conjectural evidence (Schulzer et al., 1992; The Parkinson Study Group, 1993), and on work with patients with motor fluctuations (Hubble et al., 1993). Measurement of treatment related changes in motor function have shown only a subclinical effect of deprenyl on rigidity, reaction time, and rate of muscular activation (Ziv et al., 1993). Since deprenyl monotherapy does have a clear beneficial effect, and is able to postpone the initiation of L-DOPA therapy (Shults, 1993; The Parkinson Study Group, 1996) an attempt to evaluate its effect on high-order motor deficits is warranted.

Changes in visuo-motor control (VMC) characterize the PD related deterioration of high-level motor functions (Hocherman and Aharon-Peretz, 1994). In the present work the effects of deprenyl treatment on VMC are evaluated in two contexts. In one, a task that provides an external movement cue (tracking) is used. In the other, a task that requires internal direction of hand movement (tracing) is employed.

Our results show that deprenyl monotherapy affects significantly performance of the second task, i.e. the task which is more impaired in PD (Hocherman and Giladi, 1995). The main behavioral change relates to the patients' ability to control the direction of hand movement, rather than to their ability to control hand velocity. Together, these findings suggest that the beneficial effects of deprenyl treatment pertain to the level of motor planning rather than to the level of motor execution.

68 S. Hocherman et al.

Clearly, early PD patients benefit from treatment with a low dose of 2.5 mg/day. However, a significantly greater improvement in VMC is attained at the full dose of 10 mg/day. In view of these findings it is suggested that a careful dose — response study on the symptomatic effects of deprenyl treatment will be carried out.

References

Carrillo MC, Kitani K, Kanai S, Sato Y, Ivy GO (1992) The ability of (−)deprenyl to increase superoxide dismutase activities in the rat is tissue and brain region selective. Life Sci 50: 1985–1992

Chia LG, Liu SP, Lee EH (1992) Differential effects of deprenyl and MPTP on catecholamines and activity in BALB/c mice. Neuroreport 3: 777–780

Hocherman S, Aharon-Peretz J (1994) Two dimensional tracing and tracking in patients with Parkinson's disease. Neurol 44: 111–116

Hocherman S, Giladi N (1995) Early diagnosis of Parkinson's disease in new neurological patients by testing of Visuo — Manual coordination. In: Ohye C, Kimura M, McKenzie SJ (eds) The basal ganglia V. Plenum, New York, pp 427–431

Hubble JP, Koller WC, Waters C (1993) Effects of selegiline dosing on motor fluctuations in Parkinson's disease. Clin Neuropharmacol 16: 83–87

Jenkins IH, Fernandez W, Playford ED, Lees AJ, Frackowiak RSJ, Passingham RE, Phil D, Brooks DJ (1992) Impaired activation of the supplementary motor area in Parkinson's disease is reversed when akinesia is treated with apomorphine. Ann Neurol 32: 749–757

Knoll J (1992) (−)Deprenyl-medication: a strategy to modulate the age-related decline of the striatal dopaminergic system. J Am Geriatr Soc 40: 839–847

Knoll J, Toth V, Kummert M, Sugar J (1992) (−)Deprenyl and (−)parafluorodeprenyl-treatment prevents age-related pigment changes in the substantia nigra. A TV-image analysis of neuromelanin. Mech Ageing Dev 63: 157–163

Okuda C, Segal DS, Kuczenski R (1992) Deprenyl alters behavior and caudate dopamine through an amphetamine-like action. Pharmacol Biochem Behav 43: 1075–1080

Parkinson Study Group (1993) Effects of tocopherol and deprenyl on the progression of disability in early Parkinson's disease. N Engl J Med 3: 176–183

Parkinson Study Group (1996) Impact of deprenyl and tocopherol treatment on Parkinson's disease in DATATOP patients requiring levodopa. Ann Neurol 39: 37–45

Shoulson I (1992) An interim report of the effect of selegiline (L-deprenyl) on the progression of disability in early Parkinson's disease. The Parkinson Study Group. Eur Neurol 32: 46–53

Shults CW (1993) Effect of selegiline (deprenyl) on the progression of disability in early Parkinson's disease. Parkinson Study Group. Acta Neurol Scand [Suppl] 146: 36–42

Schulzer M, Mak E, Calne DB (1992) The antiparkinson efficacy of Deprenyl derives from transient improvement that is likely to be symptomatic. Ann Neurol 32: 795–798

Tardos L, Janvarine Kanyo E (1992) Pharmacology of selegiline (recent considerations). Acta Pharm Hung 62: 237–241

Tatton WG, Greenwood CE (1991) Rescue of dying neurons: a new action of deprenyl in MPTP parkinsonism. J Neurosci Res 30: 666–672

Vezina P, Mohr E, Grimes D (1992) Deprenyl in Parkinson's disease: mechanisms, neuroprotective effect, indications and adverse effects. Can J Neurol Sci 19: 142–146

Vrana SL, Azzaro AJ, Vrana KE (1992) Chronic selegiline administration transiently decreases tyrosine hydroxylase activity and mRNA in the rat nigrostriatal pathway. Mol Pharmacol 41: 839–844

Yasar S, Schindler CW, Thorndike EB, Szelenyi I, Goldberg SR (1993) Evaluation of the stereoisomers of deprenyl for amphetamine-like discriminative stimulus effects in rats. J Pharmacol Exp Ther 265: 1–6

Ziv I, Achiron A, Djaldetti R, Dressler R, Melamed E (1993) Short-term beneficial effect of deprenyl monotherapy in early Parkinson's disease: a quantitative assessment. Clin Neuropharmacol 16: 54–60

Authors' address: Prof. S. Hocherman, Department of Physiology, Faculty of Medicine, Technion, Haifa, P.O.B. 9649, Israel

Endogenous monoamine oxidase·A inhibitory activity (tribulin), measured in saliva, is related to cardiovascular reactivity in normal individuals

A. Clow[1], A. Doyle[1], F. Hucklebridge[1], D. Carroll[2], C. Ring[2], J. Shrimpton[2], G. Willemsen[2], and P. D. Evans[1]

[1] Psychophysiology and Stress Research Group, University of Westminster, London, and [2] School of Sport and Exercise Sciences, University of Birmingham, United Kingdom

Summary. Salivary monoamine oxidase A inhibitory activity (MAO-AI), mean arterial blood pressure (MAP) and heart rate (HR) were determined simultaneously in healthy male students (n = 13) at rest, before a mild psychological stressor, twice during the task and 18 minutes after the end of the task. The sample as a whole showed significant differences in MAP and HR across occasions (respectively, $p < 0.001$ for both). Salivary MAO-AI could distinguish novice and experienced game players ($p < 0.02$) and was consistently positively correlated with MAP ($r = 0.58$, $p < 0.05$ on occasion 2). Pre-task measures of MAO-AI for an increased sample (n = 18) were associated with higher MAP (but not HR) throughout the experiment ($p < 0.05$). Those subjects with falling MAO-AI profiles from task to recovery showed significantly greater simultaneous decline in HR than those with a rising MAO-AI profile ($p < 0.05$).

Introduction

Since the discovery (Glover et al., 1980) of endogenous MAO inhibitory activity (MAO-I), also known as tribulin (Sandler, 1982), many studies have shown it to be related to pathological anxiety (human studies) and severe stress (animal studies). The early studies provided robust evidence of this relationship by measurement of MAO-I in urine and brain (see Glover, 1996 for review). More recently our group has focused on the role of tribulin (distinguishing between its MAO A and B inhibitory components: MAO-AI and MAO-BI respectively) in the normal physiological regulation of the monoamines. Urinary MAO-AI and MAO-BI, in healthy students prior to an important assessed oral seminar presentation, were shown to correlate with the elevation in arousal experienced during the task. Furthermore MAO-AI correlated with cortisol measured in the same urine samples (Doyle et al., 1996a). In another study normal individuals were again investigated but this time over a five day period with no stress challenge. Mean MAO-AI and MAO-BI (across days) were found to be positively correlated with self-reported stress scores over the same time period (Doyle et al., 1996b). This was the first study to demonstrate a relationship between tribulin and everyday stress. We concluded that tribulin output may be indicative of mild enduring state stress in normal people and that these levels could be raised even further by an acute stressful episode (as reported in the early literature). As such tribulin could be an endogenous modulator of brain monoamine levels both during resting (everyday) stress and during acute, severe stress. Raised MAO inhibitory activity in reducing monoamine metabolism would increase availability of the monoamines.

To facilitate more detailed examination of this hypothesis we have developed a sensitive technique to determine MAO-I in human saliva. This technique allows frequent and convenient sampling and overcomes the inherent problems associated with vastly fluctuating diuresis (Doyle et al. 1997). Using this technique we have again demonstrated an association between MAO-AI and MAO-BI and self-reported stress in normal people (Doyle et al., 1997).

To date our investigation into endogenous MAO-I in normal people has concentrated on its correlation with self-reported feelings of stress. Although useful, self-reported measures are inherently subjective contributing "noise" to data collection especially in between subject experimental paradigms. In the light of the known sympathomimetic activity of anti-depressant MAO A inhibitory drugs we decided to investigate, for the first time, endogenous MAO inhibitory activity and cardiovascular reactivity. Hence in the present study we report the relationship between MAP and HR (objective physiological measures of reactivity) and MAO A inhibitory activity (unfortunately insufficient saliva was available to determine both MAO-AI and MAO-BI).

Methods

This study was part of a larger investigation (using 26 subjects) into the effects of an acute psychological challenge on cardiovascular responses and secretory immunity (Carroll et al., 1996). All subjects were healthy male volunteer students (mean age ± SD: 21.0 ± 1.2 years) and involved use of the computer game Doom. Only 13 of these subjects collected sufficient saliva for analysis on all 4 occasions (although eighteen subjects collected a saliva sample on occasion 1) as outlined below, of which seven of the subjects had never played the game before and six were experienced players. Subjects sat in a comfortable chair facing a computer monitor. The computer game Doom (in which players are required to navigate a series of rooms killing hostile creatures in a bid to survive) was introduced to them and they were allowed a five minute practice session where the level of difficulty was set. The study began with a 6 minute rest period followed by the 30 minute computer game task and a 20 minute recovery period. Heart rate and blood pressure was recorded continuously during the session. Saliva was collected on 4 occasions: 4 minutes into the initial rest period, 6 minutes into the task, 24 minutes into the task and 18 minutes into the recovery period. Immediately before the start of the computer game subjects were asked to rate, on a 7-point scale, how tense they felt.

Impedance cardiography (ICG) and electrocardiography (ECG) were recorded using the VU-AMD system (Vrije Universiteit Amsterdam), (see Willemsen et al., 1996). The ICG and ECG signals were sampled at 250 Hz and 1,000 Hz respectively. Ensemble averages were obtained over 60 second intervals and scored using an interactive software programme. HR (bpm) was obtained from the R-R interval using the formula: HR = 60,000 / R-R interval. Systolic blood pressure (SBP) and diastolic blood pressure (DBP) were recorded using an Accutraker II ausculatory monitor (Suntech, Model 104). Mean arterial pressure (MAP) (mmHg) was calculated using the formula: MAP = DBP + ((SBP-DBP)/3). Mean MAP and HR were calculated for the corresponding times (4 occasions) when saliva had been collected.

Unstimulated saliva was collected using Salivettes (Sarstedt Ltd, Leicester). Subjects were asked not to eat or drink for 1 hour prior to the experiment. The cotton swab, contained in the Salivette, was placed under the tongue for exactly 2 minutes and passively absorbed all the saliva collected there. The swab plus saliva was then removed from the mouth returned to the Salivette and frozen at $-20°C$ until analysis.

Saliva was recovered, once thawed, by centrifugation at 1,000 g for 10 minutes. The MAO-A inhibitory activities were determined using a method based on that of Glover et al. (1980). Aliquots (250 µl) of saliva were acidified (50 µl 2 M HCl) prior to extraction into 2 volumes of ethyl acetate (analytical grade). After vortexing and centrifugation the organic layer was decanted and reduced to dryness under nitrogen. Water blanks were extracted in the same way. The residue was reconstituted in half the original volume (125 µl) 100 mM sodium phosphate buffer pH 7.4. MAO A inhibitory activity was determined by incubation (75 minutes at 37°C) of aliquots (50 µl) with 10 µl ^{14}C-5-hydroxytryptamine (final concentration 140 µM, specific activity, 55 mCi/mmol), and 10 µl 0.5% w/v rat liver homogenate. Assays were carried out in duplicate and the coefficient of variance between duplicates was always less than 10%.

The sample consisted of novice and experienced game players and 2 way ANOVAs (group $\times$ occasion) were used to analyse initial effects.

Results

Salivary MAO-AI and cardiovascular measures were collected from 13 subjects on 4 occasions before, during and after a mild, acute psychological stressor. Background information revealed 6 out of the 13 subjects had previous experience of the computer game whereas 7 were novices. It is worth noting that the novices rated their tension as higher than the experienced players at the start of the computer game task (mean $\pm$ SD: 2.57 $\pm$ 0.79 and 1.67 $\pm$ 0.52, respectively, t = 2.4, df = 11, p < 0.05).

ANOVA revealed a significant group $\times$ occasion interaction for MAO-AI (F = 4.19, df = 1,11, p < 0.02, see Fig. 1). More detailed comparisons showed that novice players had higher salivary MAO-AI pre-task and early on in the task than later on in the task and post-task (p at least <0.05 for all relevant comparisons). There was also a strong tendency (p < 0.10) for the novice players to have higher levels of MAO-AI than experienced players at the beginning of the task. Experienced players did not in fact differ significantly in any comparison of occasions, despite the appearance of a rise towards the end of the experiment apparent in Figure 1. Similar ANOVAs for MAP and HR revealed no group differences or group $\times$ occasion interactions, but main effects for occasions were highly significant (F = 7.61, df = 3,36, p < 0.001, and F = 8.32, df = 3,36, p < 0.001 for MAR and HR respectively). MAP

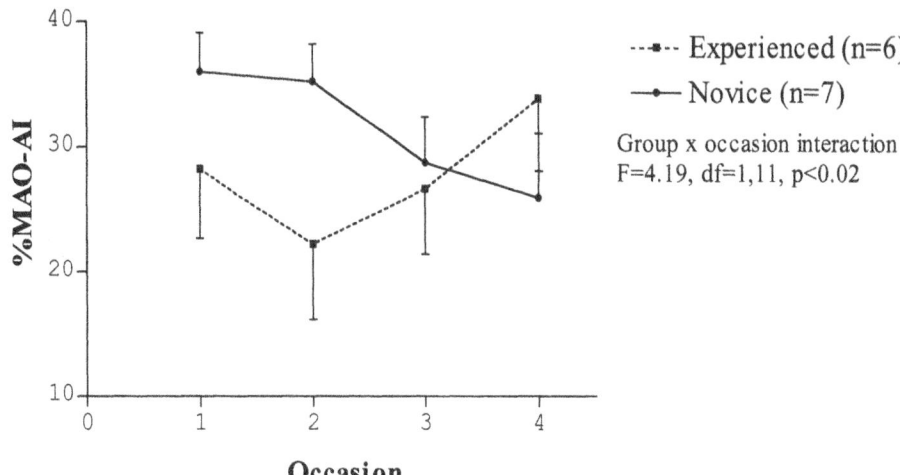

Fig. 1. Mean ± SEM salivary monoamine oxidase A inhibitory activity (MAO-AI; % inhibition) in experienced (n = 6) and novice (n = 7) game players — 4 min. into the initial rest period (occasion 1), 6 min. into the game (occasion 2), 24 min. into the game (occasion 3) and 18 min. into the recovery period (occasion 4). There was a significant group × occasion intraction: $F = 4.19$, $df = 1,11$, $p < 0.02$ (ANOVA)

showed a classic pressor response to challenge, rising distinctly from pre-task to task and then fell back again in the recovery period (p at least <0.05 for all comparisons). HR showed little movement pre-task and on task but did show a steep decline from all three prior occasions to recovery (p at least <0.05 for all comparisons). Means illustrating the significant occasions effects for MAP and HR are plotted in Fig. 2.

There was a consistent pattern of positive correlation between MAP and salivary MAO-AI. Out of sixteen possible correlation coefficients, including simultaneous and lagged effects, fifteen were positively signed (median $r = 0.24$), and on occasion 2, at the peak of the MAP pressor effect, the correlation was statistically significant ($r = 0.53$, $p < 0.05$). Similar correlational analysis revealed no consistent or significant effects for MAO-AI and HR.

In order to investigate the relationship between MAP and MAO-AI in more detail, we looked at the MAP profiles across occasions for a slightly increased sample of eighteen subjects for whom we had initial pre-task MAO-AI and full cardiovascular data across occasions. High and low MAO-AI groups were formed by median split (above and below 33.5% inhibition: mean ± SD MAO-AI for the high and low groups were 40 ± 5.2 and 25 ± 8.3 respectively), and a two-way (group × occasion) ANOVA was run on MAP scores. There was a significant main effect of group such that those high on initial MAO-AI manifested higher MAP throughout the experiment ($F = 4.5$, $df = 1,16$, $p < 0.05$). A significant occasions factor ($F = 9.4$; $df = 3,48$, $p < 0.001$) confirmed the same rise and fall in MAP reported earlier for thirteen of the eighteen subjects. Means are plotted in Fig. 3.

A similar ANOVA was performed with HR as the dependent variable but no effects of MAO-AI group were found. However one relationship

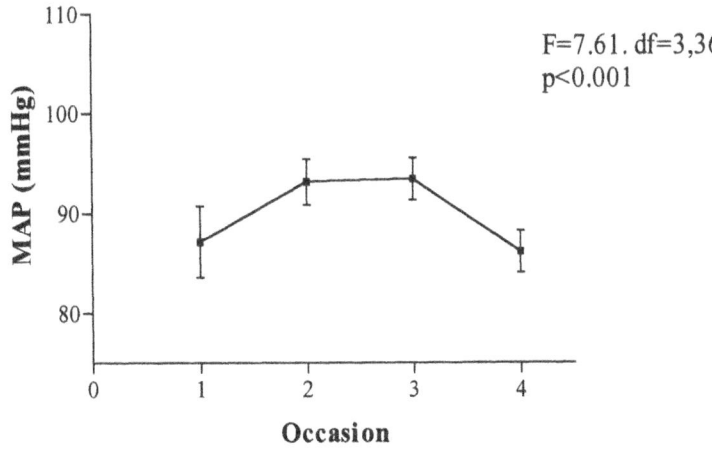

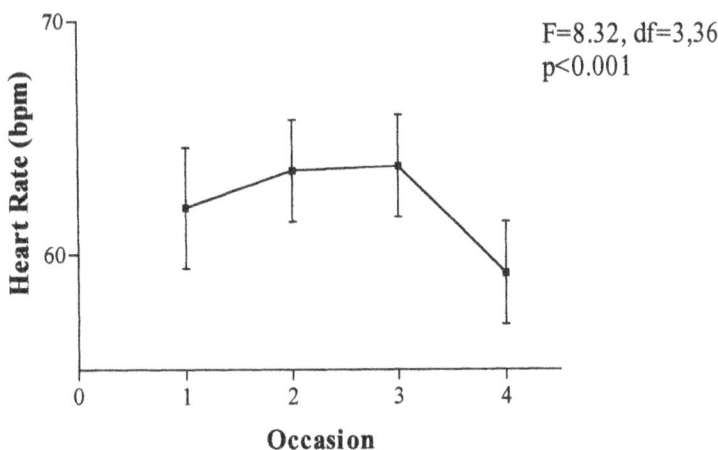

Fig. 2. Mean ± SEM mean arterial blood pressure (MAP; mmHg) and heart rate (HR; beats per minute) in subjects (n = 13) — 4 min. into the initial rest period (occasion 1), 6 min. into the game (occasion 2), 24 min. into the game (occasion 3) and 18 min. into the recovery period (occasion 4). There was a significant effect against occasions: F = 7.61, df = 3,36, p < 0.001 and F = 8.32, df = 3,36, p < 0.001 for MAP and HR respectively (ANOVA)

between HR and MAO-AI was observed in a specific analysis of the distinct fall in HR from occasion 3 to 4 (see Fig. 2) for the original n = 13 sample. Subjects were classified as either showing a rise or fall in MAO-AI across the same interval and a two-way (group × occasion) ANOVA was used to examine HR effects for occasions 3–4. The falling MAO-AI group showed a fall in HR (mean ± SD) of 6.7 ± 2.9 beats per minute (bpm) whereas the rising group showed a less marked fall in HR 2.9 ± 3.3 bpm. The relevant group × occasion interaction in the ANOVA was statistically significant (F = 4.66, df = 1,11, p < 0.05).

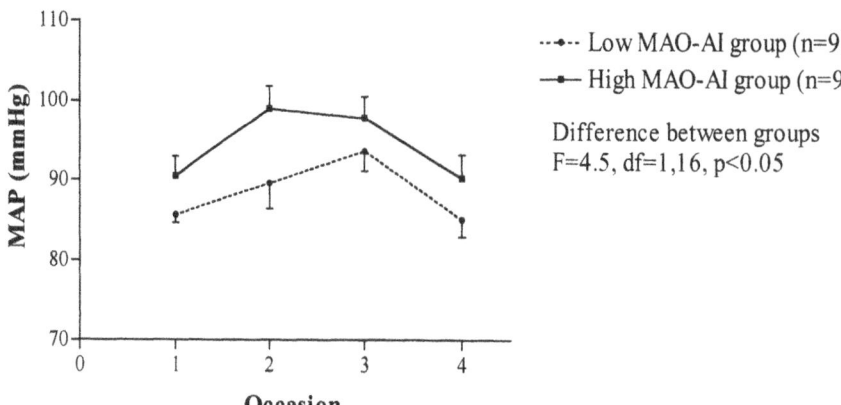

Fig. 3. Mean ± SEM mean arterial blood pressure (MAP; mmHg) in subjects (n = 18) divided by median split according to their pre-task salivary monoamine oxidase A inhibitory activity (MAO-AI; % inhibition) — 4 min. into the initial rest period (occasion 1), 6 min. into the game (occasion 2), 24 min. into the game (occasion 3) and 18 min. into the recovery period (occasion 4). There was a significant difference between the groups: F = 4.5, df = 1,16, p < 0.05 (ANOVA)

Discussion

As it is well known that MAO-A inhibitory drugs, used in the treatment of depression, have marked effects on cardiovascular activity we decided to investigate the relationship between stress-related endogenous MAO-AI (see Glover, 1997; Doyle et al., 1996a,b) and CV measures. This is the first study to investigate these relationships. All previous research on human subjects has focused on relating MAO-I (tribulin) to pathological anxiety states or self-reported stress in normal individuals. It is known that computer games can produce substantial cardiodynamic changes in normal individuals (Miller and Ditto, 1989; Turner and Carroll, 1985) so we choose this as a suitable method for investigating these relations in a controlled environment. The newly developed salivary assay for MAO-I (Doyle et al., 1997) enabled frequent sampling at matched intervals. In addition detailed analysis for the CV responses of the larger group (n = 26) had already revealed significant blood pressor responses in these subjects (Carroll et al., 1996).

There were marked individual differences in the profile of MAO-AI response to the computer game task which related to the novice versus experienced status of the game players. In summary it was evident that novice players had distinctly higher MAO-AI while anticipating the task and early on in the performance of the task, compared to later on in the task and after it had finished. Moreover the high levels of the novices early in the task were also high compared to experienced players at this point, although this between group comparison just eluded conventional significance. The fact that novice status was itself associated with more self-reported tension at the beginning of the task echoes one of our previous studies which showed high urinary MAO-AI, pre-stressor, to predict arousal during the task (Doyle et al., 1996a).

What is new about this particular investigation is the finding that high endogenous MAO-AI is also linked to cardiovascular reactivity in a challenging situation. There was a significant positive correlation between MAP and salivary MAO-AI on occasion 2 at the peak of the MAP pressor effect ($r = 0.53$, $p < 0.05$) and subjects high on pre-task MAO-AI were significantly higher on MAP throughout the experiment. Thus there is strong evidence that MAO-AI is linked to blood pressure. This study provides the first evidence of a relationship between generation of the stress-linked endogenous modulator of monoamine activity (tribulin) and cardiovascular dynamics (MAP).

Effects for HR were more elusive, but it should be remembered that HR had similar values pre-task and task, and only fell in the final interval from task to recovery period. When just this period of change is looked at, simultaneous movement in HR and MAO-AI are in the predicted direction: on aggregate HR is falling but it is falling more steeply in those whose MAO-AI levels are also falling.

What is clear from this small but interesting study is that salivary MAO-AI is a useful and sensitive measure that can distinguish between novice and experienced groups in a mild stress paradigm and that it is related to cardiovascular reactivity.

References

Carroll D, Ring C, Shrimpton J, Evans P, Gonneke HM, Hucklebridge F (1996) Secretory immunoglobulin A and cardiovascular responses to acute psychological challenge. Int J Behav Med 3: 266–279

Doyle A, Pang F-Y, Bristow M, Hucklebridge F, Evans P, Clow A (1996a) Urinary cortisol and endogenous monoamine oxidase inhibitors(s), but not isatin, are raised in anticipation of stress and/or arousal in normal individuals. Stress Med 12: 43–49

Doyle A, Hucklebridge F, Evans P, Clow A (1996b) Urinary output of endogenous monoamine oxidase inhibitory activity is related to everyday stress. Life Sci 58: 1723–1730

Doyle A, Hucklebridge F, Evans P, Clow A (1997) Salivary monoamine oxidase A and B inhibitory activities correlate with stress. Life Sci 59: 1357–1362

Glover V (1996) Function of endogenous monoamine oxidase inhibitors. J Neural Transm (this volume)

Glover V, Reveley MA, Sandler M (1980) A monoamine oxidase inhibitor in human urine. Biochem Pharmacol 29: 467–470

Miller SB, Ditto B (1989) Individual differences in heart rate and peripheral vascular responses to an extended stress task. Psychophysiology 26: 506–513

Sandler M (1982) The emergence of tribulin. Trends Pharmacol Sci 3: 471–472

Turner JR, Carroll D (1985) Heart rate and oxygen consumption during mental arithmetic, a video game and graded excerise: further evidence of metabolically-exaggerated cardiac adjustments? Psychophysiology 22: 261–267

Willemsen GHM, de Geus EJC, Klaver CHAM, van Doornen LJP, Carroll D (1996) Ambulatory monitoring of the impedance cardiogram. Psychophysiology 33: 184–193

Authors' address: Dr.A. Clow, Psychophysiology and Stress research Group, Division of Physiology, University of Westminster, 115 New Cavendish Street, London W1M 8JS, United Kingdom

MAO inhibitory side effects of neuroleptics and platelet serotonin content in schizophrenic patients

Z. Mészáros[1], **D. Borcsiczky**[1], **M. Máté**[2], **J. Tarcali**[1], **K. Tekes**[1], and **K. Magyar**[1]

[1] Department of Pharmacodynamics, Semmelweis University of Medicine, and
[2] IVth Department of Psychiatry, Jahn Ferenc Hospital, Budapest, Hungary

Summary. In order to study the putative monoamine oxidase (MAO) inhibitory side effect of neuroleptics and simultaneous changes in platelet serotonin content both MAO-B activity and serotonin (5-HT) content in platelets of 30 healthy volunteers and 50 schizophrenic patients treated with neuroleptics were investigated. Our results have shown significantly lower MAO-B activity (15.26 ± 6.81 S.D. vs. 8.63 ± 3.82 mmol/hour/10^9 platelets) and higher platelet 5-HT content (906.19 ± 285.33 vs. $1,727.85 \pm 947.40$ ng/10^9 platelets) in the schizophrenic group. Platelet MAO-B activity was considerably lower in paranoid and residual schizophrenics compared with other patients, however, no difference was found in platelet 5-HT content between different subtypes of schizophrenia. Various neuroleptic treatments did not produce different effects either on platelet serotonin content or platelet MAO-B activity.

Introduction

Human blood platelets are accessible and reliable models for several aspects of the central serotonergic neurones. Uptake, storage and release of serotonin (5-HT) in platelets are kinetically and pharmacologically similar to those in central nervous system (Stahl, 1977). Therefore, this homogeneous preparation of human blood elements is suitable for monitoring pharmacological effects supposed to interfere with central serotonergic mechanisms. The utility of platelets as a biological marker for well-established psychopathological states is still controversial (Da Prada, 1988).

Human platelets possess mitochondrial monoamine oxidase type B (MAO-B) exclusively. 5-HT is not a preferred substrate of this enzyme form. Therefore, changes in platelet MAO-B activity are not likely to alter platelet serotonin content (Youdim, 1988). The main catabolic pathway of plasma 5-HT is deamination by MAO-A in the liver and the lungs. A decrease in MAO-A activity may possibly contribute to an increase in plasma 5-HT level (Celada et al., 1990). However, up till now there is no adequate procedure for direct monitoring of human MAO-A activity. We hypothetise that platelet

MAO-B activity and platelet serotonin content are independent biological markers and, thus, they may be modified to a various extent by different neuroleptics.

Studies on platelet serotonin content of schizophrenic patients treated with neuroleptics provided controversial results so far. Several authors reported significantly elevated platelet 5-HT concentration in medicated and untreated schizophrenic patients compared with normal controls (Garelis et al., 1975; Freedman et al., 1981). Other investigators, however, have reported no differences in platelet 5-HT levels between schizophrenics and controls (Joseph et al., 1977). After 28 days of treatment with neuroleptics there was no change in 5-HT content or uptake of 5-HT into platelets, and hyperserotoninaemia of unknown origin was observed especially in chronic cases (Muck-Seler et al., 1988).

Based on the results of Murphy et al. (1977) it has been confirmed that platelet MAO-B activity decreases in chronic schizophrenia. The greatest mean percentage decrease in platelet MAO activity was reported in paranoid schizophrenic patients (Zureick and Meltzer, 1988). The effect of neuroleptic treatment on MAO-B activity is not clearly defined. Platelet MAO-B activity in neuroleptic-naive schizophrenics was similar to healthy controls (Marcolin et al., 1992), however, it gradually decreased in proportion to doses of haloperidol during treatment (DeLisi et al., 1981). It has been reported that haloperidol is able to block MAO-A activity in fibroblast cell culture, but hardly reduces platelet MAO-B activity itself (Giller et al., 1984). Certain metabolites of haloperidol are also capable of decreasing human platelet MAO-B activity and human placental MAO-A activity in vitro (Fang et al., 1995). These observations indicate that haloperidol and its metabolites may contribute to a decrease in platelet MAO-B activity of schizophrenic patients treated with neuroleptics. However, only a few studies have analysed the effect of other neuroleptic treatments (Spivak et al., 1994). We believe that any additional data in this field may contribute to clarify the extent of utility of platelets as pharmacological or biological markers in neuropsychiatric diseases.

The aim of the present work was to study the putative MAO inhibitory side effect of neuroleptics and simultaneous changes in platelet serotonin content. Different neuroleptic treatments and different diagnostic subtypes of schizophrenia were investigated, respectively.

Materials and methods

Patients

A control group of 30 healthy volunteers who were not treated with any neuroleptic, or agents that may alter serotonin metabolism, MAO activity or platelet function, and a group of 50 patients who fulfilled the criteria for schizophrenia in DSM-IIIR classification, treated with neuroleptics and nursed in the IVth Psychiatric Dept. of Jahn Ferenc Hospital have been selected. The control group was age and gender-matched with patients. The following parameters have been determined: clinical diagnosis according to

DSM-IIIR criteria (hebephrenic, catatonic, paranoid, residual and non differentiated subtypes), applied medication, platelet count, platelet rich plasma serotonin level (ng/10^9 platelets) and platelet MAO-B activity (mmol/hour/10^9 platelets). Data were analysed using one-way analysis of variance between groups. Probability levels of p less than 0.05 were considered significant.

Microdetermination of platelet MAO-B activity

To determine platelet MAO-B activity a radiometric method was adapted using ^{14}C-Tyramine as substrate, based on the method of Watanabe et al. (1990). This micro-determination has several advantages over other assays, for instance there is no need to centrifuge the samples. Centrifugation may destroy fragile platelets and may lead to loss of high density platelets with high MAO-B activity. Careful platelet preparation de-scribed above may reduce methodical error, however alterations in platelet size have been neglected. Another main feature of our methodology is using semicarbazide to inhibit semicarbazide sensitive amine oxidase (SSAO) in platelet rich plasma. This en-zyme may be responsible for degradation of at least 10% of biogenic amines found in human plasma (Yu, 1990).

Venous blood samples were withdrawn between 7 and 8 hour a.m. into 2 ml Vacu-tainer tubes containing 0.17 M EDTA-K_3 anticoagulant. Samples were stored in a cooler at $+4°C$ for two hours. Supernatant (platelet rich plasma: PRP) was used for MAO-B and serotonin determination. After a 20 minute preincubation period at room temperature the reaction was started by 80 μM ^{14}C-Tyramine substrate added to the incubation mixture containing 20 μl PRP, 0.05 M phosphate buffer (pH = 7.4), 1 mM semicarbazide in 100 μl final volume. After a 20 minute incubation period at 37°C the reaction was stopped by 40 μl 2 M citric acid. The formed product was extracted using 1 ml of toluene : ethylacetate 1 : 1 mixture and the samples were centrifuged (3,000 g, 1 min). An aliquot of 0.5 ml supernatant was added to 5 ml liquid scintillation cocktail and radioactivity was measured using a BECKMAN LS9000 counter. Platelet count was determined by CELL-DYN900 immediately after sedimentation. MAO-B activity was defined as mmol substrate de-graded/hour/10^9 platelets.

Determination of platelet serotonin content

Withdrawal and storage of samples were the same as above. Platelet count was deter-mined before and after sonication. Aliquots were deproteinated by 0.8 N perchloric acid containing N-methyl-serotonin as internal standard for HPLC-EC determination, then centrifuged again (8,000 g, 15 min). Supernatant was injected onto a HPLC column (Inertsyl 100 GL-4 ODS I 10/5). Serotonin content was expressed in ng/10^9 platelets.

Results

In this study we found that platelet serotonin content was significantly higher (mean ± S.D.: 906.19 ± 285.33 vs. 1,727.85 ± 947.40 ng/10^9 platelets, $p < 0.05$) and MAO-B activity was lower (15.26 ± 6.81 vs. 8.63 ± 3.82 mmol/hour/10^9 platelets, $p < 0.05$) in schizophrenics (n = 50) than in healthy controls (n = 30). There was no correlation found between these two parameters in the schizophrenic group (r = 0.217).

Table 1. Platelet MAO-B activity (mmol/hour/10^9 platelets: mean $\pm$ S.D.)

Treatment \ Type	Hebephrenic	Paranoid	Residual	Other type	Sum
Haloperidol	14.31 (n = 1)	8.21 ± 3.31 (n = 10)	9.10 ± 3.45 (n = 5)	11.60 (n = 1)	9.03 ± 3.42 (n = 17)
Clozapine	7.81 ± 3.25 (n = 3)	7.94 ± 4.43 (n = 8)	5.92 (n = 1)	11.70 (n = 1)	8.04 ± 3.84 (n = 13)
Other or combined neuroleptic therapy	13.11 ± 6.21 (n = 4)	6.94 ± 2.81 (n = 10)	7.34 ± 2.92 (n = 3)	9.84 ± 3.43 (n = 3)	8.67 ± 4.26 (n = 20)
Sum	11.27 ± 5.28 (n = 8)	7.68 ± 3.42 (n = 28)	8.16 ± 3.21 (n = 9)	10.56 ± 2.29 (n = 5)	8.63 ± 3.82 (n = 50)

Control: 15.26 ± 6.81 (n = 30)

To study whether these changes were induced only by haloperidol and its metabolites we examined the effect of various neuroleptic treatments. Table 1 shows MAO-B activity in schizophrenic patients treated with haloperidol (n = 17), clozapine (n = 13), thioridazine (n = 4), depot preparations (n = 3) and a combination of the above-mentioned neuroleptics (n = 13). No difference was experienced either in MAO-B activity or in platelet serotonin content between groups of schizophrenic patients treated with haloperidol, clozapine or other neuroleptics (see Table 2 for platelet serotonin content in the above-mentioned groups).

Finally, the patients were arranged according to DSM-IIIR diagnostic criteria. Platelet MAO-B activity was considerably lower in paranoid (7.68 ± 3.42 mmol/hour/10^9 platelets, n = 28) and residual schizophrenics (8.16 ± 3.21 mmol/hour/10^9 platelets, n = 9) compared with hebephrenic (n = 8), cataton (n = 2) and non differentiated (n = 3) subtypes. However, this difference was not significant. Platelet serotonin content was high in each diagnostic group, compared with healthy controls, but there was no significant difference between diagnostic subtypes of schizophrenia.

Discussion

In this study we determined both MAO-B activity and 5-HT content in platelets of schizophrenic patients and healthy controls to study the putative MAO inhibitory side effect of neuroleptics and simultaneous changes in platelet serotonin content.

The significantly lower MAO-B activity experienced in schizophrenics treated with neuroleptics compared with healthy controls corresponds to some previous findings reviewed by Zureick and Meltzer (1988). We have similarly found that paranoid schizophrenics had the greatest mean percentage decrease in platelet MAO-B activity, approximately 49.6%. However, MAO-B activity was not significantly different in diagnostic subtypes of

Table 2. Platelet serotonin content (ng/10⁹ platelets: mean ± S.D.)

Type Treatment	Hebephrenic	Paranoid	Residual	Other type	Sum
Haloperidol	763.44 (n = 1)	2,033.79 ± 773.80 (n = 9)	1,442.98 ± 657.77 (n = 5)	355.70 (n = 1)	1,664.89 ± 837.39 (n = 16)
Clozapine	1,275.32 ± 433.39 (n = 3)	1,852.26 ± 1,318.78 (n = 8)	—	—	1,694.91 ± 1,152.22 (n = 11)
Other or combined	1,925.14 ± 1,374.35 (n = 4)	1,636.06 ± 1,116.59 (n = 9)	2,441.49 (n = 1)	1,958.92 ± 455.68 (n = 3)	1,808.43 ± 1,026.11 (n = 17)
Sum	1,536.25 ± 1,031.54 (n = 8)	1,840.26 ± 1,051.70 (n = 26)	1,609.40 ± 715.75 (n = 6)	1,558.11 ± 883.75 (n = 4)	1,727.85 ± 974,40 (n = 44)

Control: 906.19 ± 285.33 (n = 26)

schizophrenia. On the basis of several observations on neuroleptic-naive patients, the experienced low MAO-B activity in schizophrenics is due to neuroleptic treatment rather than the disease itself (Marcolin et al., 1992). However, impact of genetic anomalies affecting MAO-B synthesis or activity may not be excluded. Other disturbing effects — i.e. non-specific platelet or mitochondrial alterations, changes in platelet turnover as a side effect of neuroleptics, duration of illness and of treatment with neuroleptics, disturbances in protein metabolism, hormone-related changes, endogenous inhibitors, smoking, stress, anaemia, alcohol abuse — may alter platelet MAO-B activity as well. According to our hypothesis, neuroleptics might accumulate in plasma and mitochondrial membranes of platelets and, thus they might reduce MAO-B activity regardless of their biochemical structure.

5-HT oxidation is hardly affected by the platelet MAO-B activity therefore negative correlation between serotonin levels and MAO-B activity was not experienced.

There may be several explanations for the significantly higher 5-HT platelet concentration experienced in schizophrenics. The concentration of platelet 5-HT may be regulated by its synthesis, uptake, storage, catabolism and/or release. Neuroleptics in vivo had no effect on platelet 5-HT level and uptake (Mück-Seler et al., 1988). Thus, the high serotonin content might be due to the illness regardless of the diagnostic subtype. Schizophrenia specific enhanced serotonin synthesis in enterochromaffin cells or a decreased serotonin release from platelets of schizophrenics may also be an explanation. We suppose that decreased degradation of plasma serotonin by MAO-A in the liver, lungs and endothelial cells may contribute to the experienced high platelet 5-HT content. Other factors — i.e. nutrition, hormone-related changes, diurnal and seasonal rhythms — may also influence the platelet serotonin content.

In summary, neither the low MAO-B activity nor the high platelet serotonin content may be specific for the diagnostic subtypes of schizophrenia. Different neuroleptic treatments have not affected these parameters differentially. The exact cause of hyperserotoninaemia in schizophrenia and the relationship between platelet MAO-B activity and neuroleptic treatment is not clearly defined till now.

Acknowledgements

We are grateful to our Ph.D. supervisors and colleagues for their indispensable help. Our research was supported by grant FEFA II–III.

References

Celada P, Sarrias MJ, Artigas F (1990) Serotonin and 5-hydroxyindoleacetic acid in plasma. Potential use as peripheral measures of MAO-A activity. J Neural Transm [Suppl] 32: 149–154
Da Prada M, Cesura AM, Launay JM, Richards JG (1988) Platelets as a model for neurones? Experientia 44: 115–126

DeLisi LE, Wise CD, Bridge TP, Rosenblatt JE, Wagner RL, Morihisa J, Karson C, Potkin SG, Wyatt RJ (1981) A probable neuroleptic effect on platelet monoamine oxidase in chronic schizophrenic patients. Psychiatry Res 4: 95–107

Fang J, Yu PH, Gorrod JW, Boulton AA (1995) Inhibition of monoamine oxidases by haloperidol and its metabolites: pharmacological implications for the chemotherapy of schizophrenia. Psychopharmacology 118: 206–212

Freedman D, Belendiuk K, Belendiuk G, Crayton J (1981) Blood tryptophan metabolism in chronic schizophrenics. Arch Gen Psychiatry 38: 655–659

Garelis E, Gillin J, Wyatt R, Neff N (1975) Elevated blood serotonin concentration in unmedicated chronic schizophrenic patients. Am J Psychiatry 132: 184–186

Giller E, Hall H, Reubens L, Wojciechoswki J (1984) Haloperidol inhibition of monoamine oxidase in vivo and in vitro. Biol Psychiatry 19: 517–523

Joseph M, Owen F, Baker H, Bourne R (1977) Platelet serotonin concentration and monoamine oxidase activity in unmedicated chronic schizophrenics and schizophrenic patients. Psychol Med 7: 159–162

Marcolin MA, Davis JM (1992) Platelet monoamine oxidase in schizophrenia: a meta-analysis. Schizophr Res 7: 249–267

Meltzer HY, Duncavage M, Arora RC, Tricou BJ, Jackman H, Young M (1982) Effects of neuroleptic drugs on platelet monoamine oxidase in psychiatric patients. Am J Psychiatry 139: 1242–1247

Murphy DL, Donnelly CH, Miller L, Wyatt RJ (1977) Platelet monoamine oxidase in chronic schizophrenia. Arch Gen Psychiatry 33: 1377–1381

Mück-Seler D, Jakovljevic M, Deanovic Z (1988) Time course of schizophrenia and platelet 5-HT level. Biol Psychiatry 23: 243–251

Siever LJ, Kahn RS, Lawlor BA, Trestman RL, Lawrence TL, Coccaro EF (1991) Critical issues in defining the role of serotonin in psychiatric disorders. Pharm Rev 43: 509–525

Spivak B, Kosower N, Zipser Y, Shreiber Schul N, Apter A, Tyano S, Weizman A (1994) Platelet monoamine oxidase activity in neuroleptic-naive schizpohrenic patients: lack of influence of chronic perphenazine treatment. Clin Neuropharmacol 17(1): 83–88

Stahl SM (1977) The human platelets. Arch Gen Psychiatry 34: 509–516

Watanabe K, Kobayashi S, Oguchi K (1990) Microdetermination of human platelet MAO activity. Showa Med Lett 44(2): 205–211

Wirz-Justice A (1988) Platelet research in psychiatry. Experientia 44: 145–151

Youdim MBH (1988) Platelet monoamine oxidase B: use and misuse. Experientia 44: 137–141

Yu PH (1990) Oxidative deamination of aliphatic amines by rat aorta semicarbazide-sensitive amine oxidase. J Pharm Pharmacol 42: 882–884

Zureick JL, Meltzer HY (1988) Platelet MAO activity in hallucinating and paranoid schizophrenics: a review and meta-analysis. Biol Psychiatry 24: 63–78

Authors' address: Dr. Z. Mészáros, Department of Pharmacodynamics, Semmelweis University of Medicine, Nagyvárad tér 4, P.O. Box 370, H-1445 Budapest, Hungary

Modulation of glutamate neurotoxicity in the transformed cell culture by monoamine oxidase inhibitors, clorgyline and deprenyl

O. Yu. Abakumova, O. V. Podobed, T. A. Tsvetkova,
I. V. Yakusheva, T. A. Moskvitina, L. I. Kondakova,
D. G. Navasardyantz, and A. E. Medvedev

Institute of Biomedical Chemistry, Academy of Medical Sciences, Moscow, Russia

Summary. Addition of 30 mM glutamate to the culture medium decreased growth of rat glioma C6 cells accompanied by a decrease of DNA synthesis and an increase of lactate dehydrogenase (LDH) detected in the conditioned medium. The presence of 1 μM deprenyl attenuated the glutamate effect on cell growth only during the first 24–48 h incubation and had a minor influence on the glutamate-induced decrease of DNA synthesis. Clorgyline (1 μM) potentiated glutamate-induced DNA synthesis during the first 24 h incubation without significant influence on the cell growth. Deprenyl slightly attenuated the glutamate-induced LDH increase during 24 h incubation but potentiated the glutamate effect at 96 h. Clorgyline decreased the glutamate influence at 24 h and especially 96 h. All these effects were observed in the absence of exogenous monoamines in the culture medium.

These results suggest that in transformed cells monoamine oxidase (MAO) inhibitors may influence processes of cell death via MAO-independent mechanisms.

Introduction

L-Depenyl, a selective inhibitor of monoamine oxidase B (MAO-B), has a positive therapeutic influence on the progression of disability in early Parkinson's disease (Tetrud and Lanston, 1989; Parkinson Study Group, 1993). Apart from the inhibition of MAO deprenyl has other actions. It increases longevity of rats (Knoll et al., 1989), induces superoxide dismutase activity in rat striata (Knoll, 1988; Carrillo et al., 1991; Clow et al., 1991). It has recently been shown that deprenyl reduced cell death without MAO inhibition (Tatton et al., 1994).

In the present report we have investigated the influence of monoamine oxidase inhibitors, clorgyline and deprenyl, on glutamate neurotoxicity in the transformed cell culture.

Materials and methods

Rat glioma C6 cells were kindly supplied by Dr. A. S. Kholansky (Institute of Human and Animal Morphology, Russian Academy of Medical Sciences, Moscow). For growth and DNA synthesis measurements cells were seeded after the treatment with 0.02% EDTA on 24 well plates. The initial cell density was 10^3 per well. Cells were routinely cultured 3 days in RPMI-1640 medium (Flow, UK) supplemented with 5% calf serum, 5% embryonic calf serum and 50 μg/ml Gentamycin. Cells were incubated unsealed at 37°C in incubators continuously flushed with a humidified atmosphere of 5% CO_2: 95% air. At the stage of dense monolayer 30 mM glutamate was added for the induction of cytotoxicity (Ratan et al., 1994). Monoamine oxidase inhibitors were added together with glutamate.

DNA synthesis was evaluated by incorporation of ^{14}C-thymidine (Isotope, Russia, sp. radioactivity 56 mCi/mmole). ^{14}C-Thymidine (5 μCi) was added to 1 ml of the culture medium 22, 70, and 94 h after glutamate (and MAO inhibitors). After the labeling for 2 h cells were washed with cold Hanks medium and fixed overnight with a cold mixture ethanol: ice-cold acetic acid (9:1) for removal of unbound ^{14}C-thymidine.

The cell monolayer was stained with 0.2% crystal violet in 2% ethanol (Medvedev et al., 1990). After the staining cells were washed with water and the dye was eluted with 10% acetic acid. Number of the cells was measured as the optical density at 595 nm (Keung et al., 1989) and OD of 0.1 at E^1_{595} corresponded to 32,500 cells (Abakumova et al., unpulished observation). Then cells were lysed with 0.3 N KOH overnight at 37°C, neutralised to pH 7.0 and the radioactivity was counted using Bray's solution. In our conditions DNA synthesis in control incubated without glutamate and MAO inhibitors was $31,100 \pm 2,920$ cmp/10^6 cells.

Lactate dehydrogenase (LDH) was measured in conditioned medium using Humalyser-2000 (Germany).

For the determination of MAO activity cells were harvested to 0.05 M phosphate buffer after the treatment with 0.02% EDTA-Na aqeous solution. Cell suspension was sonicated during 1 min at 0°C (400 Wt, 22 Hz) and 0.1 mg of the cell protein was used for the determination of MAO activity. The activity of MAO was assayed radiometrically (Tipton and Youdim, 1976) with minor modifications (Medvedev et al., 1994) using 0.1 mM ^{14}C-5-hydroxytryptamine (5HT) and 10 μM ^{14}C-2-phenylethylamine (PEA). Specific radioactivity in both cases was 10 Ci/mole.

In all the cases data represent mean of 4 independent experiments (each in quadruplicate). The statistical significance was evaluated by paired Student's t test.

Results and discussion

Sonicated cells slowly deaminated radiolabelled 5HT and PEA. These reactions were detected only at relatively high specific radioactivity of the substrates. Deamination of serotonin was inhibited by tetrindole ($IC_{50} = 5$ nM), the selective MAO-A inhibitor (Medvedev et al., 1994), whereas PEA deaminating activity exhibited lower sensitivity to deprenyl ($IC_{50} = 0.5$ μM).

Addition of 30 mM glutamate to the culture medium decreased growth of glioma cells (Fig. 1). The effect became evident after 24 h incubation and developed to 96 h. The presence of 1 μM deprenyl attenuated glutamate effect during 24–72 h incubation ($p < 0.05$). This influence disappeared to 96 h incubation. Clorgyline (1 μM) also slightly attenuated the glutamate effect but only during 24 h incubation.

A

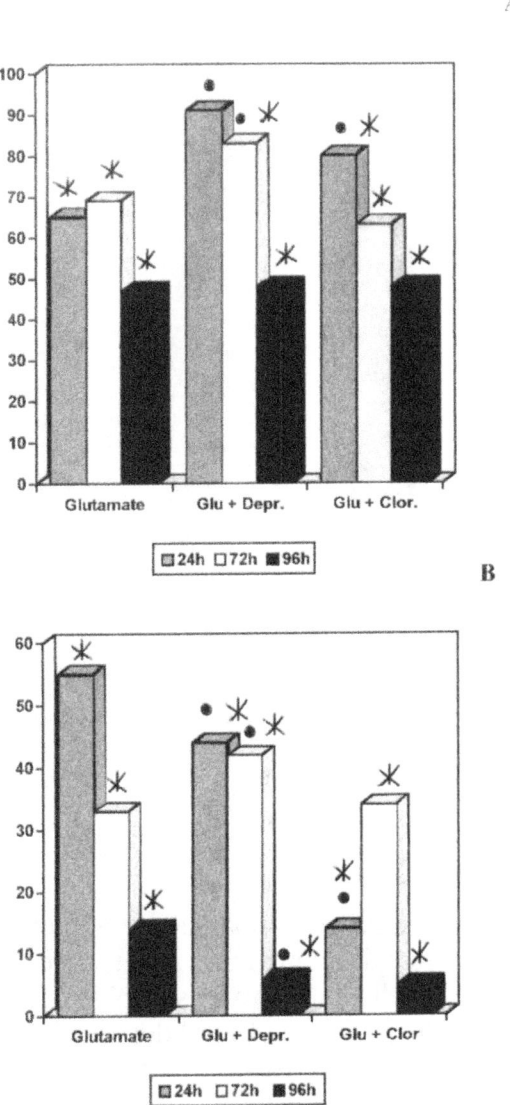

B

Fig. 1. The influence of $1 \mu M$ deprenyl and $1 \mu M$ clorgyline on the cell growth (**A**) and DNA synthesis (**B**) in the presence of 30mM glutamate. Data represent % of control without glutamate. Asterisk shows the significance of differences compared with control, incubated without glutamate and MAO inhibitors ($p < 0.05$). Closed circle shows the significance of the influence of MAO inhibitors ($p < 0.05$) on the glutamate effect

Glutamate addition resulted in a gradual decrease of DNA synthesis from 45% (after 24h) to 86% (after 96h). Deprenyl had a dual effect on the influence of glutamate on DNA synthesis. During the first 24h incubation deprenyl even potentiated glutamate-induced decrease in the DNA synthesis. However after 72h deprenyl slightly (but significantly, $p < 0.05$) attenuated the glutamate effect. In these conditions clorgyline did not protect DNA synthesis against the influence of glutamate.

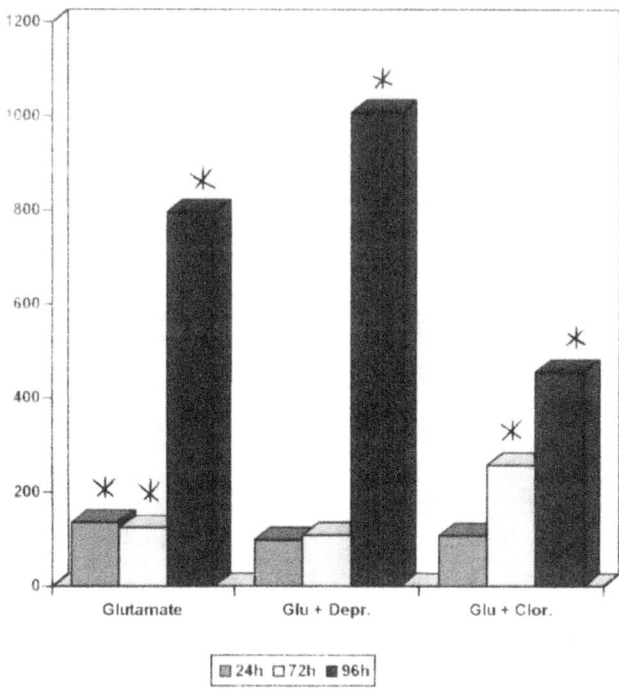

Fig. 2. The influence of 1 μM deprenyl and 1 μM clorgyline on glutamate effect on LDH release to the conditioned medium (% of control). Asterisk shows the significance of differences compared with control, incubated without glutamate and MAO inhibitors (p < 0.05)

Glutamate neurotoxicity was accompanied by an increase of lactate dehydrogenase in the condensed culture medium. After some augmentation (+20–40%, p < 0.05) during 24–72 h LDH activity sharply increased (+695%) after 96 h incubation (Fig. 2). The presence of 1 μM deprenyl abolished the increase of LDH activity during 24–72 h. However after 96 h incubation depreynl potentiated the glutamate effect on LDH release. In contrast to deprenyl, clorgyline potentiated the glutamate effect on LDH release after 72 h incubation but attenuated augmentation of LDH activity after 96 h.

It is known that glutamate may cause an oxidative stress (Murphy et al., 1990), leading to apoptosis in embryonic cortical neurons (Ratan et al., 1994). The MAO inhibitors deprenyl and to a lesser extent clorgyline antagonise the glutamate effect mainly during 24–72 h incubation. Since biogenic amines MAO substrates, were not included in the medium, the results of the present report give further evidence that in transformed cells MAO inhibitors may influence processes of cell death via MAO-independent mechanism(s).

Further experiments are required to clarify these mechanisms.

Acknowledgements

This work was supported by a grant from Russian Academy of Medical Sciences.

References

Carillo MC, Kanai S, Nokubu M, Kitani K (1991) (−)Deprenyl induces activities of both superoxide dismutase and catalase but not glutathione peroxidase in the striatum of young male rats. Life Sci 48: 517–521

Clow A, Hussain T, Glover V, Sandler M, Dexter DT, Walker M (1991) (−)Deprenyl can induce soluble superoxide dismutase in rat striata. J Neural Transm 86: 77–80

Keung W, Sieber E, Eppenberger U (1989) Quantitation of cells cultured on 96 well plates. Anal Biochem 182: 16–19

Knoll J (1988) The striatal dopamine dependency of life span in male rats. Longevity study with (−)deprenyl. Mech Ageing Dev 46: 237–262

Knoll J, Dallo J, Yen TT (1989) Striatal dopamine, sexual activity and life span. Longevity of rats treated with (−)deprenyl. Life Sci 45: 525–531

Medvedev AE, Fuchs BB, Rakhmilevich AI (1990) A study of the action of immunosupressive factors from tumor cells on lymphocytes and macrophages in vitro and on the graft-versus-host reaction in mice. Biomed Sci 1: 261–266

Medvedev AE, Kirkel AZ, Kamyshanskaya NS, Axenova LN, Moskvitina TA, Gorkin VZ, Andreeva NI, Mashkovsky MD (1994) Monoamine oxidase inhibition by novel antidepressant tetrindole. Biochem Pharmacol 47: 303–308

Murphy TH, Miyamoto M, Sastre A, Schnaar R, Coyle JT (1989) Glutamate toxicity in a neuronal cell line involves inhibition of cysteine transport leading to oxidative stress. Neuron 2: 1547–1558

Parkinson Study Group (1993) Effects of tocopherol and deprenyl on the progression of disability in early Parkinson's disease. N Engl J Med 328: 176–183

Ratan RR, Murphy TH, Baraban JM (1994) Oxidative stress induced apoptosis in embryotic cortical neurons. J Neurochem 62: 376–379

Tatton WG, Seniuk NA, Ju WYH, Ansari KS (1994) Reduction of nerve cell death by deprenyl without monoamine oxidase inhibition. In: Lieberman A, Olanow O, Youdim MBH, Tipton KF (eds) Monoamine oxidase inhibitors in neurological diseases. Raven Press, New York, pp 217–248

Tetrud VW, Lanston JW (1989) The effect of deprenyl (selegiline) on the natural history of Parkinson's disease. Science (Wahsington DC) 245: 519–522

Tipton KF, Youdim MBH (1976) Assay of monoamine oxidase. In: Wolstenholm GEW, Knight J (eds) Monoamine oxidase and its inhibition. Elsevier, Amsterdam, pp 393–403

Authors' address: A. E. Medvedev, Institute of Biomedical Chemistry, Russian Academy of Medical Sciences, 10 Pogodinskaya str., Moscow, 119832 Russia

A cell culture model of cerebral ischemia as a convenient system to screen for neuroprotective drugs

J. Ekblom[1], **H. Garpenstrand**[1], **O. Tottmar**[2], **J. A. Prince**[1], and **L. Oreland**[1]

Departments of [1] Medical Pharmacology and [2] Animal Physiology, Uppsala University, Biomedical Center, Uppsala, Sweden

Summary.   Aggregation cultures of rat brain were exposed to a combination of anoxia and hypoglycaemia for 30 minutes. Thereafter, the release of lactate dehydrogenase into the cell culture medium was monitored up to 4 days as a measure of cell damage after the ischemic insult. Some cultures were treated with different concentrations of deprenyl or tolcapone, selective inhibitors of monoamine oxidase B and catechol-O-methyltransferase, respectively. After 1 day in culture, the release of lactate dehydrogenase was significantly reduced in cultures treated with deprenyl (at 1 nM, 100 nM, and 10 μM), as well as in cultures treated with 1 nM or 100 nM tolcapone; 10 μM of tolcapone, on the other hand, resulted in a toxic effect on the cell aggregates. No differences in the release of lactate dehydrogenase into the medium was observed in the aggregates treated with drugs as compared with the control cultures after 2 or 4 days post-ischemia.

Introduction

Cerebral ischemia is the main cause of permanent brain damage after traumatic or vascular disease. Development of drugs that could reduce the damage following ischemic insults would dramatically improve the treatment of disorders such as stroke and neurotrauma. Different animal models of transient ischemia have been developed and a variety of biochemical mechanisms involved in the development of ischemic damage have been described, such as the effects of free radical reactions, failing energy metabolism with excessive lactate accumulation giving rise to acidosis, and neurotoxic effects of excitatory aminoacids.

Today, a large number of drugs with potential neuroprotective capacity are evaluated in animal models for brain damage. Current rodent models of cerebral ischemia are associated with several disadvantages, e.g. the inter- and intra-assay variation is often considerable and the study of drug-induced effects is complicated by peripheral metabolism. Therefore, the use of an *in vitro* model seems very attractive. The technique of rotation-mediated aggregating cell cultures was originally introduced by Moscona and collaborators

(1960). They demonstrated that freshly isolated immature cells from any organ possess the ability to re-associate spontaneously *in vitro*, giving rise to three-dimensional, organotypic cultures (Moscona et al., 1960). In the present report we describe the use of fetal rat brain aggregation cultures (Honegger et al., 1979), a system with multiple advantages as compared with other cell culture models. For instance, the aggregates are composed of multiple cell types similar with the mature CNS and they may be maintained in a serum-free medium. In a recent investigation it has been shown that these cultures respond to oxygen- and glucose-deprivation in a fashion that is very similar to the process observed after transient ischemia in humans and experimantal animals (Tottmar et al., unpublished observation). In the present study, we have therefore used this system to screen for drugs that may be neuroprotective after ischemia.

The L-form of deprenyl (selegiline) has been shown to have neuroprotective and/or neurorescuing capacity in several different paradigms for neuronal damage (for a review, see Tipton, 1995). This was first demonstrated more than 10 years ago in a retrospective study of patients suffering from Parkinson's disease (Birkmayer et al., 1985). It was shown that patients who received deprenyl in conjunct with L-dopa had a longer average lifespan as compared with patients that received L-dopa only. However, today, a decade after the study of Birkmayer and coworkers was published, the mechanism is still enigmatic. It has been postulated that the mechanism is independent on the capacity of this drug to cause inhibition of MAO-B (for a review, see Tipton, 1995). For instance, it has been shown that deprenyl enhances rescue of motor neurons in the nucleus facialis after axotomy at very low doses (5–10μg/kg/day) (Ansari et al., 1993). These doses do not cause any measurable inhibition MAO.

The effects of catechol-O-methyltransferase (COMT) inhibitors are even less known. In a recent work by Khromova and collaborators (1995) it was shown that chronic treatment with moderate doses of tolcapone, which is a reversible COMT inhibitor, enhanced the behavioural recovery of rats with toxic lesions in the forebrain.

Materials and methods

Cell cultures

The cell cultures were prepared and maintained according to the protocol of Honegger (1985). Briefly, pregnant Sprague-Dawley rats were purchased from ALAB (Stockholm, Sweden) and at gestational day 15–16 the brains of the embryos were dissected and the forebrains (telencephlon + diencephalon) were dissociated through nylon nets and thereafter diluted in a deflned serum-free DMEM-based medium. The cell suspension was kept under continous gyratory rotation (80rpm) in a humidifled atmosphere consisting of 10% CO_2 and 90% air. Every second day 5ml of old medium was removed and the same volume of fresh medium was added. On day 40, aggregates were exposed to the anoxic and hypoglycemic insult. Aggregates were washed once in phosphate-buffered saline (PBS) containing $CaCl_2$ and $MgCl_2$ and thereafter exposed to PBS that was "deoxygenated" by bubbling with sterile nitrogen. The nitrogen exposure was maintained for 30 min. The treatment was terminated by transfer of the aggregates to fresh medium. Control cells were washed in PBS in parallell with the treated cells but were thereafter transferred back to fresh cell culture medium. After the treatment the cells were kept in 6-well plates (Falcon, Cedex, France) and the test substances were included in the medium. The medium was completely changed every day after the treatment. Samples of medium and aggregates were collected after 1, 2, and 4 days, repsectively.

Enzyme activity assays

For the enzyme activity assays, the medium samples were centrifuged at 10,000 g for 5 min and thereafter stored at $-70°$ until analysis. The aggregate samples were washed twice in PBS and thereafter sonicated in ice-cold 10 mM Tris-HCl-buffer, PH 7.4 for 3 × 5 s. The homogenates were stored at $-70°C$ in aliquots until analyzed at a concentration of approximately 5% (tissue w/v). Enzyme activities estimated in homogenates were standardized to protein concentration which was assayed according to the procedure of Lowry and coworkers (1951). Lactate dehydrogenase (LDH) was estimated according to the protocol of Martinek (1972). The LDH activiy in the medium was expressed as a ratio LDH_{med}/ LDH_{agg}.

Catalytic activities of MAO and COMT were determined in order to reveal the level of enzyme inhibition caused by the the two drugs, according to the procedure described by Ekblom and coworkers (1993). For analysis of MAO A and B (MAO-A and MAO-B; E.C. 1.4.3.4) catalytic activity, the $[^{14}C]$-serotonin and $[^{14}C]$-phenylethylamine were used as substates for the two subtypes, respectively. COMT (E.C. 2.1.1.6) catalytic activities were analyzed using $[^{3}H]$methyl-S-adenosylmethionine as methyl donor and catechol as substrate.

Statistical analysis

Statistical signiflcance of differences between different groups was analysed using Students unpaired t-test.

Results

LDH release

Detectable amounts of LDH were released into the medium at all the time points studied after the ischemic insult. In both control cultures and cultures treated with drugs, the LDH-release was most prominent during the first 24 hours. Thereafter, the release gradually decreased. In the control cultures, the fraction of LDH released into the medium at day 1 was 9.9%. At the two later time points (day 2 and 4) the relative release was 2.9% and 2.1%, respectively. Treatment with deprenyl and tolcapone significantly reduced the LDH-release during the first 24 hours (Fig. 1a). There were no statistically significant effects of the drugs after 2 and 4 days post-ischemia. With regard to deprenyl, all three concentrations used, resulted in a cytoprotection after 24 h that was statistically significant. The MAO-B catalytic activity was approximately 45% as compared

LDH-release (%)

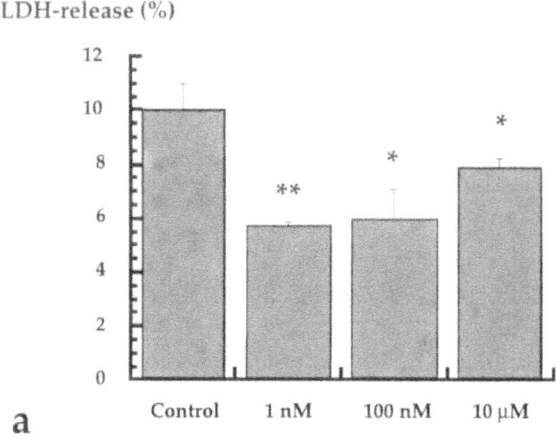

a

LDH-release (%)

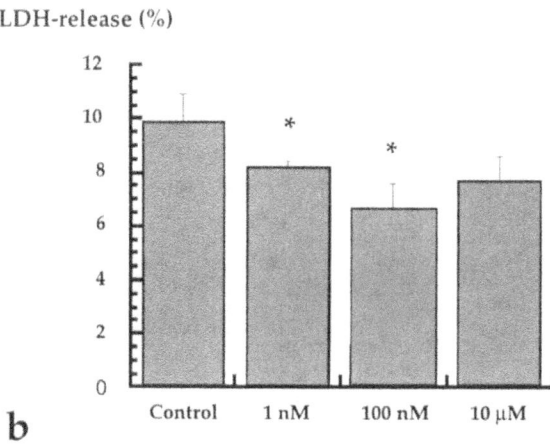

b

Fig. 1. The graphs shows the fraction of tissue-bound LDH that was released into the cell culture medium during the first 24 hours after 30 min ischemia. The cultures were treated with **a** deprenyl or **b** tolcapone. Statistical analysis was performed using Students t-test ($*P < 0.05$, $**P < 0.01$)

with control cultures after treatment with 10^{-9}M of deprenyl, which was the lowest concentration tested. The two higher concentrations caused a complete inhibition of both MAO-A and -B in accordance with published data. Three different concentrations of the reversible COMT inhibitor tolcapone were also investigated in the system. The two lower of these concentrations, 10^{-9} and 10^{-7}M, respectively, caused a significant reduction in the LDH-release into the medium, suggesting that also this drug may have cytoprotective capacity (Fig. 1b). In contrast, when the aggregates were exposed to 10^{-5}M of tolcapone, no change in the LDH-release was observed, as compared with the control aggregates. However, on day 2 and 4, the LDH-release from the aggregates treated with 10^{-5} M tolcapone was significantly higher, suggesting that there was a toxic effect of the drug at this concentration. The colour of the culture medium revealed that the pH was lower than in the control cultures. The toxic effect of tolcapone concentrations above 10^{-5} M was confirmed by treatment of non-ischemic control cultures (data not shown).

Discussion

It has previously been demonstrated that fetal rat brain aggregation cultures respond to ischemia in a very "*in vivo*-like" fashion (Tottmar et al., unpublished observations). For example, neuronal cell-death, both in the early and late phase of the ischemic response as well as the occurence of reactive gliosis, has been reported (Tottmar et al., 1996). Thus, the system seems excellent for screening of neuroprotective drugs.

In the present investigation we observed a high level of LDH release from the aggregates after the ischemic insults. The level of the LDH release was highest during the first 24 hours. However, we also observed a significant LDH release at later time points, possibly reflecting a component of delayed cell death. The control aggregates also showed LDH release, although significantly lower. These findings are all in accordance with the previous report (Tottmar et al., 1996). This model has many advantages, e.g., it offers a high level of reproducibility, that makes it suitable for initial (first-stage) screening of drug candidates that may have neuroprotective/neurorescuing properties.

In a recent paper it was shown that deprenyl has a neuroprotective effect in a rat model of transient ischemia (Lahtinen et al., 1995). The present finding, of an effect of deprenyl on cell survival after oxygen deprivation is therefore in accordance with previous investigations. It did not seem as if the protective effect was dependent upon the level of MAO-B inhibition, since the effect of the lowest dose, with modest MAO inhibition was similar to the higher doses where complete inhibition of MAO was achieved.

Interestingly, tolcapone treatment also resulted in a protective effect. To our knowledge, this is the first study where it has been shown that a COMT inhibitor has neuroprotective properties in an *in vitro* paradigm. It is interesting to note that it has been shown that COMT catalytic activity is increased in CNS tissue in a neurodegenerative disorder such as amyotrophic lateral sclerosis (Ekblom et al., 1993). The present study suggests that COMT inhibitors may be evaluated with regard to potential therapeutic effects in neurodegenerative disorders such as Alzheimer's disease and amyotrophic lateral sclerosis.

In conclusion, the present study describes the use of brain aggregation cultures as a system to screen for neuroprotective drugs. It was shown that both deprenyl and tolcapone could reduce the extent of the cell damage occuring during the first 24 hour after an ischemic insult.

Acknowledgements

The authors would like to thank Prof. P. Männistö for kindly providing tolcapone. This work was supported by grants from The Swedish Medical Research Council, no. 4145.

References

Ansari KS, Yu P, Kruck TPA, Tatton WG (1993) Rescue of axotomized immature rat facial motoneurons by R(−)-deprenyl: stereospecificity and independence from monoamine oxidase inhibition. J Neurosci 13: 4042–4053

Ekblom J, Aquilonius S-M, Jossan SS (1993) Differential increases in catecholamine metabolizing enzymes in amyotrophic lateral sclerosis. Exp Neurol 123: 289–294

Honegger P, Lenoir D, Favrod P (1979) Growth and differentiation of aggregating fetal brain cells in a serum-free defined medium. Nature 282: 305–308

Honegger P (1985) Biochemical differentiation in serum-free aggregating brain cell cultures. In: Bottenstein JE, Sato G (eds) Cell cultures in the neurosciences. Plenum Press, New York, pp 223–243

Khromoval I, Rauhala P, Zolotov N, Männistö P (1995) Tolcapone, an inhibitor of catechol-O-methyltransferase counteracts memory deflcits caused by bilatral cholinotoxin lesions of the basal nuclei of Meynert. Neuroreport 6: 1219–1222

Lahtinen H, Koistinaho J, Kauppinen R, Haapalinna A, Keinanen R, Sivenius J (1997) Selegiline treatment after transient global ischemia in gerbils enhances the survival of CA1 pyramidal cells in the hippocampus. Brain Res 757: 260–267

Lowry O, Rosebrough N, Farr A, Randall R (1951) Protein measurement with the Folin phenol reagent. J Biol Chem 193: 265–275

Martinek RG (1972) A rapid ultraviolet spectrophotometric lactic dehydrogenase assay. Clin Chim Acta 40: 91

Moscona AA (1960) Patterns and mechanisms of tissue reconstruction from dissociated cells. In: Rudnick R (ed) Developing cell system and their control. Ronald Press, New York, pp 45–70

Tipton K (1995) What is it that l-deprenyl (selegeline) might do? Clin Pharmacol Ther 56: 781–796

Authors' address: Dr. J. Ekblom, Department of Medical Pharmacology, Uppsala University, Biomedical Center, Box 593, S-751 24 Uppsala, Sweden

Neuroprotection by selegiline and other MAO inhibitors

G. Stern

Department of Clinical Neurology, University College, London Hospitals,
United Kingdom

Summary. A proposal for the nomination of the father of monoamine oxidase inhibitors is presented. A brief history of the human clinical pharmacology of selegiline is considered including the results of two major prospective ongoing clinical trials and recent evidence on the effects of sustained selegiline therapy on postural blood pressures in parkinsonians is discussed.

Introduction

As a naive clinician among distinguished pharmacologists, I would have assumed that "the mother of amine oxidases" was Miss Hare (1928) who discovered tyramine oxidase in the liver. Although Miss Hare did not see any oxidation of adrenaline, when nine years later Blaschko clearly established the actions of MAO on adrenaline, noradrenaline and dopamine, he and his colleagues acknowledged Miss Hare's priority.

As for a "father" of monoamine oxidase inhibitors I propose the name of Richard Spruce (1908). An intrepid British botanist, Spruce explored the Amazon and its tributaries during the years 1849–1864 and in his "Notes of botanists on the Amazon and Andes" provides a detailed account of his discovery indicating that field work in the Amazon Valley was as interesting as in the laboratory, but potentially more dangerous.

"In the accounts given by travellers of the festivities of the South American Indians and of the incantations of their medicine men, frequent mention is made of powerful drugs used to produce intoxication or even temporary delirium . . . having had the good fortune to see the two most famous narcotics in use and to obtain specimens of the plants that afford them sufficiently perfect to be determined botanically I propose to record my observations on them, made on the spot. The first of these narcotics is afforded by a climbing plant called caapi . . . I drew up the following description in November 1853. Woody twiner . . . petals five on longish thick claws . . . lamina pentagonal fimbriate . . . habitat on the river Yuapes and other upper tributaries of the Rio Negro where it is commonly planted in mandicca plots . . . the lower part of the stem is the part used (caapi — the Portuguese have made it cappim — is the tupi or lingoa Geral name for grass. It simply means thin leaf).

In November 1852 I was present at a feast of gifts held in a village house . . . we reached the Malloca at nightfall just as the sacred trumpets began to boom lugubriously within the margin of the forest . . . at that sound every female outside makes a rush into the house before the sacred trumpets emerge on the open, for to merely see one of them would be to her a sentence of death ... about three hundred people are assembled and the dance is at once commenced ... in the course of the night young men that took part took of caapi five or six times in the intervals between the dances, but only a few of them at a time and very few drank it twice. The cup bearer — who must be a man for no woman can touch or taste caapi — starts at a short run from the opposite end of the house with a small calabash containing about a teacupful of caapi in each hand ... in two minutes or less after drinking it its effects begin to be apparent. The Indian turns deadly pale, trembles in every limb, and horror is in his aspect. Suddenly contrary symptoms succeed: he burst into a perspiration and seems possessed with reckless fury, seizes whatever arms are at hand, his bow and arrows or cutlasses, and rushes to the doorway where he inflicts violent blows on the ground or the doorpost calling out all the while "Thus would I do to mine enemy were this he!" In about ten minutes excitement has passed off the Indian grows calm but appears exhausted. . . .

I had gone with the full intention of experimenting with the caapi myself, but I had scarcely despatched one cup of the nauseous beverage, which is but half a dose, when the ruler of the feast — desirous apparently that I should taste all his delicacies at once — came up with a woman bearing a large calabash of mendicca beer of which I must needs take a copious draft and as I knew from the mode of its preparation was gulped down with secret loathing. Scarcely had I accomplished this feat when a large cigar, two feet long and as thick as the wrist, was put lighted into my hand and etiquette demanded that I should take a few whiffs of it — I, who had never in my life smoked a cigar or a pipe of tobacco. Above all this I must drink a large cup of palm wine and it will readily be understood that the effects of such a complex dose was a strong inclination to vomit . . .

White men who have partaken of caapi in the proper way concur in the accounts of the sensations under its influence. They feel alternations of cold and heat, fear and boldness. The sight is disturbed, and visions pass rapidly before the eyes wherein everything gorgeous and magnificent they have heard or read seems combined; presently the scene changes to things uncouth and horrible ... a Brazilian friend said that when he once took a full dose of caapi he saw all the marvels that he had read of in the Arabian nights pass rapidly before his eyes as in a panorama, but the final sensations and sights were horrible . . .".

Thus Spruce merited his nomination. Early chemical investigations of the plant indicated the presence of an alkaloid given the mysterious name telepathine and later Beringer obtained a crystalline.

In Berlin, Louis Lewin (1928) isolated an alkaloid which he named banisterine. German chemists later considered banisterine to be identical with the base harmine an alkaloid isolated eighty-seven years earlier from the plant Peganum harmala known throughout the Middle East as an intoxicant. All of the bases isolated from b. Caapi had a beta carboline structure with different degrees of hydrogenation of the pyridine group identical to those of the harmala alkaloids.

Returning to the story of Louis Lewin, he recommended banisterine for the treatment of diseases of the nervous system. Kurt Beringer (1928) pursued Lewin's suggestion and gave banisterine to fifteen patients with post-encephalitic parkinsonism. He described dramatic positive effects on rigidity and akinesia when banisterine was given subcutaneously, orally and even rectally although little benefit was seen on tremor. A year later Lewin and Schuster (1929) presented a paper at the Berlin Medical Association describing the dramatic beneficial effects of banisterine in eighteen parkinsonians when given subcutaneously in doses of 20–40 mg. A film of the effects of the drug was shown at the conference and caused considerable popular interest in the media of that time. Side-effects were reported to be slight. Others confirmed the beneficial effects. One courageous investigator, L. Halpern (1930), ingested up to 40 mg by mouth and 30 mg subcutaneously and described an immediate sensation of excitement and restlessness.

"The impression was felt as if the consciousness was packed in ether. When lying on a sofa, the lightness increased to a feeling of fleeting sensation and the weight of the body was subjectively less. These clinical observations should be compared to the state of levitation frequently reported to occur with the crude drug ayahauasca or caapi . . . with higher doses excitation was increased even in a belligerent way . . . the author who is normally not belligerent started a fight with a man in the street where she was the one who attacked, even though according to the circumstances the prospect for the attacker was unfavourable".

Enthusiasm for the caapi and harmala alkaloids waned after the mid-1930s, but pharmacological interest was rekindled when Udenfriend (1958) demonstrated that these drugs were inhibitors of MAO. From that point harmine and harmaline were used experimentally for their capacity to reversibly inhibit MAO. If the reader sees any parallel between the media-generated frenzy stemming from caapi and that from events related to selegiline then the writer cannot resist repeating the adage "If the caapi fits wear it".

The selegiline story

When synthesised over thirty years ago the drug was intended to be a "psychic energiser" combining the pharmacological properties of amphetamine with those of a monoamine oxidase inhibitor (Knoll, 1965). It was shown to have a modest antidepressant effect (Mann, 1978), that it possessed a cocaine-like inhibition of dopamine uptake (Knoll, 1978), and was found to be metabolised to met-amphetamine and amphetamine (Reynolds, 1978).

After a decade selegiline was accepted to be a selective inhibitor of MAO-Type B and was the first MAO inhibitor to be free from the hypertensive hazards following the ingestion of tyramine-rich foods — "the cheese effect" (Elsworth, 1978). It was soon combined with levodopa in the treatment of Parkinson's disease. Several studies showed that selegiline had a useful anti-parkinsonian symptomatic benefit and also a significant levodopa-sparing effect (Birkmayer, 1975; Lees, 1977).

Meanwhile laboratory evidence was accumulating that selegiline had a protective effect at least as far as MPTP toxicity was concerned (Heikkila, 1984). Thus, when Birkmayer et al. (1985) reported "Increased life expectancy resulting from addition of l-deprenyl to Madopar treatment in Parkinson's disease: a long-term study" great interest among neurologists, the public at large and of course inevitably the media was generated. The authors interpreted their results "as indicating l-deprenyl's ability to prevent or retard the degeneration of striatal dopaminergic neurons".

The study was impressive in that it compared the effects of treatment with Madopar alone (77) or in combination with selegiline (564) in a large group of patients and survival analysis revealed a significant increase of life expectancy in those taking selegiline. Any trial of patients involving large numbers and studied over many years, is likely to be imperfect. This influential study was open, uncontrolled and was analysed retrospectively. The daily dose of selegiline was 5–10 mg and the Madopar dosage was adjusted according to patients' needs. The average length of selegiline treatment was 3.92 years which was started between 1974 and 1983. The time from the onset of Parkinsonism to the start of levodopa substitution was significantly longer (3.9 years) in the group with combined treatment than in the Madopar only group (2.7 years). There were differences between the male/female ratios and baseline disability scores. The daily l-dopa dose averaged for the whole period of substitution therapy was lower in the Madopar group (524 mg) than in the combined treatment group (627 mg). Statistical analysis revealed a significant difference between the two treatments in favour of the combined medications, the estimated survival time was 129 months in the Madopar and 145 in the Madopar/selegiline patients.

It is noteworthy that no causes of death were mentioned in the study. The authors were aware of the limitations of their trial methodology and were appropriately cautious about explaining how the addition of selegiline might possibly increase life expectancy.

It was clearly desirable to confirm or refute this potentially important study and conclusions. While the concept of "neuroprotection" was understandably attractive to patients, clinicians and scientists, confirmatory long-term studies in homo sapiens are considerably more arduous than those in laboratory animals. Notwithstanding these considerable difficulties at least two major, long-term randomised, prospective studies were initiated — one in the United States and the other in the United Kingdom. Interim reports have been published. These will be briefly reviewed and some recent as yet unpublished information concerning selegiline in parkinsonians will be presented.

Ongoing long-term trials

a) In the Parkinson Study Group (1989) controlled trial of deprenyl and tocopherol antioxidative therapy of Parkinsonism (DATATOP), 800 hitherto untreated patients were enrolled. The initial results appeared to demonstrate that selegiline (but not tocopherol) delayed the progress of Parkinson's disease. However, this decision was primarily determined by the interval of time that elapsed between commencing "treatment" versus placebo, and the time when the patient's deterioration was deemed to be sufficient to introduce conventional levodopa therapy. It seems that the possibility of symptomatic benefit from selegiline monotherapy — a conclusion which could have been reasonably drawn from reported studies prior to the commencement of DATATOP — was not taken into consideration in the trial design. Slowly, initial enthusiasm for the selegiline-neuroprotection hypothesis began to wane and the two most recent publications from the ongoing DATATOP trial reflect decreasing assertiveness concerning the original claim. In the interim DATATOP report (1996a) concerning "the impact of deprenyl and tocopherol treatment on Parkinson's disease insect.... prior treatment with deprenyl did not lead to superior survival with respect to end point of disability requiring levodopa, suggesting that the initial advantages of deprenyl were not sustained". In those patients requiring levodopa the rate of development of levodopa-associated side-effects did not differ among the original treatment groups (early v. late deprenyl and tocopherol v. non-tocopherol). At the end of the study the groups were similarly disabled on standard rating scales and taking similar amounts of levodopa. The authors concluded that "prior treatment with deprenyl or tocopherol did not reduce the occurrence of subsequent levodopa-associated effects in this population" (1996b).

The only wholly persuasive evidence from this trial concerning demonstration of "neuroprotection" will not emerge until mortality figures are available.

ii) The Parkinson's Disease Research Group of the United Kingdom open long-term prospective randomised trial involved 782 patients with early Parkinson's disease who had not previously received dopaminergic treatment. They were allocated to Arm 1 (conventional treatment with levodopa and dopa/decarboxylase inhibitor), Arm 2 (selegiline combined with the treatment in Arm 1), and Arm 3 — neat bromocriptine. It was not thought ethically advisable or indeed practical — although it would have been interesting — to include a placebo arm. The reason for the bromocriptine arm was that by the mid-1980s evidence was accumulating from other trials that those patients who were able to tolerate an adequate dose of bromocriptine could benefit at least as well as from a comparable dose of levodopa and later showed fewer of the long-term side-effects associated with levodopa therapy.

The first interim report (PDRG-UK, 1993) showed no marked differences in functional improvement between the progress of the patients in the three arms, suggesting that the choice of treatment in the early stages of the disease might not be as critical as others had claimed. There were insufficient deaths at that stage of the trial to study mortality.

The second interim report (PDRG-UK, 1995) was published after an average follow-up period of 5.6 years. The mortality ratio in Arm 2 compared with Arm 1 was 1.57 (95% confidence interval 1.09–2.30) and differences in survival between the two arms was significant (p = 0.015). Hazard ratio adjusted for age and sex was 1.49 (1.02–2.16) and after adjustment for other baseline factors it increased to 1.57 (1.07–2.31). Patients in Arm 1 had slightly worse disability scores than those in Arm 2, but the differences were not significant. Peak dose dyskinesia and on-off oscillations were more frequent in Arm 2 than Arm 1. During the trial the dose of levodopa required to produce optimal clinical control increased in Arm 1 (mean daily dose 375 mg at one year, 625 mg at four years), but the median dose in Arm 2 did not change (375 mg). The latter would be compatible with the previously demonstrated levodopa-sparing effect of selegiline.

The unexpected mortality figures indicated that mortality among patients in Arm 2 was about 60% higher than in Arm 1 and there was no evidence that this was affected by sex or age. Possible bias was considered — for example, that a physician's knowledge of a particular patient's progress could influence a decision as to whether to withdraw that patient from the trial, but if this was the case "on-treatment" analysis would produce different results whereas it gave a similar hazard ratio. In fact the number of patients withdrawn because of deteriorating responses in the two arms did not differ greatly and in fact more patients were withdrawn from treatment with levodopa and selegiline because of adverse reactions. Furthermore, "we had no a priori reason to believe that the patients receiving levodopa and selegiline would fare worse than those receiving levodopa monotherapy; the opposite would have been predicted from previously reported studies". As in most European countries the number of autopsies is not large.

This paper provoked vigorous correspondence and in response the PDRG-UK (1996) emphasised that a definite cause or relationship between selegiline 10 mg a day and increased mortality in Arm 2 had not been established and no firm and final conclusions could be drawn. However, it should be recalled that there is no study truly comparable with respect to size, duration of follow-up and the use of mortality as a primary end point. The PDRG-UK does not of course exclude the possibility that selegiline may have a mild neuroprotective effect, but the question of clinical relevance against other emerging results must be carefully considered. After careful deliberation the PDRG-UK felt that all patients receiving selegiline in the trial should be notified of these results, that selegiline should be withdrawn and the consequences of this decision on disability and subsequent levodopa requirements will be reported together with further mortality figures in about six months' time. Meanwhile, an independent team of assessors are conducting a cause of death examination study of the deaths which have occurred so far.

Tests of autonomic function in parkinsonians on sustained selegiline therapy

Twenty-five outpatients attending the neurology clinics at University College London Hospitals, matched for age, disease and duration of disease severity on conventional therapy (levodopa plus or minus anticholinergics, tricyclics and dopamine agonists) underwent autonomic studies. Nine had never taken selegiline and sixteen had taken selegiline for a minimum period of three years. Standard tests of autonomic function were performed in a department exclusively devoted to these assessments. Tests included recordings of blood pressures, catechols (adrenaline, noradrenaline and dopamine), and heart rate; all three assays were determined in the reclining and standing position for two minutes and supine and tilted at 45 degrees for 2 and 10 minutes. In addition, tests were made of the heart rate response to deep breathing for two minutes, mental arithmetic for two minutes, cold face pack for 45 seconds, cold pressor test for 40 seconds and Valsalva ratio (15 seconds; minimum pressure 20 mmHg). Selegiline was then withdrawn and patients re-tested three months later — this empirical time was determined as thrice the half period of MAO-B inhibition in the central nervous system as determined by positron emission tomography studies elsewhere. There was no selection of patients with respect to autonomic symptoms from their histories.

The full details of this study will be published elsewhere, but the main findings were as follows. Sustained selegiline therapy was associated with a selective and often severe systolic hypotension which was most evident on tilting and to a lesser degree on standing. The systolic blood pressure was affected more often and more severely than the diastolic. Those with severe postural hypotension showed a poor and variable tachycardia. A history suggestive of postural hypotension such as dizziness was in fact a poor indicator of severe hypotensive response on tilting.

Retesting after withdrawal of selegiline resulted in normal postural blood pressure responses; resolution of postural dizziness occurred in 63% of the previously symptomatic patients. Selegiline did not appear to affect autonomic function in tests other than postural control of blood pressure. Withdrawal of selegiline resulted in a severe decline in motor function in the majority of patients (69%) necessitating an increase in dopaminergic medication and clearly illustrating the marked levodopa-sparing effect of selegiline. The absence of significant postural hypotension in the non-selegiline-treated patients or in those who had previously taken selegiline suggests that significant underlying autonomic dysfunction as a feature of their Parkinson's disease was not a contributing factor.

These figures do not adequately illustrate the potential severity and gravity of the induced postural hypotension. Two of the 16 patients completely lost consciousness and had fits when the blood pressure dropped to zero (only one of these had a previous history of postural dizziness) and an additional four with marked postural hypotension became very dizzy with impairment of consciousness. All six patients with significant provoked postural hypotension were symptomatic within one minute of tilting, no patient who had symptomatic hypotension during tilting was symptomatic on standing even when there was demonstrable hypotension, and the consequent severe hypotension on tilting did not appear to be related to a low supine blood pressure. All symptomatic patients recovered promptly and completely when normal posture was regained. No postural dizziness was seen in those who had never taken selegiline.

Thus it seems reasonable to conclude that sustained selegiline therapy can cause a reversible selective and often severe postural hypotension, of considerable potential clinical significance, although the exact mechanisms invoking postural hypotension remains uncertain. What relationship these effects might have on long-term mortality figures in those who take selegiline remains to be elucidated.

References

Beringer K (1928) Über ein neues, auf das extrapyramidal-motorische System wirkendes Alkaloid (Banisterin). Nervenarzt 1: 265–275

Birkmayer W, Riederer P, Youdim MBH, Linauer W (1975) The potentiation of the anti-akinetic effect after l-dopa treatment by an inhibitor of MAO-B, Deprenyl. J Neural Transm 36: 303–326

Birkmayer W, Knoll J, Riederer P, Youdim MBH, Harts V, Marton J (1985) Increased life expectancy resulting from addition of L-deprenyl to Madopar treatment in Parkinson's disease: a long-term study. J Neural Transm 64: 113–127

Blaschko H, Richter D, Schlossmann H (1937) The inactivation of adrenaline. J Physiol (Lond) 90: 1–15

Blaschko H, Richter D, Schlossmann H (1937) The oxidation of adrenaline and other amines. Biochem J 31: 2187–2196

Elsworth JD, Glover V, Reynolds GP, Sandler M, Lees AJ, Phuapradit P, Shaw KM, Stern GM, Kumar P (1978) Deprenyl administration in man: a selective mono-amine oxidase B inhibitor without the "cheese effect". Psychopharmacology 57: 33–38

Halpern L (1930) Ueber die Harminwirkung im Selbstversuch. Dtsch Med Wochenschr 56: 1252–1254

Hare MLC (1928) Tyramine oxidase. 1. A new enzyme system in the liver. Biochem J 22: 968–979

Heikkila RE, Manzino L, Cabbat FS, Duvoisin RC (1984) Protection against the dopam-inergic neurotoxicity of l-methyl-4-phenyl-1,2,3,6-tetrahydropyridine by monoamine oxidase inhibitors. Nature 311: 467–469

Knoll J (1978) The possible mechanisms of action of (−)-deprenyl in Parkinson's disease. J Neural Transm 43: 177–198

Knoll J, Ecsery Z, Kelemen K, Nievel JG, Knoll B (1965) Phenylpropylmethyl-propinylamine (E-250): a new spectrum psychic energizer. Arch Int Pharmacodyn 155: 154–164

Lees AJ, Shaw KM, Kohout L, Stern GM, Elsworth JD, Sandler M, Youdim MBH (1977) Deprenyl in Parkinson's disease. Lancet ii: 791–795

Lewin L (1928) Untersuchungen uber Banisterin caapi Spr. Arch Exp Pathol Pharmakol 129: 133–149

Lewin L, Schuster P (1929) Ergebnisse von Banisterinversuchen an Kranken. Dtsch Med Wochenschr 55: 149

Mann J, Gershon S (1980) L-deprenyl, a selective monoamine oxidase inhibitor in endog-enous depression. Life Sci 26: 877–882

The Parkinson's Disease Research Group of the UK (1993) Comparison of therapeutic effects of levodopa, levodopa and selegiline, and bromocriptine in patients with early, mild Parkinson's disease: three year interim report. BMJ 307: 469–472

The Parkinson's Disease Research Group of the UK (1995) Comparison of therapeutic effects and mortality data of levodopa and levodopa combined with selegiline in patients with early, mild Parkinson's disease. BMJ 311: 1602–1607

The Parkinson's Disease Research Group of the UK (1996) BMJ 312: 704–705 (letter)

The Parkinson Study Group (1989) Effect of deprenyl on the progression of disability in early Parkinson's disease. N Engl J Med 321: 1364–1371

The Parkinson Study Group (1996) Impact of Deprenyl and tocopherol treatment on Parkinson's disease in DATATOP subjects not requiring levodopa. Ann Neurol 39: 29–36

The Parkinson Study Group (1996) Impact of Deprenyl and tocopherol treatment on Parkinson's disease in DATATOP subjects requiring levodopa. Ann Neurol 39: 37–45

Reynolds GP, Elsworth JD, Blau K, Sandler M, Lees AJ, Stern GM (1978) Deprenyl is metabolised to methamphetamine and amphetamine in man. Br J Clin Pharmacol 6: 542–544

Spruce R (1908) Notes of a Botanist on the Amazon and Ande, vol 11. Macmillan, London

Udenfriend S, Witkop B, Redfield BG, Wiessbach H (1958) Studies with reversible inhibitors of monoamine oxidase: harmaline and related compounds. Biochem Pharmacol 1: 160–165

Author's address: Dr. G. Stern, Department of Clinical Neurology, University College, London Hospitals, United Kingdom

The neuroprotective and neuronal rescue effects of (−)-deprenyl

K. Magyar[1], **B. Szende**[2], **J. Lengyel**[3], **J. Tarczali**[1], and **I. Szatmáry**[4]

Departments of [1] Pharmacodynamics, and [2] Pathology and Experimental Cancer Research, and [3] Central Isotope Laboratory, Semmelweis University of Medicine, and [4] Chinoin Pharmaceutical Works, Budapest, Hungary

Summary. The pharmacological effects of (−)-deprenyl is multi-fold in its nature (dopamine sparing activity, neuroprotective and neuronal rescue effects), which cannot be explained solely by the irreversible MAO-B inhibitory action of the substance. Deprenyl slightly inhibits the re-uptake of noradrenaline and dopamine, but methylamphetamine, the metabolite of the inhibitor, by one order of magnitude is more potent in this respect, than the parent compound. Neither the metabolite nor (−)-deprenyl acts on the uptake of serotonin. The inhibitor has an intensive first pass metabolism after oral treatment. The in vivo pharmacokinetic studies with (−)-deprenyl, using the double labelled radioisotope technique (1.5 mg/kg; orally) in rats revealed that the molar concentration of methylamphetamine can reach the level suitable to induce a significant inhibition of amine uptake. Deprenyl, but especially methylamphetamine pre-treatment can prevent the noradrenaline release induced by the noradrenergic neurotoxin DSP-4. The uptake inhibitory effect of (−)-deprenyl and the metabolites is reversible. After repeated administration of (−)-deprenyl (1.5 mg/kg daily, for 8 days) sustained concentration of its metabolites was detected, compared to that of the acute studies. This can at least partly explain why (−)deprenyl should be administered daily to evoke therapeutic effects in Parkinson's disease. Administration of (−)-deprenyl in a low dose, following the toxic insult, can rescue the damaged neurones. The neuronal rescue effect of the drug was studied on M-1 human melanoma cells in tissue culture. The inhibitor reduced the apoptosis of serum-deprived M-1 cells, but the (+)-isomer failed to exert this effect. The (±)-desmethyl-deprenyl almost lacks the property to inhibit apoptosis. For neuroprotection and neuronal rescue an optimal dose of (−)-deprenyl should be administered, because to reach a well balanced concentration of the metabolites in tissues is critical.

Introduction

Deprenyl (phenyl-isopropyl-methyl-propargylamine) has been synthesised by Z. Ecseri, the chemist of the Chinoin Pharmaceutical Works in Budapest. The first paper regarding the pharmacological activity of the drug was published in 1965 by Knoll and his co-workers (1965). Deprenyl irreversibly inhibits monoamine oxidase (MAO; E.C.1.4.3.4.) and its ($-$)-isomer (selegiline) is a more potent inhibitor, than its ($+$)-enantiomer (Magyar et al., 1967). The irreversible inhibition induced by the substance is preceded by a competitive reversible phase and during this time the presence of the specific substrate can prevent the irreversible inactivation of the enzyme (Tipton, 1980). Two forms of MAO are existing, as type A and B. The distinction is based on the substrate specificities and the inhibitor sensitivities of the enzyme (Johnston, 1968). ($-$)-deprenyl inhibits selectively MAO-B and still it is the most widely used MAO-B inhibitor in clinics. Its pharmacological effects are rather complex. It is beyond controversy that the dopamine sparing activity, as well as the neuroprotective and neuronal rescue effects of ($-$)-deprenyl cannot be explained solely by its irreversible enzyme inhibitory action. In addition, the reversible inhibition of the biogenic amine re-uptake by ($-$)-deprenyl may also play role in the complex pharmacological activity of the drug, including neuroprotection.

Neuroprotection

Both forms of MAO play a protective role by saving the central and peripheral nerves from the harmful effects of the exogenous food derived amines. The activation of the pretoxin 1-methyl-4-phenyl-1,2,3,6-tetrahydropyridine (MPTP) by MAO-B to toxin, 1-methyl-4-phenylpyridinium (MPP^+), is an opposite mechanism of the enzyme's normal function. The role of the exo- and endotoxins in the pathogenesis of the neurodegenerative diseases, in spite of the intensive research in this field is not clear. Attempts to identify the existence of the natural environmental or a possible endogenously-generated neurotoxin which could cause an idiopathic neurodegenerative illness as Parkinson's disease or Alzheimer type of dementia, were still unsuccessful. MPTP does not exists in natural environment and it is rather unlikely the cause of the idiopathic Parkinson's disease.

Nagatsu and Hirato (1987) suspected tetrahydro-isoquinoline (TIQ) as a candidate to produce parkinsonism. It is an endogenous substance formed in the body which can be N-methylated by human brain homogenate. The product, N-methyl-TIQ is a substrate for MAO, which catalyses the formation of the positively charged N-methyl-isoquinoline$^+$ ($NMIQ^+$), a close chemical analogue of MPP^+. Both forms of MAO catalyse the activation of TIQ, therefore the inhibition of MAO-B by ($-$)-deprenyl cannot prevent the activation of the pretoxin. Nevertheless, a convincing evidence proving the role of TIQ in the pathogenesis of the known neurodegenerative disorders does not exist.

Birkmayer and his colleagues were the first who have indicated the neuroprotective effect of ($-$)-deprenyl based on a retrospective evaluation of a seven year long clinical application of the drug (1985). The neuroprotective action of ($-$)-deprenyl, is also multi-fold in nature. There are at least four accepted mechanisms by which ($-$)-deprenyl could prevent neurodegeneration. Firstly, it may decrease the free radical formation (generation of H_2O_2) from the normal metabolism of the biogenic amines, mainly dopamine, by inhibition of MAO-B in the central nervous system (Cohen and Spina, 1989). Hydrogen peroxide oxidises Fe^{++} ion and generates hydroxyl radical formation by the so-called metal-catalysed Haber-Weiss reaction. It should be considered that MAO-B activity increases with age, which leads to the rise in H_2O_2 formation. The over-production of H_2O_2 may contribute to the neural damage (Strolin-Benedetti and Dostert, 1989). In addition to the oxidative stress caused by the age dependent increase of MAO-B activity, further reactions between endogenous amines and aldehydes formed by MAO-B, can also play a role in neurodegeneration (Glover et al., 1986). Secondly, according to some authors the inhibitor may increase the free radical scavenging capacity of the brain by an elevation of superoxide dismutase (SOD) activity (Knoll, 1988; Carrillo et al., 1991), but others failed to observe any increase in SOD function (Lay et al., 1994). Thirdly, due to MAO-B inhibition, (−)-deprenyl may prevent the activation of the environmental pre-toxins (Langston, 1990). Finally, due to the uptake inhibitory properties of the substance it can prevent the selective uptake of neurotoxins into the nerve endings, thereby obviating the neuronal damage. Concerning the inhibition of uptake the metabolites of (−)-deprenyl [(−)-methyl-amphetamine and (−)-amphetamine] are even more potent, than the parent compound (Magyar, 1994). This paper is dealing mainly with the pharmacokinetics of the formation of the metabolites, including the role of metabolism and metabolites of (−)-deprenyl in neuroprotection, by using animal model of neurodegeneration induced by the synthetic neurotoxin, DSP-4 (Ross and Renyi, 1976).

Metabolism of (−)-deprenyl

The metabolic conversion of (−)-deprenyl was widely studied in both animals and human (Magyar, 1992). It has been confirmed in many laboratories, among them in ours, that (−)-deprenyl is metabolised to methylamphetamine (MA) and amphetamine (A). The biotransformation of (−)-deprenyl in man was firstly published by Reynolds and his co-workers. They proved that both MA and A can be detected post-mortem in brain tissue (1978). Heinonen confirmed similar pattern of metabolism of the inhibitor also in man (Heinonen et al., 1989). In addition to MA and A, (−)-deprenyl is converted to desmethyl-deprenyl in rats without producing any remarkable signs of psychostimulant activity, when it was administered in doses sufficient to elicit selective inhibition of MAO-B (Magyar and Szüts, 1982; Magyar and Tóthfalusi, 1984). In spite of the formation of amphetamines, (−)-deprenyl did not prove to be a potent releaser of biogenic amines (Knoll and Magyar, 1972).

The lack of releasing potency and psychostimulant activity could partly be due to the fact that from (−)-deprenyl, (−)-enantiomers of the metabolites are formed (Szökö and Magyar, 1995, 1996) which have less psychostimulant activity than the (+)-amphetamines. It was proved earlier in parkinsonian patients that racemase did not convert the (−)-amphetamines to their (+)-forms (Schachter et al., 1980).

(−)-deprenyl is metabolised by the microsomal enzymes, which can be induced by phenobarbital and inhibited by SKF-525A pre-treatment. Phenobarbital pre-treatment (80 mg/kg daily for 3 days) intraperitoneally decreased the inhibition of phenylethylamine (PEA) oxidation in rats, elicited by 5 mg/kg of (−)-deprenyl oral administration, both in brain and in liver. This finding indicates that only the parent compound is responsible for the inhibition of MAO-B. When rats were pre-treated 1 h before (−)-deprenyl administration with SKF-525A (50 mg/kg; intraperitoneally), the oxidative deamination of PEA evoked by 0.25 mg/kg of (−)-deprenyl oral treatment, was increased in the nuclei free homogenate both of the brain and the liver (Magyar, 1996a,b).

In recent studies we investigated the in vivo pharmacokinetics of (−)-deprenyl in rats by using two radioisomers of the compound: (−)-deprenyl-propargyl-^{14}C and (−)-deprenyl-phenyl-^{3}H. The animals (three rats in a group at every time of determinations) were treated orally with the mixture of the two radioisomers and the radioactivity of both labels was measured simultaneously in the plasma and 15 brain regions. The actual concentrations of (−)-deprenyl were calculated from the amount of radioactivities measured in tissues and the specific activities of the radioisomers being present in the dose (1.5 mg/kg) administered. The alternate labels (^{3}H and ^{14}C) also provided us a good estimation of the metabolic disintegration of the molecule (Magyar, 1994; Magyar et al., 1995).

From the data obtained from these studies around $1-2 \times 10^{-6}$ M concentration of amphetamines could be estimated in brain tissues shortly after the oral intake ($t_{max} = 0.5$ h) of 1.5 mg/kg of (−)-deprenyl (Fig. 3). The above concentration range of MA, which can inhibit the amine uptake was observed only for a short duration (about 0.5–1 h). We have to take into account that MA is more potent to inhibit the uptake, than the parent compound (Magyar, 1994). The data obtained by the aid of this double labelled technique indicated an intensive "first pass" metabolism of (−)-deprenyl after oral administration, which means that only a minor portion of the orally administered substance takes part in inducing the irreversible MAO-B inhibition. The integrity of the molecule is essential for the inhibition of the enzyme.

In an other series of experiments rats were orally treated daily for 8 subsequent days (subacute experiments) with the mixture of the same tracers and the dose, as in the above acute experiments (Magyar et al., 1995). The time related changes of (−)-deprenyl concentrations in the plasma and 15 brain regions were followed, using the technique as in the acute experiment.

One of our aims was to determine the equilibrium concentration of (−)-deprenyl in the tissues during repeated treatment. The radioactivities (^{3}H and ^{14}C) were measured 24 h after each drug administration (one h before the next treatment), but following the last treatment a detailed pharmacokinetic study was performed for a 72 h period. The selected time points of measuring radioactivity after the last treatment were equivalent with those of the acute treatment.

On the 6–7th day an equilibrium has been reached in the case of the ^{3}H tracer, while with ^{14}C-radioactivity the equilibrium was set earlier on the 4th day, both in the plasma and in the striatal tissue (Figs. 1, 2). Corpus striatum was selected not only as a typical example from the 15 brain regions, but as a special part of the brain playing essential role in the pathogenesis of the parkinsonian syndrome.

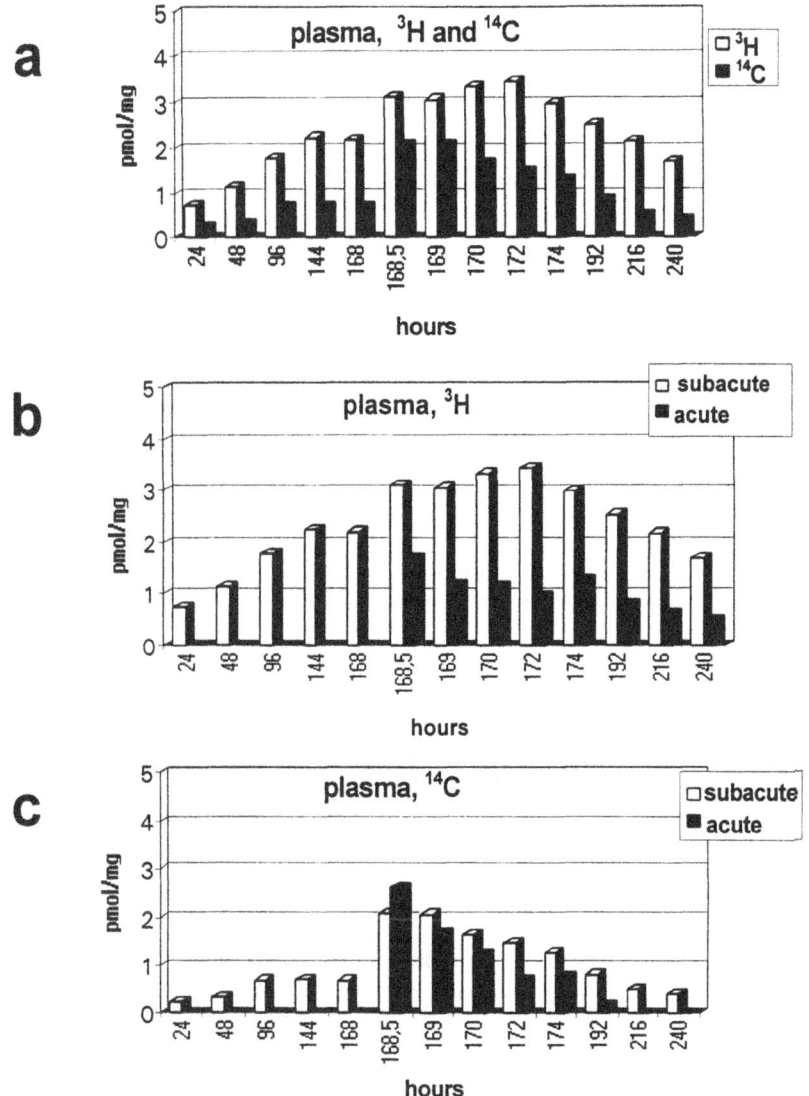

Fig. 1. Time dependent changes in plasma concentrations of (−)-deprenyl in rats after subacute treatment. **a** Detected by ^{3}H- and ^{14}C-labels; **b** detected by ^{3}H-label and compared with the corresponding values of the acute experiments; **c** detected by ^{14}C-label and compared with the corresponding values of the acute experiments. Rats were treated orally with the mixture of alternatively labelled (^{3}H or ^{14}C) (−)-deprenyl (1.5 mg/kg/day, for 1 to 8 days). The radioactivity of both labels was simultaneously detected at the times registered on the abscissa

In addition to the equilibrium state Fig. 1a also shows the concentrations of (−)-deprenyl following the very last treatment (8th treatment at 168h) calculated on the basis of the amount of radioactivities measured in the plasma and the specific activities of the two labels in the dose administered. The figure shows that the ^{3}H-related concentrations of (−)-deprenyl are beyond controversy higher compared to that of ^{14}C-radioactivity. Since the amount of non-metabolised (−)-deprenyl cannot exceed that calculated from ^{14}C-radioactivity, the surplus of the ^{3}H-related concentrations of the inhibitor should indicate the formation of metabolites free from the propargyl-^{14}C group.

Figure 1b and c show the concentrations of (−)-deprenyl in the plasma calculated from both labels, respectively, and the results are compared with those obtained in acute studies. The comparison of the results obtained after

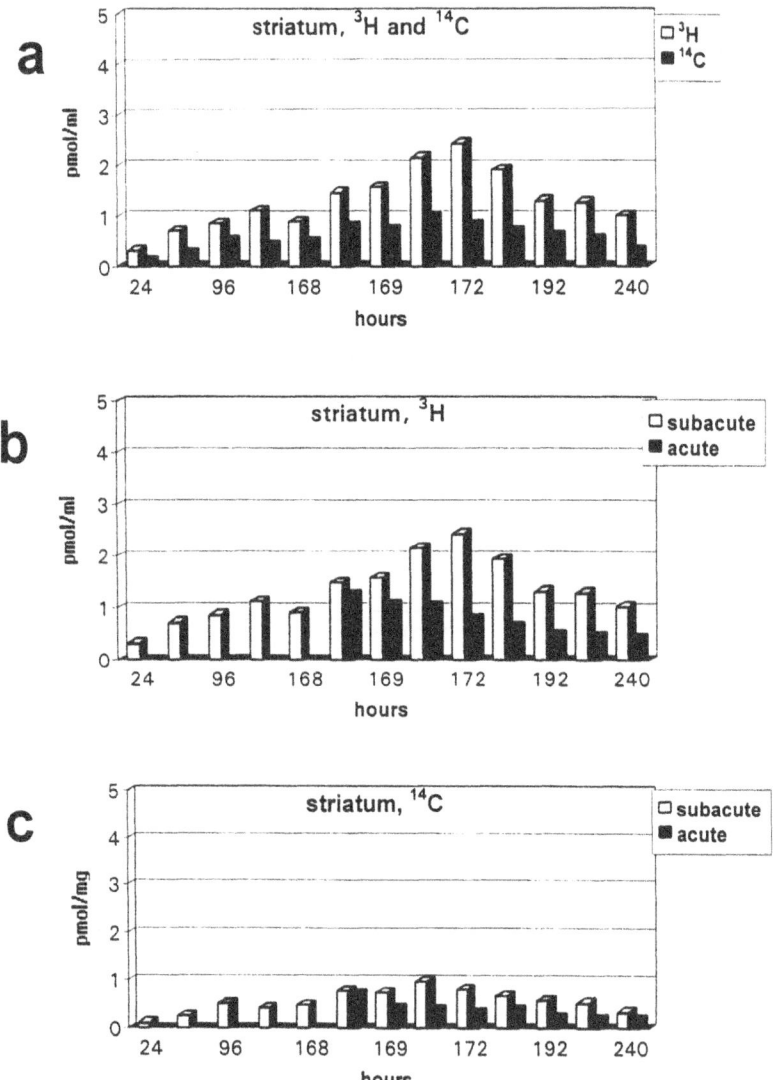

Fig. 2. Time dependent changes of the concentrations of (−)-deprenyl in the rat striatum after subacute treatment. Legends for Fig. 2a,b,c are presented in Fig. 1

168 h from the subacute experiments with those of the acute studies, lead us to the conclusion, that in the case of ^{3}H label the plasma concentration-time curve covers much bigger area ($AUC_{168-240}$) than that seen in the acute studies. In the case of ^{14}C-radioactivity, less difference was noticed between the two values in this respect (Tables 1 and 2).

The tissue concentration-time curves of the corpus striatum obtained in acute and subacute studies are presented in Fig. 2a,b,c. The method of the presentation is identical with Fig. 1. The results are similar to that presented in Fig. 1 regarding the plasma.

In addition to the higher AUC values in subacute studies, Figs. 1 and 2 also show that the highest tissue levels (C_{max}) were reached earlier in acute than in subacute studies. Nevertheless, the values of the C_{max} in the tissues are higher in subacute experiments. The higher AUC values are mostly resulted from the lower rate of elimination of the radioactivies rather than from the increase of the C_{max} values, as compared to the acute experiments. The lower rate of elimination of the radioactivity found in subacute studies might be due to the slower rate of excretion of the metabolites from the tissues compared to that of the parent compound.

In the two selected organs (plasma and corpus striatum) the time dependent changes of the tissue concentrations calculated from the ^{14}C-label in acute and subacute studies are subtracted from that of the ^{3}H-radioactivity and the results are presented in Fig. 3. The difference arising from the surplus of ^{3}H-label, could represent the actual concentrations of the metabolites. As it can be seen, the plasma concentrations calculated from ^{14}C in the first hour, for unknown reason, is higher than that of the ^{3}H. We suppose that the distribution volume of the propargyl radical carrying the ^{14}C-tracer is smaller, than that of the ^{3}H-one, indicating the core of the molecule. The former finding was not observed in the case of the corpus striatum.

Table 1. The main pharmacokinetic parameters in the plasma and the pooled brain tissues after acute and subacute treatment

Plasma	C_{max} nmol/ml	t_{max} h	AUC_{0-72} nmol/ml·h	k_e h
acute ^{3}H	1.69 ± 0.19	0.5	107.89	0.02
acute ^{14}C	2.61 ± 0.11	0.5	15.79	0.22
subacute ^{3}H	3.44 ± 0.01	4	336.61*	0.09
subacute ^{14}C	2.08 ± 0.34	0.5	89.92*	0.03
Brain				
acute ^{3}H	3.36 ± 1.12	0.62 ± 0.15	132.9 ± 34.2	0.03**
acute ^{14}C	1.18 ± 0.29	0.83 ± 1.02	41.1 ± 9.4	0.03**
subacute ^{3}H	3.75 ± 0.92	1.42 ± 0.99	83.4 ± 57.9*	0.01**
subacute ^{14}C	1.73 ± 0.51	1.88 ± 1.02	182.6 ± 89.3*	0.01**

* $AUC_{168-240}$, ** frontal cortex

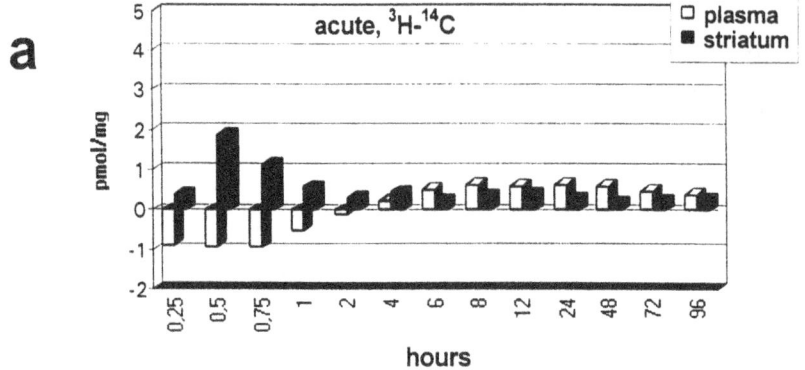

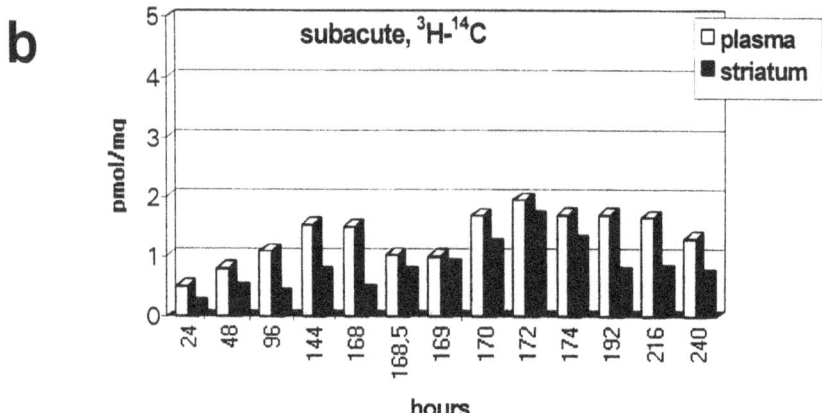

Fig. 3. Time dependent changes of the concentrations of the possible metabolites of (−)-deprenyl calculated from the surplus of ^{3}H-label over ^{14}C radioactivity in the plasma and striatum in acute (**a**) and subacute (**b**) experiments in rats. The values of (−)-deprenyl concentrations calculated from the specific activity of the ^{14}C-label are subtracted from that of ^{3}H-radioactivity

In the acute studies the peak concentration of the metabolites was reached at 0.5 h in the corpus striatum, which was amounted to 2 μmol equ/L for a short duration. In some of the other brain areas the calculated concentrations of the metabolites can exceed the level found in the corpus striatum. The results presented in the Fig. 3 indicate that the estimated concentrations of the metabolites are much lower after a single dose (Fig. 3a), than following repeated treatments (Fig. 3b).

Figure 3b shows that the metabolites are not cumulated either in the plasma or in the striatum, but their level in both tissues are over 1 μmol equ/L for hours (for 6 h in the striatum, but till the end of the experiment in the plasma). The concentrations of the metabolites in the striatum are in every case lower than that found in the plasma. It is worth to mention that the values before 168 h represent the tissue concentrations of the metabolites measured 24 h after the daily administration.

The main pharmacokinetic parameters calculated from the values of the acute and subacute treatments in the plasma and the brain tissue are presented in Table 1. The mean values regarding the brain tissue represent the average measures of 15 brain areas ± S.D. In case of both labels the AUC and C_{max} values of the plasma and the cerebral tissue after repeated treatments are higher than those obtained in the acute experiments. The data demonstrate the time, needed to reach the C_{max} values (t_{max}), which is shorter in the acute than in the subacute experiments.

The most striking difference documented in Table 2 is the higher AUC values obtained in subacute studies both in the plasma and in the brain tissues. The table presenting the AUC values of 15 brain areas also shows that (−)-deprenyl and its metabolites are unevenly distributed in the brain tissue. The ratio of the AUC subacute and AUC acute values in the plasma and the brain with ^{3}H-label are 3.39 and 2.13 while with ^{14}C-tracer 5.69 and 4.44, respectively. The mean values of the brain areas were used for the calculation. If we consider the difference between the AUC values for ^{3}H and ^{14}C as metabolites of (−)-deprenyl in the plasma, in that case 85% of the dose are metabolised during the equilibrium state in subacute studies.

Table 2. AUC values of (−)-deprenyl in the plasma and the brain regions of rats, obtained in acute and subacute studies

Tissue	^{3}H acute	^{3}H subacute	^{14}C acute	^{14}C subacute
Plasma	107.89	336.61	15.79	89.92
corpus pineale	133.70	282.50	0	0
bulbus olf.	99.11	205.10	41.71	159.34
hypophysis	55.25	276.16	33.27	348.52
hypothal.	166.31	327.73	48.92	176.46
tuberculum olf.	144.47	306.27	53.70	219.74
subst. nigra	131.88	220.41	0	0
nucl. mamill.	116.56	330.59	0	0
frontal cortex	159.89	401.63	34.26	181.05
pariet. cortex	122.64	296.51	37.45	224.38
striatum	74.98	245.16	24.96	59.31
hippocamp.	123.01	251.09	33.91	36.19
collic. sup.	186.15	317.01	54.79	259.74
cerebellum	161.27	360.99	47.45	211.34
pons + col.inf.	141.60	262.33	37.71	251.39
medulla obl.	142.12	234.05	36.45	149.91
Brain regions mean ± S.D.	132.99 ± 34.22	283.36 ± 57.92	41.08 ± 9.36	182.65 ± 89.33

The values in the subacute experiments were calculated between 168 and 240 h from the ^{3}H- and ^{14}C-counts and the specific activity of the two radioisomers (1.5 mg/kg (−)-deprenyl orally; pmol/mg tissue·h; n = 3)

(−)-Deprenyl induced neuroprotection against DSP-4 toxicity

All of the MAO-B inhibitors are protective against MPTP toxicity, but (−)-deprenyl and not the MDL 72974 compound, in spite of being a potent MAO-B inhibitor, was found by Finnegan to be effective against DSP-4 toxicity (Finnegan et al., 1990). It was published recently that (−)-deprenyl also protected the effect of the cholinergic neurotoxin AF64A (Ricci et al., 1992).

DSP-4, originally described by Ross and Renyi (1976), is a beta-haloethylamine derivative of benzylamine which has the ability to induce a long-term depletion of noradrenaline (NA) stores. The irreversible damage of the presynaptic nerve ending caused by the covalently bound toxin can be blocked by the co-administration of uptake inhibitors, like desipramine. The recovery from the irreversible effect of DSP-4 is based on the synthesis of new transport protein. The selectivity of the compound to the noradrenergic synapse is not fully known (Dudley et al., 1990), but most likely it is based on the selective uptake of the positively charged aziridinium ion, formed from DSP-4 in solution, into the noradrenergic nerve endings. Because of this the protective mechanism is also requiring a functional transporter system which can be competitively and reversibly inhibited by the protective agents.

We reported as early as in 1972 that deprenyl and its optical isomers inhibit ^{3}H-NA uptake into the cerebral cortex slices of mice (Knoll and Magyar, 1972) and in the synaptosomal fraction of the rat brain (Tekes et al., 1988). Recent experiments revealed that not only the parent compound but also it metabolites are responsible for the inhibition of the synaptosomal uptake of NA and dopamine (Magyar, 1994). Neither deprenyl nor its metabolites inhibit the synaptosomal uptake of serotonin. These studies also revealed that the metabolites by one order of magnitude are more potent to inhibit the uptake of NA than the parent compound. In respect of the release of transmitter amines the positive isomers of the metabolites are more potent than the (−)-variants, but we did not find significant difference between the uptake inhibitory potencies of the stereoisomers.

The protective role of (−)-deprenyl against DSP-4 induced NA release was studied under the following experimental conditions: Rats were treated with 50 mg/kg of DSP-4 intraperitoneally, and the NA content of the hippocampus was measured by HPLC technique on the 7th day of treatment. NA content of the hippocampus was dramatically reduced by this time. When the animals were pre-treated orally 1 h before DSP-4 administration with 5 mg/kg of (−)-deprenyl, a remarkable protection against NA depletion was detected. SKF-525A pre-treatment decreased the protective effect of (−)-deprenyl against NA depletion induced by the noradrenergic neurotoxin, while SKF-525A treatment in itself did not influence NA level of the hippocampus (Magyar, 1996a,b). From these data we can conclude that the inhibition of the metabolism of (−)-deprenyl decreases the protective capacity of the inhibitor against NA release, suggesting that the metabolites of (−)-deprenyl in respect of uptake inhibition might be more effective, than the parent compound. In these series of experiments 5 mg/kg of orally administered (−)-deprenyl induced a degree of protection against DSP-4 toxicity comparable to that caused by 10 mg/kg of (−)-deprenyl given intraperitoneally (Finnegan et al., 1990). In these studies even a dose of 1 mg/kg of (−)-deprenyl administered orally was found to be protective against DSP-4 toxicity. This finding might be due to the more intensive ("first pass") metabolism of (−)-deprenyl, occurring after oral administration than after intraperitoneal injection, which process is resulted in the formation of more MA.

In accordance with the uptake studies both optical isomers of MA prevented the NA release from the hippocampus by 60% in rats, when the enantiomers were injected intraperitoneally in a dose of 1 mg/kg, one hour before DSP-4 treatment (50 mg/kg, intraperitoneally; Magyar, 1996a,b). In contrast, (−)-deprenyl was effective in a dose of 10 mg/kg in Finnegan's experiments (Finnegan et al., 1990).

We wanted to reach a 100% protection of the NA release by rising the dose of (−)-MA up to 5 mg/kg, preserving the former experimental conditions. The combination [DSP-4 50 mg/kg + (−)-MA 5 mg/kg i.p.] became so toxic that all of the rats died. From these studies we concluded that the concentration of (−)-MA for neuroprotection is critical, because MA in one respect is more potent inhibitor of the uptake, but in an other one over the optimal concentration its higher toxicity will dominate. For neuroprotection a proper concentration of MA should be established which, as a self-regulating process, depends on the dose of (−)-deprenyl administered.

Although it is widely accepted that the inhibition of the carrier mediated re-uptake process of noradrenaline plays an essential role in the prevention of DSP-4 induced neurotoxicity, some contradictory data are also cumulating in the literature, suggesting that the inhibition of NA re-uptake cannot be responsible alone for the protection (Berry et al., 1994). Gibson reported that clorgyline, which has similar inhibitory properties to (−)-deprenyl on NA re-uptake process, does not protect against DSP-4 toxicity (1987). On the contrary, MDL 72145 is also able to protect DSP-4 toxicity without having uptake inhibitory potency (Bey et al., 1984). It has been published that some relatively short chain aliphatic compounds, as N-2-hexyl-N–methylpropargylamine (2-HxMP) synthesised recently, is a potent selective inhibitor of MAO-B, but with the lack of uptake inhibitory potency, was able to protect DSP-4 toxicity (Yu et al., 1994). It is apparent that the toxicity induced by DSP-4 is more complex than it has been thought, but the role of the uptake inhibition in the protection cannot be ruled out. We provided direct evidence that (−)-MA the metabolite of (−)-deprenyl is able to protect DSP-4 induced neurotoxicity and in this respect it is more effective than the parent compound. In addition, the results of our subacute studies suggest that the higher and more prolonged concentration of MA seen in that condition should improve the protective capacity of (−)-deprenyl during repeated treatment. Nevertheless, the above findings indicate the need for further studies to clarify a yet unidentified action of (−)-deprenyl, and some but not all other inhibitors of MAO-B, which might play a role in the protection of DSP-4 toxicity.

Neural rescue effect of (−)-deprenyl

It has become apparent that (−)-deprenyl administration following the toxic insults can rescue the damaged neurones. The administration of (−)-deprenyl in a small selective dose (0.25 mg/kg) three times a week, starting three days after MPTP treatment, increased neuronal survival (Tatton and Greenwood, 1991). This type of post-treatment cannot prevent activation of MPTP by MAO-B. The mechanism of protection is not yet known, but it was supposed that during post-insult treatment (−)-deprenyl increases the trophic support of the damaged neurones, which results in their longer survival time. It was also published that when rats were treated with 0.005 to 0.01 mg/kg of (−)-deprenyl after facial motoneuron axotomy, the treatment increased the neural survival without inhibition of brainstem MAO-B activity (Ansari et al., 1993). The low dose of the inhibitor in these studies suggests that the neuronal rescue effect represents a novel, hitherto unknown mechanism which is presumably independent of the MAO-B inhibition and particularly of the blockade of amine re-uptake. The latter process requires higher concentration of (−)-deprenyl than that of the inhibition of MAO-B. It was also published that (−)-deprenyl is capable to rescue hippocampal neurones in acute cerebral ischemia by a complex mechanism (Barber et al., 1993).

It has also been shown that (−)-deprenyl increased the neuronal survival of PC12 cells in tissue culture. The withdrawal of serum and nerve growth factor induced apoptosis in PC12 cells, but (−)-deprenyl inhibited the programmed cell death in a concentration of less than 10^{-9} M. Concerning the concentration of the inhibitor, the rescue effect in tissue culture should also be independent of MAO-B inhibition (Tatton et al., 1994). The (+)-enantiomer of deprenyl lacks this property.

In order to study the neuronal rescuing effect of (−)-deprenyl we used M-1 human melanoma cells (Ladányi et al., 1990) which were plated in 6-well Grainer (Kremsmünster, Germany) plates containing glass cover slips. The culture medium was RPMI (GIBCO), supplemented with 10% fetal calf serum (Bioproduct, Gödöllö, Hungary). Cell number at plating was 1.5×10^5/ well. The medium was changed 48 hours after plating and no more serum supplementation was given since that time. Five days after changing the medium, (−)-deprenyl was given to the cell cultures to reach the final concentrations of 10^{-3}, 10^{-7} and 10^{-13} M. Samples were taken 24, 48 and 72 hours after treatment. Duplicate cover slip cultures per dose and per day, including 2-2 untreated controls were stained with Hematoxylin and Eosin. The percentage of apoptosis was determined taking into consideration the morphological signs of apoptosis (Wyllie et al., 1986).

The ratio of apoptotic cells increased gradually in the untreated cell cultures from the 24th to 72nd hours of the treatment period. At 72 hours practically only apoptotic cells and cell debris were found. (−)-deprenyl treatment significantly decreased the number of apoptotic cells even in the lowest concentration (10^{-13} M). At 72 hours the cultures appeared to be viable. (+)-deprenyl, however, did not prevent the high incidence of apoptosis (Magyar et al., 1996b). Our findings on serum-deprived M-1 cells support the results of Tatton who found that $(-)$-deprenyl reduced apoptosis, but the $(+)$-isomer failed to exert such effect (Tatton et al., 1994).

During the present studies we prolonged the observation period for apoptosis to 120 h. At a concentration of 10^{-3} M of $(-)$-deprenyl no protection was observed, but at 10^{-7}, 10^{-9}, 10^{-13} M concentrations apoptosis was reduced by 77, 44, 44%, respectively. Like Tatton's studies revealed, our results also suggest that an optimal concentration range of $(-)$-deprenyl should exist for neuroprotection.

We are continuing our studies on the M-1 human melanoma cell line which derives from melanocytes of neuroectodermal origin. In order to get some insight into the mode of action of $(-)$-deprenyl on neuronal rescue a few chemical analogues of the inhibitor are involved into our present studies. Preliminary observations indicate that the $(\pm)$-desmethyl-deprenyl almost lacks rescuing potency.

Acknowledgement

The experimental part of the paper was supported by the grant OTKA T 017749 of the Hungarian Academy of Sciences.

References

Ansari KS, Tatton WG, Yu PH, Kruck TPA (1993) Rescue of axotomised immature rat facial motoneurons by R(−)-deprenyl: stereospecificity and independence from monoamine oxidase inhibition. J Neurosci 13: 4042–4053

Barber AJ, Paterson IA, Gelowitz DL, Voll CL (1993) Deprenyl protects rat hippocampal pyramidal cells from ischaemic insult. Soc Neurosci Abstr 19: 1646

Berry MD, Juorio AV, Paterson IA (1994) The functional role of monoamine oxidases A and B in the mammalian central nervous system. Prog Neurobiol 42: 375–391

Bey P, Fozard J, McDonald I, Palfreyman MG, Zreika M (1984) MDL 72145: a potent and selective inhibitor of MAO type B. Br J Pharmacol 81: 50P

Birkmayer W, Knoll J, Riederer P, Youdim MBH, Hars V, Martin J (1985) Increased life expectancy resulting from addition of L-Deprenyl to Madopar^R treatment in Parkinson's disease: a long-term study. J Neural Transm 64: 113–128

Carrillo MC, Kanai S, Nokubo M, Kitani K (1991) (−)-Deprenyl induces activities of both superoxide dismutase and catalase but not of glutathione peroxidase in the striatum of young male rats. Life Sci 48: 517–521

Cohen G, Spina MB (1989) Deprenyl suppresses the oxidant stress associated with increased dopamine turn-over. Ann Neurol 26: 689–690

Dudley MW, Howard BD, Cho AK (1990) The interaction of the beta-haloethyl benzylamines, xylamine and DSP-4 with catecholaminergic neurones. Annu Rev Pharmacol Toxicol 30: 387–403

Finnegan KT, Skratt JS, Irwin I, DeLanney LE, Langston JW (1990) Protection against DSP-4-induced neurotoxicity by deprenyl is not related to its inhibition of MAO-B. Eur J Pharmacol 184: 119–126

Gibson CJ (1987) Inhibition of MAO-B, but not MAO-A blocks DSP-4 toxicity on central NE neurones. Eur J Pharmacol 141: 135–138

Glover V, Gibb C, Sandler M (1986) The role of MAO in MPTP toxicity — a review. J Neural Transm [Suppl 20]: 65–76

Heinonen EH, Myllyla V, Sotaniemi K, Lammintausta R, Salonen JS, Anttila M, et al (1989) Pharmacokinetics and metabolism of selegiline. Acta Neurol Scand 126: 93–99

Johnston JP (1968) Some observations upon a new inhibitor of monoamine oxidase in brain tissue. Biochem Pharmacol 17: 1285–1297

Knoll J (1988) The striatal dopamine dependency of life span in male rats. Longevity study with (−)-deprenyl. Mech Aging Dev 46: 237–262

Knoll J, Magyar K (1972) Some puzzling pharmacological effects of monoamine oxidase inhibitors. Adv Biochem Psychopharmacol 5: 393–408

Knoll J, Ecseri Z, Kelemen K, Nievel J, Knoll B (1965) Phenylisopropylmethyl-propinylamine (E-250), a new spectrum psychic energizer. Arch Int Pharmacodyn Ther 155: 154–164

Ladányi A, Timár J, Paku S, Molnár G, Lapis K (1990) Selection and characterization of human melanoma lines with different liver-colonizing capacity. Int J Cancer 46: 456–461

Lai CT, Zuo DM, Yu PH (1994) Is brain superoxide dismutase activity increased following chronic treatment with L-deprenyl? J Neural Transm 41: 221–229

Langston JW (1990) (−)-deprenyl as neuroprotective therapy in Parkinson's disease: concepts and controversies. Neurology [Suppl 3] 40: 61–66

Magyar K (1992) Pharmacology of monoamine oxidase type B inhibitors. In: Szelenyi I (ed) Inhibitors of monoamine oxidase B. Birkhäuser, Basel, pp 125–143

Magyar K (1994) Behaviour of (−)-deprenyl and its analogues. J Neural Transm 41: 167–175

Magyar K (1996a) The role of the metabolism of (−)-deprenyl in neuroprotection. J Neurochem 66 [Suppl 2]: S20 D

Magyar K (1996b) The role of the metabolism of (−)-deprenyl in neuroprotection. J Neurochem (in press)

Magyar K, Szüts T (1982) The fate of (−)-deprenyl in the body. Preclinical studies. In: Proceedings of the International Symposium on (−)-Deprenyl, Jumex, Szombathely, Hungary, pp 25–31

Magyar K, Tóthfalusi L (1984) Pharmacokinetic aspects of deprenyl effects. Pol J Pharmacol Pharm 36: 373–384

Magyar K, Vizi ES, Ecseri Z, Knoll J (1967) Comparative pharmacological analysis of the optical isomers of phenyl-isopropyl-methyl-propinylamine (E-250). Acta Physiol Hung 32: 377–387

Magyar K, Lengyel J, Szatmári I, Gaál J (1995) The distribution of orally administered (−)-deprenyl-propargyl-^{14}C and (−)-deprenyl-phenyl-^{3}H in rat brain. Prog Brain Res 106: 143–153

Magyar K, Szende B, Lengyel J, Tekes K (1996) The pharmacology of B-type selective monoamine oxidase inhibitors; milestones in (−)-deprenyl research. J Neural Transm 48: 29–43

Nagatsu T, Hirata Y (1987) Inhibition of the tyrosine hydroxylase system by MPTP, 1-methyl-4-phenylpiridinium ion (MPP$^+$) and the structurally related compounds in vitro and in vivo. Eur Neurol 26 [Suppl 1]: 11

Reynolds GP, Elsworth JD, Blau K, Sandler M, Lees AJ, Stern GM (1978) Deprenyl is metabolized to methamphetamine and amphetamine in man. Br J Clin Pharmacol 6: 542–544

Ross SB, Renyi AL (1976) On the long-lasting inhibitory effect of N-(2-chloroethyl)-N-ethyl-2-bromobenzylamine (DSP-4) on the active uptake of adrenaline. J Pharm Pharmacol 28: 458–459

Schachter M, Marsden CD, Parkes JD, Jenner P, Testa B (1980) Deprenyl in the management of response fluctuations in patients with Parkinson's disease on levodopa. J Neurol Neurosurg Psychiatry 43: 1016–1021

Strolin-Benedetti M, Dostert P (1989) Monoamine oxidase, brain ageing and degenerative diseases. Biochem Pharmacol 38: 555–561

Szökö É, Magyar K (1995) Chiral separation of deprenyl and its major metabolites using cyclodextrin-modified capillary zone electrophoresis. J Chromatogr A 709: 157–162

Szökö É, Magyar K (1996) Enantiomer identification of the major metabolites of (−)-deprenyl in rat urine by capillary electrophoresis. Int J Pharm Adv 1: 320–328

Tatton WG, Greenwood CE (1991) Rescue of dying neurones: a new action for deprenyl in MPTP parkinsonism. J Neurosci Res 30: 666–672

Tatton WG, Ju WYL, Holland DP, Tai C, Kwan M (1994) (−)-Deprenyl reduces PC12 cell apoptosis by inducing new protein synthesis. J Neurochem 63: 1572–1575

Tekes K, Tóthfalusi L, Gaál J, Magyar K (1988) Effect of MAO inhibitors on the uptake and metabolism of dopamine in rat and human brain. Pol J Pharmacol Pharm 40: 653–658

Tipton KF (1980) Kinetics and enzyme inhibition studies. In: Sandler M (ed) Enzyme inhibitors and drugs. McMillen, London, pp 1–23

Wyllie AH, Kerr JFR, Currie AR (1986) Cell death: the significance of apoptosis. Int Rev Cytol 68: 251–306

Yu PH, Davis BA, Fang J, Boulton AA (1994) Neuroprotective effects of some monoamine oxidase-B inhibitors against DSP-4 induced noradrenaline depletion in the mouse hippocampus. J Neurochem 63: 1820–1828

Authors' address: K. Magyar, Department of Pharmacodynamics, Semmelweis University of Medicine, Nagyvárad tér 4., Budapest, 1089, Hungary

Oxidation of N-methyl(R)salsolinol: involvement to neurotoxicity and neuroprotection by endogenous catechol isoquinolines

M. Naoi[1], **W. Maruyama**[2], **T. Kasamatsu**[3], and **P. Dostert**[4]

[1] Department of Biosciences, Nagoya Institute of Technology, Nagoya, and
[2] Laboratory of Biochemistry and Metabolism, Department of Basic Gerontology,
National Institute for Longevity Sciences, Obu, Japan
[3] Faculty of Pharmaceutical Sciences, Nagoya City University, Nagoya, Japan,
[4] Pharmacia-Upjohn, Milan, Italy

Summary. 1(R), 2(N)-Dimethyl-6,7-dihydroxy-1,2,3,4-tetrahydroisoquinoline, N-methyl(R)salsolinol, is a potent dopaminergic neurotoxin to induce parkinsonism in rats. The cytotoxicity of N-methyl(R)salsolinol proved to be ascribed to its oxidation into cytotoxic 1,2-dimethyl-6,7-dihydroxyisoquinolinium ion with generation of hydroxyl radical. The isoquinolinium ion caused massive necrosis in the striatum, whereas N-methyl(R)salsolinol depleted selectively dopaminergic neurons in the substantia nigra without necrotic tissue reaction. N-Methyl(R)salsolinol induced DNA damage to human neuroblastoma SH-SY5Y cells, which could be prevented by anti-oxidants and cycloheximide. These results suggest that oxidative stress through oxidation of N-methyl(R)salsolinol induces apoptotic cell death. On the other hand, (R)salsolinol proved to scavenge hydroxyl radical produced by oxidation of dopamine. The neurotoxicity and neuroprotection of catechol isoquinolines may be ascribed to their oxidation and scavenging of radicals.

Introduction

"Oxidative stress" is a hypothesis commonly proposed for the etiology of cell death in some neurodegenerative disorders, such as Parkinson's disease. The characteristic pathological finding in Parkinson's disease is the selective death of dopaminergic neurons in the nigro-striatal system. Clinical data (Riederer et al., 1992; Stanley and Gerald, 1992) and experimental results with a potent dopaminergic neurotoxin, 1-methyl-4-phenyl-1,2,3,6-tetrahydropyridine (MPTP) (Chiueh et al., 1992) suggest that oxidative stress may be involved in the pathogenesis of this disease. In the case of Parkinson's disease, where a specific type of neurons degenerate in a selective brain region, the mechanism of the specificity remains to be elucidated. Recently, endogenous 1(R),2(N)-dimethyl-6,7-dihydroxy-1,2,3,4-tetrahydroisoquinoline [N-methyl-(R)salsolinol, NM(R)Sal] was found to induce parkinsonism in rodents (Naoi et al., 1996a,b). After injection in the rat striatum, the observed biochemical, behavioral and histopathological changes were very similar to those detected in parkinsonian patients. NM(R)Sal is produced in the brain by N-methylation of 1(R)-methyl-6,7-dihydroxy-1,2,3,4-tetrahydroisoquinoline [(R)salsolinol, (R)Sal] (Maruyama et al., 1992). (R)Sal is the only enantiomer detected in human brain (Deng et al., 1995), as suggested by previous observation that (R)Sal is the predominant enantiomer in human urine (Doster et al., 1990). Recently (R)Sal was confirmed to be synthesized from dopamine and acetaldehyde by a novel enzyme (Naoi et al., 1996b). NM(R)Sal was found to be oxidized into 1,2(N)-dimethyl-6,7-dihydroxyisoquinolinium ion (DMDHIQ$^+$) non-enzymatically (Maruyama et al., 1995a), or enzymatically (Naoi et al., 1995b) with concomitant formation of hydroxyl radicals. Histopathological studies on the rat model of Parkinson's disease revealed that after injection of DMDHIQ$^+$ massive necrosis was induced in the striatum. However, continuous administration of NM(R)Sal selectively depleted dopamine neurons in the substantia nigra, without necrotic tissue reaction. These results suggest that the oxidized and reduced catechol isoquinolines may induce different types of cell death, necrosis and apoptosis, in dopamine neurons.

There have been increasing evidences to suggest that apoptotic cell death is involved in development of the brain system (Johnson and Deckwerth, 1993) and neurodegeneration. 1-Methyl-4-phenylpyridinium ion (MPP$^+$) produced from MPTP by type B monoamine oxidase [monoamine: oxygen oxidoreductase (deaminating) EC 1.4.3.4, MAO] was reported to induce apoptosis to cultured cerebellar granule neurons (Dipasquale et al., 1991), cultured rat ventral mesencephalic and striatal cells (Mochizuki et al., 1994), and to pheochromocytoma PC12 cells (Mutoh et al., 1994). In addition, recently apoptotic cell death was detected in the brain of parkinsonian patients (Anglade et al., 1995). These results suggest that apoptosis may be mechanism underlying the selective cell death of dopamine neurons in the substantia nigra of parkinsonian brain.

In this paper, the mechanism of cytotoxicity of NM(R)Sal and related compounds was examined by in vitro and in vitro experiments. In vivo microdialysis studies at the rat striatum revealed that the generation or scavenging of hydroxyl radical may account for neurotoxic and neuroprotective potency of catechol isoquinolines (Maruyama et al., 1995a,b). NM(R)Sal could induce DNA damage to dopamine neurons. To assess DNA damage, single cell gel electrophoresis (comet) assay (Singh et al., 1988; Östling and Johanson, 1984) was applied to human dopaminergic neuroblastoma SH-SY5Y cells. The involvement of oxidation of NM(R)Sal was discussed in relation to the cell death process of dopamine neurons of the nigra-striatal system in Parkinson's disease.

Materials and methods

The (R)- and (S)enantiomers of Sal and NMSal were synthesized according to Teitel et al. (1972). 1,2-Dimethyl-6,7-dihydroxyisoquinolinium ion (DMDHIQ$^+$) was obtained by quaternalization of 1-methyl-6,7-dihydroxyquinoline (Bembenek et al., 1990). Hydroxyl radical was quantitatively determined by trapping as 2,3- and 2,5-dihydroxybenzoic acid (DHBA) by reaction with salicylic acid (Floyd et al., 1984) using high-performance liquid chromatography (HPLC) with electrochemical detection (ECD) and DMDHIQ$^+$ produced was measured fluorimetrically (Maruyama et al., 1995a).

In vivo microdialysis study for hydroxyl radical generation in the rat striatum was carried out as described previously (Maruyama et al., 1992, 1993). Hydroxyl radicals were trapped with salicylic acid as 2,3- and 2,5-DHBA (Chiueh et al., 1992). Monoamines, their metabolites, and 2,3- and 2,5-DHBA were quantitatively analyzed by HPLC and multi-electrochemical detection (ECD) (CEAS, ESA, Chelmsford, MA). The conditions of the analysis were reported previously (Naoi et al., 1993; Maruyama et al., 1992, 1993).

An animal model of Parkinson's disease was prepared using these catechol isoquinolines (Naoi et al., 1996a,b) by injection into the striatum of male Wistar rats. For chronic continuous infusion, NM(R)Sal was administrated in the striatum by attachment of a cannula to a mini-osmotic pump.

DNA fragmentation was assayed by single cell gel electrophoresis (comet assay), using SH-SY5Y (Singh et al., 1988; Östling and Johanson, 1984), and the length (the nucleus plus migrated DNA tail) of comet image was measured. The cells with the length longer than 40 μm were classified to be positive for DNA damage.

Results

Auto-oxidation of N-methyl(R)salsolinol with hydroxyl radical production

After incubation of NM(R)Sal, DMDHIQ$^+$ and 2,3- and 2,5-DHBA were identified in the reaction mixture. The amount of DMDHIQ$^+$ produced from NM(R)Sal was found to depend on the NM(R)Sal amount added to the reaction mixture (Fig. 1A, correlation = 0.921). Production of hydroxyl radical expressed as the sum of the 2,3- and 2,5-DHBA (total DHBA), was also linearly dependent on the NM(R)Sal amount (Fig. 1B, correlation = 0.931). The total amount of produced DHBAs was about 1/3 of that of DMDHIQ$^+$ (Fig. 1C).

The comparative ability of naturally-occurring isoquinolines to produce hydroxyl radicals was examined. Among all compounds examined, dopamine produced the largest amount of hydroxyl radical and the DHBA amount was obtained to be 32.1 ± 6.2 pmol/20 min after the incubation of 20 nmol of dopamine. (R)Sal and NM(R)Sal produced hydroxyl radicals; 11.4 ± 2.3 and 7.2 ± 1.0 pmol/20 min, respectively. The oxidized catechol isoquinolinium ion, DMDHIQ$^+$, did not produce hydroxyl radicals.

The effects of the isoquinolines on hydroxyl radical produced by dopamine autooxidation were examined. The amount of hydroxyl radical produced from dopamine in the presence of each of these three catechol isoquinoline derivatives is shown in Fig. 2A. (R)Sal and DMDHIQ$^+$ reduced hydroxyl radical production compared with that by dopamine alone. The formation of DHBAs generated from dopamine was reduced to 48.6 ± 20.4% of the control value by addition of (R)Sal, and to 63.2 ± 3.4% with DMDHIQ$^+$ (Fig. 2B). With NM(R)Sal, the amount of DHBAs produced from dopamine was not affected (99.3 ± 3.2% of the control value).

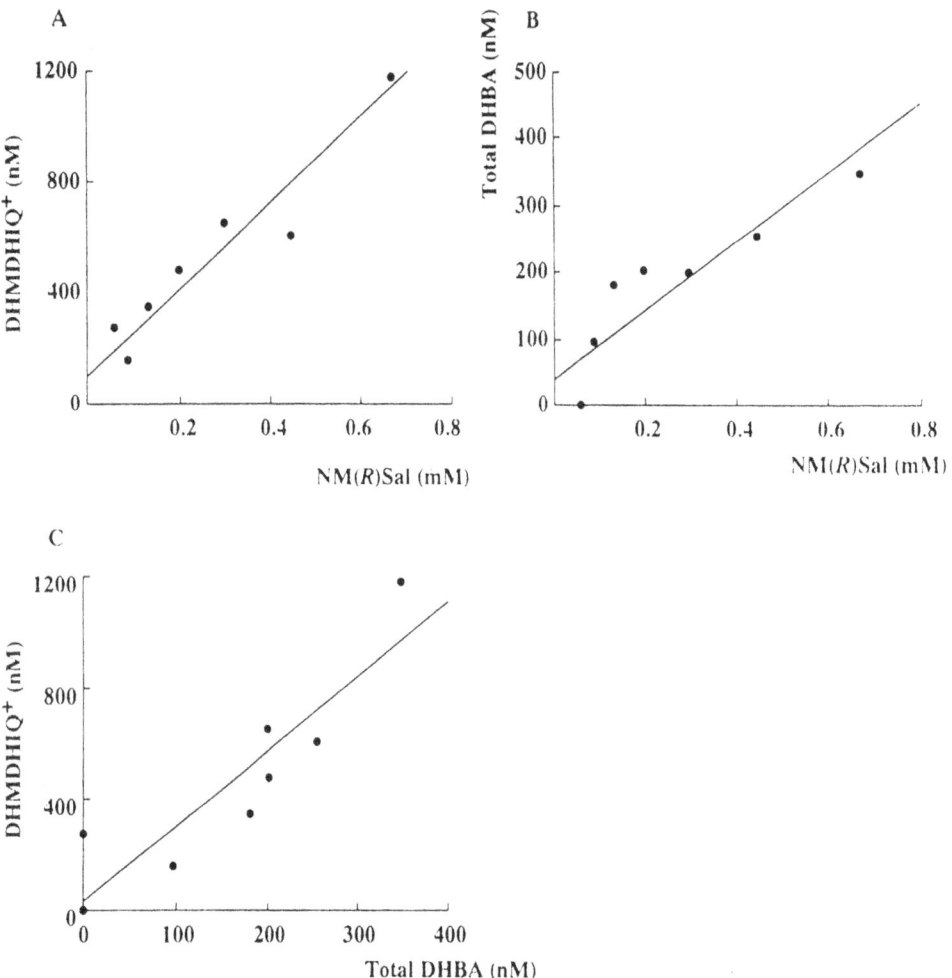

Fig. 1. The effects of NM(R)Sal amount on productions of DMDHIQ⁺ and hydroxyl radicals. Different concentrations of NM(R)Sal were added in the reaction mixture and DMDHIQ⁺ (**A**) and 2,3- and 2,5-DHBA (total DHBA) (**B**) were analyzed in the same sample, and the product concentrations were plotted against that of NM(R)Sal. **C** The concentrations of DMDHIQ⁺ produced were plotted against those of hydroxyl radicals. Each point represents mean of 2 experiments

These results indicate that oxidation of catechol isoquinolines generates hydroxyl radical, whereas they scavenge hydroxyl radical generated from oxidation of dopamine.

In vivo generation of hydroxyl radical in rat striatum:
In vivo microdialysis study

When Ringer's solution containing salicylic acid was perfused through the rat striatum, 2,3-DHBA and 2,5-DHBA were detected in the dialysate. The basal levels of 2,3- and 2,5-DHBA and dopamine metabolites were 26.9 ± 5.7 nM (mean $\pm$ SD) for 2,3-DHBA, 40.9 ± 8.9 nM for 2,5-DHBA, 1.9 ± 1.2 nM for

A

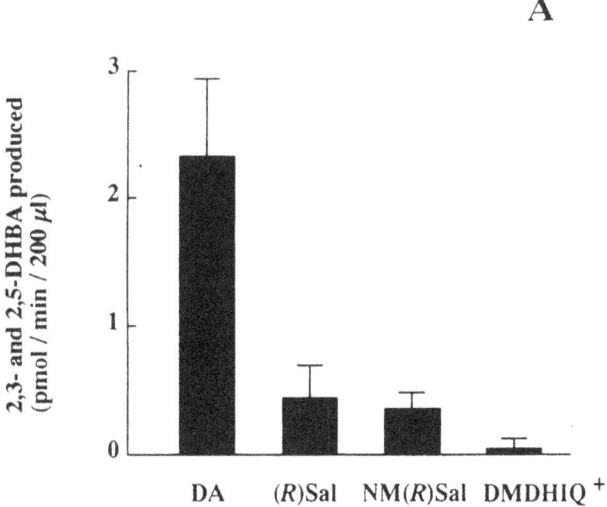

B

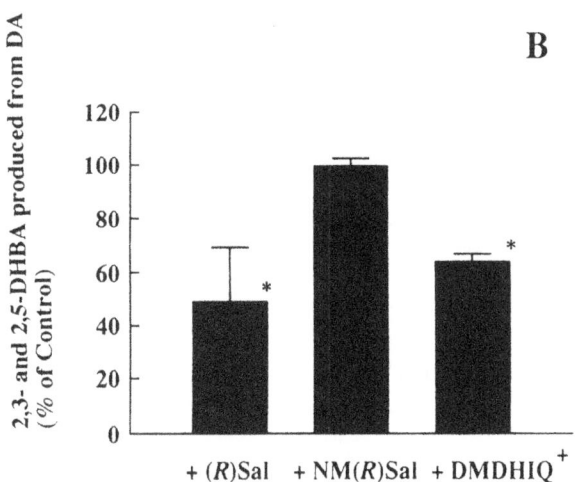

Fig. 2. The production of hydroxyl radicals from dopamine and the isoquinolines, (R)Sal, NM(R)Sal and DMDHIQ⁺. **A** Twenty nmoles of dopamine, (R)Sal, NM(R)Sal, or DMDHIQ⁺ were incubated with 1.4 µmoles of salicylic acid at 37°C for 20min. DHBAs produced were analyzed by HPLC-ECD. **B** Twenty nmoles of (R)Sal, NM(R)Sal or DMDHIO⁺ were incubated with 20nmoles of dopamine with 1.4 µmoles of salicylic acid. The hydroxyl radical production from dopamine was calculated by [(total DHBA amounts detected) − (DHBA amounts producted by the isoquinoline derivative alone)] and expressed as the percentage of DHBA amounts produced by dopamine alone. *p < 0.05 by ANOVA

dopamine, 612 ± 205 nM for 3,4-dihydroxyphenylacetic acid (DOPAC), and 284 ± 68 nM for homovanillic acid (HVA). Then, the catechol isoquinolines (40 µM) were perfused. (R)Sal, NM(R)Sal, and DMDHIQ⁺ were found to affect the level of 2,3-DHBA in the dialysate, and the time-dependent change in its level is shown in Fig. 3. The 2,3-DHBA concentrations after perfusion of the three isoquinolines were compared with those before perfusion. (R)Sal and DMDHIQ⁺ reduced the 2,3-DHBA levels significantly, whereas NM(R)Sal increased the 2,3-DHBA level significantly. The effects of the

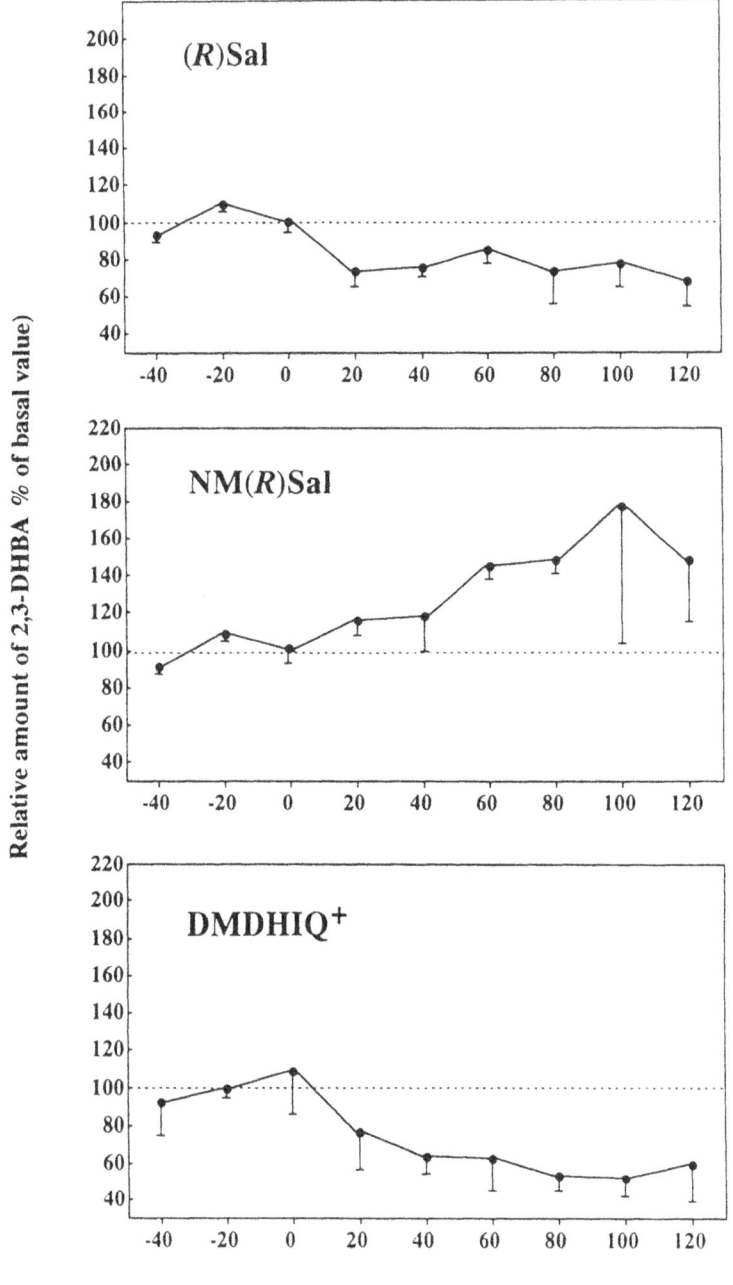

Time of catechol isoquinoline perfusion (min)

Fig. 3. Effects of perfusion of (R)Sal, NM(R)Sal and DMDHIQ$^+$ on 2,3-DHBA concentration in the striatal dialysate. The Krebs-Ringer solution containing 1 mM of salicylic acid was perfused in the rat striatum and hydroxyl radical was trapped as 2,3-DHBA and analyzed by HPLC-ECD. The relative amount of 2,3-DHBA was expressed as the percentage of the basal values: 26.9 ± 5.7 nM. The rat striatum was perfused with 40 μM of (R)Sal, NM(R)Sal, or DMDHIQ$^+$ for 120 min. Each point and bar represent the mean and SD of 4 experiments.

Table 1. Effect of (R)salsolinol, N-methyl(R)salsolinol and 1,2-dimethyl-6,7-dihydroxyisoquinolinium ion on dopamine catabolism in the rat striatum

Isoquinolines	DOPAC (% of the basal value)	HVA (% of the basal value)	Isoquinoline recovered (μM)
(R)Sal	83.4 $\pm$ 11.1	74.5 $\pm$ 10.6	25.2 $\pm$ 2.2
NM(R)Sal	101 $\pm$ 4	94.1 $\pm$ 17.1	37.1 $\pm$ 1.4
DMDHIQ$^+$	73.9 $\pm$ 6.5	20.3 $\pm$ 0.5	22.3 $\pm$ 4.5

The isoquinolines ($40\,\mu$M) were perfused through the rat striatum by the in vivo microdialysis technique. The concentrations of 3,4-dihydroxyphenylacetic acid (DOPAC) and homovanillic acid (HVA) in the dialysate were analyzed by HPLC-ECD. The concentrations after 100 min of perfusion are expressed as percent of the basal values before the perfusion. The basal value of DOPAC and HVA were $612 \pm 206\,$nM and $284 \pm 68\,$nM, respectively. Each have represents the mean and SE of 4 experiments

isoquinolines on dopamine catabolism in the rat striatum in vivo were examined, and the relative values of dopamine and its metabolites, DOPAC and HVA, in the dialysate after a 100-min perfusion are summarized in Table 1. DMDHIQ$^+$ reduced dopamine metabolite levels most potentially, followed by (R)Sal.

The results obtained by in vivo microdialysis clearly show that NM(R)Sal produces hydroxyl radical, whereas (R)Sal and DMDHIQ$^+$ reduce the radical produced in situ, which may be produced through oxidation of biogenic monoamines non-enzymatically or enzymatically by MAO.

Effect of the catechol isoquinoline derivatives on MAO activity

The effects of these catechol isoquinolines on the activity of type A and B MAO were examined. (R)Sal, NM(R)Sal and DMDHIQ$^+$ inhibited the activity of type A MAO competitively with respect to the substrate kynuramine. In contrast they were found to be non-competitive inhibitors of type B MAO. The values of the inhibitor constant, K_i, are given in Table 2. All three catechol isoquinolines inhibited type A MAO more strongly than type B. DMDHIQ$^+$ was the most potent inhibitor, with a K_i value of $8.0\,\mu$M for type A MAO, followed by (R)Sal and NM(R)Sal.

In vitro data suggest that the inhibition of MAO activity may account for the reduction of hydroxyl radical level in the striatum, as shown by in vivo microdialysis.

Animal model of Parkinson's disease

After a single injection of NM(R)Sal into the left striatum, rats exhibited postural abnormality with the head and trunk deviating toward the lesions site, lateral extension of the right hind limb and stiffness of the tail elevated

Table 2. Inhibition of type A and B monoamine oxidase by catechol isoquinolines

Catechol isoquinoline	Type A	Type B	Type A	Type B
	K_i	(μM)	Type of inhibition	
(R)Sal	52.0	181	Competitive	Noncompetitive
NM(R)Sal	85.9	463	Competitive	Noncompetitive
DMDHIQ$^+$	8.0	150	Competitive	Noncompetitive

Monoamine oxidase sample was prepared from human brain synaptosomes, and the activity was measured fluorimetrically with kynuramine as a substrate. Type A and B activity were differentiated by pre-incubation of the sample with 1 μM clorgyline or deprenyl. The K_i (inhibitor constant) and type of inhibition were obtained by plotting the data according to Lineweaver and Burk

above the ground. Some rats showed fine regular twitching of right limbs at rest. During spontaneous activity, the rats showed ipsilateral circulation toward the injection site. Injection of NM(R)Sal into both side of the striatum induced akinesia to rats. A single injection of DMDHIQ$^+$ in the left striatum induced hypokinesia in the right limbs, but no involuntary movement was detected. Other catechol isoquinolines, NM(S)Sal, (R)- and (S)-Sal, 6,7-dihydroxy-1,2,3,4-tetrahydroisoquinoline (norsalsolinol) and N-methyl-norsalsolinol, did not induce any behavioral changes to rats.

Biochemical analyses of monoamines and their metabolites are summarized in Fig. 4. In the brain slices containing the striatum and the substantia nigra, marked reduction in dopamine was observed after injection of NM(R)Sal and DMDHIQ$^+$. The reduction was more manifest with NM(R)Sal than with DMDHIQ$^+$ in the striatum ($p < 0.05$), and in the substantia nigra dopamine was lower than the detection limit. Noradrenaline was reduced in the striatum and the substantia nigra after NM(R)-Sal injection. By contrast, serotonin was not reduced after injection of NM(R)Sal. In other brain regions, no significant change in the contents of dopamine, noradrenaline, serotonin and their metabolites was observed after injection of NM(R)Sal or DMDHIQ$^+$. The activity of a rate-limiting enzyme of catecholamines synthesis, tyrosine hydroxylase [tyrosine, tetrahydropteridine: oxygen oxidoreductase (3-hydroxylating); EC 1.4.16.2, TH) was reduced in the substantia nigra after injection of NM(R)Sal, whereas the activity did not change in the striatum and after injection of DMDHIQ$^+$ (Fig. 5).

In the striatum NM(R)Sal induced only mild necrosis around the injected site, whereas DMDHIQ$^+$ caused massive necrosis and numerous macrophages were observed. Klüver-Barrera staining showed also massive destruction of myelin structure after injection of DMDHIQ$^+$, whereas NM(R)Sal did not cause significant change in the structure. After 1 week continuous injection of NM(R)Sal into the left striatum, the number of tyrosine hydroxylase-positive

Striatum

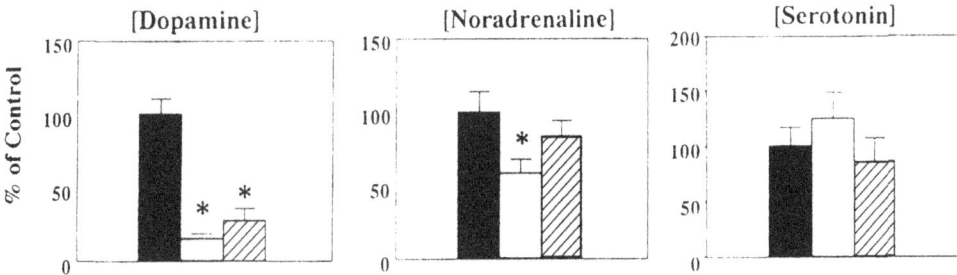

Substantia nigra

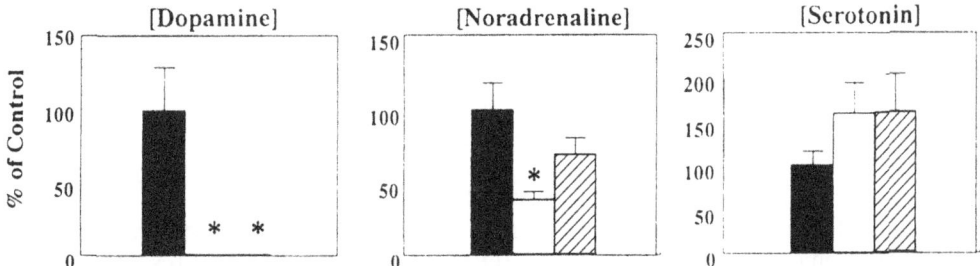

Fig. 4. Effects of the injection of NM(R)Sal and DMDHIQ⁺ on the levels of catecholamines and indoleamines in rat striatum and substantia nigra. The column and bar represent the mean and SD of duplicate measurements of 6 rats for each compound. An asterisk shows that difference from controls is statistically significant ($p < 0.05$). The filled column represents control, the hollow column after NM(R)Sal injection, and the hatched column after DMDHIQ⁺ injection. The rats were sacrificed 3 days after a single injection of 100 nmoles of each compound into the left striatum. *DOPAC* 3,4-dihydroxyphenylacetic acid, *HVA* 3-methoxy-4-hydroxyphenylacetic acid, homovanillic acid, *MHPG* 3-methoxy-4-hydroxyphenylethylene glycol, *5-HIAA* 5-hydroxyindoleacetic acid

Striatum

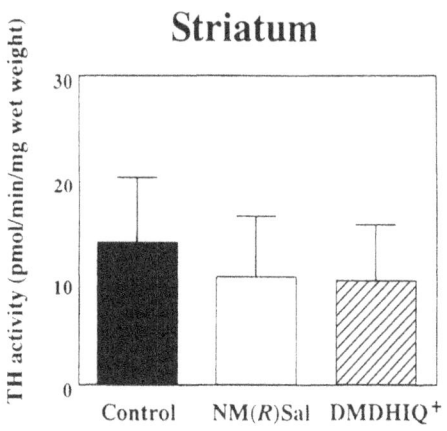

Substantia nigra

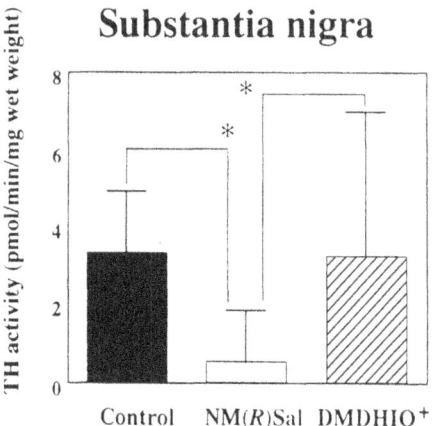

Fig. 5. The activity of tyrosine hydroxylase (TH) in the striatum and substantia nigra after a single injection of NM(R)Sal and DMDHIQ⁺. The column and bar represent the mean and SD of duplicate measurements of 6 rats for each compound. An asterisk shows that difference from control is statistically significant ($p < 0.05$). The filled column represents control, the hollow column after NM(R)Sal injection, and the hatched column after DMDHIQ⁺ injection. The rats were sacrificed 3 days after a single injection of 100 nmoles of each compound into the left striatum

neurons was significantly reduced in the substantia nigra of the injected side ($3.09 \pm 0.32/225\,\mu m^2$), in comparison with that in the opposite side ($6.69 \pm 0.62/225\,\mu m^2$) ($p < 0.05$).

The biochemical and histological results on the animal model of Parkinson's disease indicate that NM(R)Sal induces selective cell death of dopaminergic neurons, which may be ascribed to the selective uptake in the striatum and the retrograde axonal transport into the substantia nigra. The oxidation product of NM(R)Sal, DMDHIQ⁺, is more cytotoxic, but the cell death is non-selective and necrotic process.

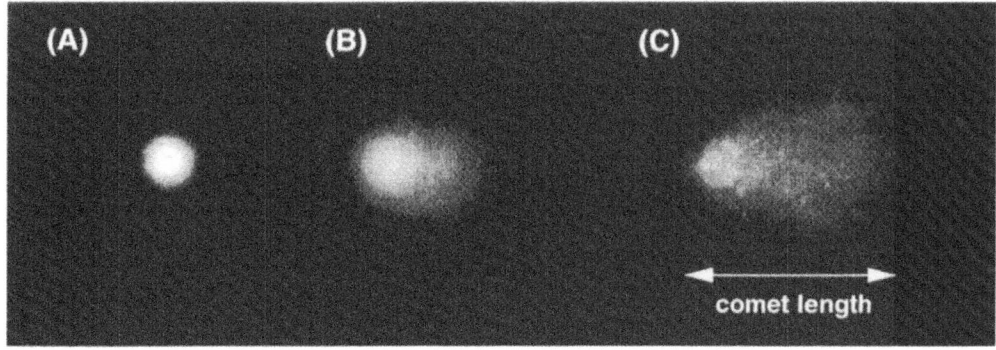

Fig. 6. Fluorescence photomicrograph of SH-SY5Y cells treated with 1 mM NM(R)
Sal for 3 hours. The cells were subjected to comet assay. **A** Control. **B** and **C** After
incubation with NM(R)Sal some cells showed partial (B) and full (A) fragmentation of
DNA. The length of the head to the end of tail of the comet image was measured to assess
DNA damage

DNA damage induced by NM(R)Sal

The selective neurotoxicity of NM(R)Sal was demonstrated further by its
induction of DNA damage. SH-SY5Y cells were incubated with catechol
isoquinolines and assessed for DNA damage using the single cell alkaline gel
electrophoresis assay (comet assay). Only after incubation of NM(R)Sal,
DNA fragmentation was detected. Other catechol isoquinolines, (R)Sal and
DMDHIQ$^+$ did not induce DNA damage. The fluorescence photomicro-
graphs (Fig. 6) showed that the typical comet image of the cells after incubation with NM(R)Sal. With 1 mM NM(R)Sal almost all the cells showed the
positive comet tails, whereas with 1 mM NM(S)Sal only about 10% cells were
positive for DNA damage. The effects of anti-oxidants and related enzymes
were confirmed. With 0.5 mM NM(R)Sal about 18% cells showed typical
Comet image of DNA damage. The presence of catalase, reduced glutathione,
deprenyl and semicarbazide protected the cells from DNA damage, suggesting the involvement of oxidative stress to the DNA damage. A protein synthesis inhibitor, cycloheximide, was proved to prevent DNA damage by
NM(R)Sal, indicating that the apoptotic cell death was induced.

The induction of DNA damage by NM(R)Sal confirms again that this
reduced catechol isoquinoline is a selective dopaminergic neurotoxin and the
cell death is apoptotic process, for which the oxidation is essentially required.

Discussion

In wine and other foods Sal is identified in the racemic form produced by
the non-enzymatic Pictet-Spengler reaction of dopamine with acetaldehyde
(Strolin-Benedetti et al., 1989). Even though both (R)- and (S)-Sal were
detected in human plasma (Sällström Baum and Rommelspacher, 1994), Sal

cannot be transported into the brain through the blood-brain barrier (Origitano et al., 1981). These results are relevant with our data on the sole presence of the (R)-enantiomers of Sal and NMSal in the brain (Deng et al., 1995), intraventricular (Maruyama et al., 1996b) and cerebrospinal fluid (Maruyama et al., 1996a). NM(R)Sal is oxidized into DMDHIQ$^+$ by an oxidase different from flavin-containing mitochondrial MAO. These catechol isoquinolines were examined in human brain and the concentration of NM(R)Sal was found to be significantly higher in the nigro-striatal system than in other brain regions, which may be due to higher activity of the N-methyltransferase in these nuclei (Maruyama et al., 1992). These results indicate that (R)Sal and N-methylated derivatives should be enzymatically synthesized from dopamine in situ in the human brain.

As shown in the substantia nigra of a rat model of Parkinson's disease, DMDHIQ$^+$ induces massive necrosis, but the cell death was non-specific. The necrosis by the ion may be due to the inhibition of ATP synthesis, as shown with a series of catechol isoquinolines (McNaught et al., 1995). The selective cell death of dopamine neurons in the substantia nigra was observed after injection of NM(R)Sal, but not by cytotoxic DMDHIQ$^+$. The depletion of dopamine neurons was not by necrotic, but by apoptotic process of cell death. Hartley et al. (1994) reported that inhibition of complex I by MPP$^+$ induced apoptosis, suggesting that ATP depletion may induce apoptotic process. On the other hand, as reported here with NM(R)Sal and SH-SY5Y cells antioxidants, reduced glutathione and catalase, could prevent DNA damage. These results suggest that oxidative stress may be primarily elicited, followed by DNA fragmentation. The selective potency of (R)-enantiomer of NMSal to induce DNA damage suggests that an enzyme should recognize the enantiometric configuration. At present, the characterization of the NM(R)Sal oxidase, which has higher affinity to the (R)enantiomer than the (S), remains to be clarified.

All these results demonstrate that oxidation is the essential process to the neural cell death by NM(R)Sal, and endogenous catechol isoquinolines, such as salsolinol, can scavenge hydroxyl radical and thus be neuroprotective to dopamine neurons.

Acknowledgements

This work was supported by a Grant-In-Aid for Scientific Research on Priority Area and (C) from the Ministry of Education, Science and Culture, Japan.

References

Anglade P, Vyas S, Javoy-Agid F, Herrero MT, Michel PP, Marquerz J, Mouatt-Prigient A, Ruberg M, Hirsch EC, Agid Y (1995) Apoptotic degeneration of nigral dopaminergic neurons in Parkinson's disease. Soci Neurosci Abstr 21: 1250
Bembenek ME, Abell CW, Chrisey LA, Rowadowska MD, Gessner W, Brossi A (1990) Ihhibition of monoamine oxidase A and B by simple isoquinoline alkaloids: racemic

and optically active 1,2,3,4-tetrahydro, 3,4-dihydro, and fully aromatic isoquinolines. J Med Chem 33: 147–152

Chiueh CC, Krishna G, Tulsi P, Obata T, Lang L, Huang S, Murphy DL (1992) Intracranial microdialysis of salicylic acid to detect hydroxyl radical generation through dopamine autooxidation in the caudate nucleus: effects of MPP$^+$. Free Radic Biol Med 13: 581–583

Deng Y, Maruyama W, Dostert P, Takahashi T, Kawai M, Naoi M (1995) Determination of the (R)- and (S)-enantiomers of salsolinol and N-methylsalsolinol by use of a chiral HPLC column. J Chromatogr B 670: 47–54

Dipasquale B, Marini AM, Youle RJ (1991) Apoptosis and DNA degradation induced by 1-methyl-4-phenylpyridinium in neurons. Biochem Biophys Res Commun 181: 1442–1448

Dostert P, Strolin Benedetti M, Bellotti V, Allievi C, Dordain G (1990) Biosynthesis of salsolinol, a tetrahydroisoquinoline alkaloid, in healthy subjects. J Neural Transm [GenSect] 81: 215–223

Floyd RA, Watson JJ, Wong PK (1984) Sensitive assay of hydroxyl free radical formation utilizing high performance liquid chromatography with electrochemical detection of phenol and salicylate hydroxylation products. J Biochem Biophys Methods 10: 221–235

Hartley A, Stone JM, Heron C, Cooper JM, Schapira AHV (1994) Complex I inhibitors induce dose-dependent apoptosis in PC12 cells: relevance to Parkinson's disease. J Neurochem 63: 1987–1990

Johnson EM Jr, Deckwerth TL (1993) Molecular mechanisms of developmental neuronal death. Ann Res Neurosci 16: 31–46

Maruyama W, Nakahara D, Ota M, Takahashi T, Takahashi A, Nagatsu T, Naoi M (1992) N-Methylation of dopamine-derived 6,7-dihydroxy-1,2,3,4-tetrahydroisoquinoline, (R)-salsolinol, in rat brains: in vivo microdialysis study. J Neurochem 59: 395–400

Maruyama W, Nakahara D, Dostert P, Hashiguchi H, Ohta S, Hirobe M, Takahashi A, Nagatsu T, Naoi M (1993) Selective release of serotonin by endogenous alkaloids, 1-methyl-6,7-dihydroxy-1,2,3,4-tetrahydroisoquinolines, (R)- and (S)-salsolinol, in the rat striatum; in vivo microdialysis study. Neurosci Lett 149: 115–118

Maruyama W, Dostert P, Matsubara K, Naoi M (1995a) N-Methyl(R)salsolinol produces hydroxyl radicals: involvement to neurotoxicity. Free Radic Biol Med 19: 67–75

Maruyama W, Dostert P, Naoi M (1995b) Dopamine-derived 1-methyl-6,7-dihydroxyisoquinolines as hydroxyl radical promoters and scavengers in the rat brain: In vivo and in vitro studies. J Neurochem 64: 2635–2643

Maruyama W, Abe T, Tohgi H, Dostert P, Naoi M (1996a) A dopaminergic neurotoxin, (R)-N-methylsalsolinol, increases in parkinsonian CSF. Ann Neurol 40: 119–122

Maruyama W, Narabayashi H, Dostert P, Naoi M (1996b) Stereospecific occurrence of a parkinsonism-inducing catechol isoquinoline, N-methyl(R)salsolinol, in the human intraventricular fluid. J Neural Transm 103: 1069–1076

McNaught K, Thull U, Carrupt P-A, Altomare C, Cellamare S, Carotti A, Testa B, Kenner P, Marsden CD (1995) Inhibition of complex I by isoquinoline derivatives structurally related to 1-methyl-4-phenyl-1,2,3,6-tetrahydropyridine (MPTP). Biochem Pharmacol 50: 1903–1911

Mochizuki H, Nakamura N, Nishi K, Mizuno Y (1994) Apoptosis in induced by 1-methyl-4-phenylpyridinium ion (MPP$^+$) in ventral mesencephalic-striatal co-culture in rat. Neurosci Lett 170: 191–194

Mutoh T, Tokuda A, Marini M, Fujiki N (1994) 1-Methyl-4-phenylpyridinium kills differentiated PC12 cells with a concomitant change in protein phosphorylation. Brain Res 661: 51–55

Naoi M, Maruyama W, Acworth IN, Nakahara D, Parvez H (1993) Multi-electrode detection system for determination of neurotransmitters. In: Parvez H, Naoi M, Nagatsu T, Parvez S (eds) Methods in neurotransmitter and neuropeptide research, vol I. Elsevier, Amsterdam, pp 1–39

Naoi M, Maruyama W, Dostert P (1995a) Dopamine-derived 6,7-dihydroxy-1,2,3,4-tetrahydroisoquinolines: oxidation and neurotoxicity. Prog Brain Res 106: 227–239

Naoi M, Maruyama W, Zhang JH, Takahashi T, Deng Y, Dostert P (1995b) Enzymatic oxidation of the dopaminergic neurotoxin, 1(R), 2(N)-dimethyl-6,7-dihydroxy-1,2,3,4-tetrahydroisoquinoline, into 1,2(N)-dimethyl-6,7-dihydroxyisoquinolinium ion. Life Sci 57: 1061–1066

Naoi M, Maruyama W, Dostert P, Hashizume Y, Takahashi T, Ota M (1996a) Dopamine-derived endogenous 1(R),2(N)-dimethyl-6,7-dihydroxy-1,2,3,4-tetrahydroisoquinoline, N-methyl-(R)-salsolinol, induced parkinsonism in rats: biochemical, pathological and behavioral studies. Brain Res 709: 285–295

Naoi M, Maruyama W, Dostert P, Hashizume Y (1996b) Animal model of Parkinson's disease induced by naturally-occurring 1(R), 2(N)-dimethyl-6,7-dihydroxy-1,2,3,4-tetrahydroisoquinoline. Biogen Amines 12: 135–147

Origitano T, Hanningen J, Collins MA (1981) Rat brain salsolinol and blood-brain barrier. Brain Res 224: 446–451

Östling O, Johanson KJ (1984) Microelectrophoretic study of radiation-induced DNA damages in individual mammalian cells. Biochem Biophys Res Commun 123: 291–298

Riederer P, Sofic E, Rausch W-D, Schmidt B, Reynolds GD, Jellinger K, Youdim MBH (1989) Transition metals, ferritin, glutathione, and ascorbic acid in parkinsonian brains. J Neurochem 52: 515–520

Sällström Baum S, Rommelspacher H (1994) Determination of total dopamine, R- and S-salsolinol in human plasma by cyclodextrin bonded-phase liquid chromatography with electrochemical detection. J Chromatogr B 660: 235–241

Singh NP, McCoy MT, Tice RR, Schneider EL (1988) A simple technique for quantitation of low levels of DNA damage in individual cells. Exp Cell Res 175: 184–191

Stanley F, Gerald C (1992) The oxidative stress hypothesis in Parkinson's disease: evidence supporting it. Ann Neurol 32: 804–812

Strolin Benedetti M, Bellotti V, Pianezola E, Moro E, Carminati P, Dostert P (1989) Ratio of the R and S enantiomers of salsolinol in food and human urine. J Neural Transm 77: 47–53

Teitel S, O'Brien J, Brossi A (1972) Alkaloids in mammalian tissue. II. Synthesis of (+) and (−) substituted-6,7-dihydroxy-1,2,3,4-tetrahydroisoquinolines. J Med Chem 15: 845–846

Authors' address: Dr. M Naoi, Department of Biosciences, Nagoya Institute of Technology, Gokiso-cho, Showa-ku, Nagoya 466, Japan

Substrate regulation of monoamine oxidases

R. R. Ramsay

School of Biomedical Sciences, University of St. Andrews, St. Andrews,
United Kingdom

Summary. The rate of oxidation by monoamine oxidase (MAO) of a particular amine in a given cell depends on the levels of MAO-A and MAO-B expressed in the mitochondrial outer membranes, on the amine concentration and the oxygen concentration. Its disposal will be slowed by the presence of competing amines or endogenous inhibitors. However, substrate binding alters the properties of MAO and influences catalytic turnover. (a) It increases the redox potential of the flavin making possible the transfer of electrons from the higher potential amine. (b) It accelerates the reactivity of the covalently bound flavin with oxygen, effectively increasing the V_m (particularly for MAO-B). (c) It bypasses the generation of free oxidised enzyme in the reaction cycle so that, at high amine concentrations, only the affinity of a substrate or inhibitor for the reduced enzyme (particularly for MAO-A) is important. These changes are induced only by substrate, not by the few stable products available nor by inhibitors suggesting a very specific interaction between a substrate ligand and the enzyme. The altered properties are very different for MAO-A and MAO-B even with the same substrate. Elucidation of the mechanisms involved must await structural information from physical studies, molecular modelling and mutational analysis.

Introduction

Monoamine oxidases (MAO) play a critical role in the degradation of neurotransmitter amines and their by-products, but are found in all tissues indicating a general function for the oxidation of bioamines. MAO is located on the mitochondrial outer membrane where it oxidises an enormous variety of amines to the corresponding imines. The general properties of the MAOs, their molecular biology, and the chemical mechanism have been reviewed (Singer, 1991; Weyler et al., 1990; Silverman, 1992).

In this article I shall try to explain the influence of substrates on the flux through MAO, without resorting to kinetic equations, in order to point out the complexities and finesse of the amine oxidising system provided by the A and B forms of MAO.

Discussion

1. Factors affecting MAO-dependent amine oxidation

MAO levels

MAO is a constitutive enzyme, with no known external regulation of its expression. The level of expression is, however, very tissue specific and the rates of synthesis and degradation determine different half-lives, varying from 3 days in liver to 30 days in brain (Weyler et al., 1990). In brain, there are large regional variations (reviewed in Weyler et al., 1990) and specific cell types have different proportions of the two enzymes. MAO-A is predominant in catecholinergic neurones and MAO-B in serotonergic neurones and in the glia. Now that the tools and techniques are available, detailed quantitative studies of the amounts and distribution of the two forms should be done. When co-ordinated with kinetic data, such information could help construct a model of the effects of MAO inhibition on cellular amine profiles in an area of interest.

Amine levels and identity

MAO is located intracellularly so the local amine levels and the composition of the pools are difficult to determine. Neurotransmitters such as serotonin, dopamine, and noradrenaline provide simple surges which presumably dominate the amine profile transiently, but the possibility of more complex mixtures, including metabolites, polyamines, β-carbolines, and isoquinolines make it difficult to predict which will be oxidised by MAO in preference to the other amines and what effect MAO inhibition would have on the local amine metabolism in the individual cell. A study of urinary amine metabolites in human deficiencies of MAO-A and MAO-B represents a limited start towards predicting the cellular pattern (Murphy DL, this volume).

The specific localisation of MAO in the mitochondria is puzzling but can be rationalised in terms of the charge properties of most amines. Neutral forms cross membranes with relative ease but positively charged (protonated) species distribute across the membrane in response to the electrochemical gradient. MAO may help prevent the build-up of amines in the matrix and also the dissipation of the gradient essential for energy production.

All amines are potential MAO substrates. Primary, secondary, tertiary amines, lipophilic or not, all compete for the active site. Lipophilic amines, in particular, tend to distribute into the membrane providing a higher local concentration in the vicinity of the active site. The neutral (non-protonated) form of the amine is the substrate for MAO. Most primary amines (e.g., dopamine) have pK_a values above 10 and secondary and tertiary amines (e.g., MPTP) have higher pK_a values. However, protonation is a rapid equilibrium and the strength of its influence on catalysis under cellular conditions has not been determined. In summary, lipophilicity, fractional ionisation and location of the amine all influence the effective concentration in the vicinity of the active site of MAO. Relative specificity values have been determined for many naturally occurring amines, although the function of some of these amines in the cell is still obscure. With techniques now available to determine absolute levels of enzymes and the assay limitations well characterized (Krueger et al., 1995) it would be useful to have a new definitive study of true specificity values (k_{cat}/K_m) for naturally occurring amines.

2. Kinetic studies

Substrate increases the rate of the oxidative half-reaction

Simple mass action considerations affect the flux through all chemical reactions. What aspects of MAO catalysis indicate that substrates alter the rate of their own disposal?

The second substrate for MAO is oxygen and chemically-reduced MAO is oxidised by oxygen at a rate of about $1 s^{-1}$ at air saturation. If the reduced MAO is pre-equilibrated with substrate before mixing with oxygen, the rate of oxidation is increased 2–100 fold (Ramsay et al., 1987; Tan and Ramsay, 1993). This means that substrate binding alters the observed rate constant for that step, because the $E_R S$ complex reacts with oxygen faster than the free reduced enzyme (E_R). The higher rate constant contributes to V_m so that, in general, the V_m is increased. In simple terms, no k_{cat} greater than $1 s^{-1}$ would be possible without this faster alternative pathway. Inspection of Table 1 shows that all of the naturally occuring amines listed there have k_{cat} values greater that. The effect of the alternate pathways on steady-state kinetics has been demonstrated with methyltetrahydrostilbazole (Ramsay et al., 1993).

Table 1. Rate constants for MAO with selected amines[a]

Substrate	Rate constant (s^{-1})					
	MAO-A			MAO-B		
	Steady state	Reduction	Oxidation	Steady State	Reduction	Oxidation
Kynuramine	2.7	3.1	120	2.7	13.6	2.2
Benzylamine	0.02	0.06	23	10	11	7.6
α-Methyl-benzylamine	0.0002	0.0002	4.3	0.0007	0.0007	21
Serotonin	2.8	2.1	5.7	0.08	0.09	1.7
MPTP	0.2	0.2	40	3.3	3.7	6.0

[a] Data are taken from Tan and Ramsay (1993). The rate constants (at 30°C) are: for the steady-state, the rate of product formation; for reduction, the rate of reduction of the flavin in the $E_{ox}S$ complex; for oxidation, the rate of oxidation of the flavin in the $E_R S$ complex at 0.24 mM oxygen

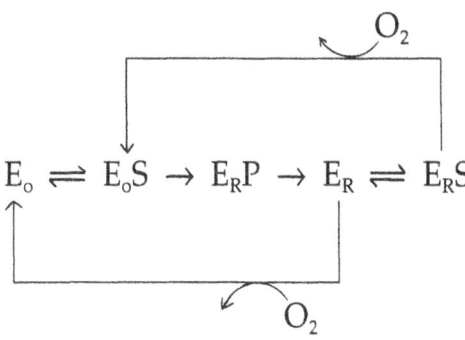

Fig. 1. Kinetic pathways for monoamine oxidases. Note that at high amine concentrations, the upper ternary complex cycle involves no free oxidised enzyme pool, whereas the lower binary mechanism generates pools of both E_R and E_{ox}

The influence of the faster oxidation of the ternary complex is obvious in MAO B where the oxidative half reaction strongly influences the steady-state rate (Table 1). For MAO-A, the rates of reduction of the enzyme by substrates are much lower and so limit the steady-state rates. However, the formation of the $E_R S$ complex is still important because it bypasses the step of binding substrate to free oxidised enzyme (Fig. 1). Substrate binding to the oxidised and reduced forms of the enzyme may be different (see below). Indirect estimates suggest that MAO-A shows a lower K_D in the reduced form (Tan and Ramsay, 1993), but MAO-B shows a much higher one (Ramsay et al., 1987).

The partition between the slow (E_R) and faster ($E_R S$) oxidation pathways is determined by the availability of E_R, the substrate concentration, and the rate of association of that substrate relative to the rate of oxidation of E_R. Using isolated enzyme, the flavin spectrum can be monitored to determine the levels of oxidised and reduced enzyme during steady state turnover. Table 2 shows that during the oxidation of serotonin, MAO-A is mostly reduced, whereas when MPTP is the substrate, MAO-A is mostly oxidised. In contrast,

Table 2. Oxidised enzyme levels during turnover

	% Oxidised enzyme	
Substrate	MAO-A	MAO-B
Kynuramine	82	75
Serotonin	3	66
MPTP	94	92

The steady-state concentration of oxidised MAO was determined by spectrophotometry during stopped-flow monitored turnover with 5 mM amine, 0.195 mM oxygen, and 18 μM MAO-A. The values are from one experiment with errors of less than ± 4%

although much the same with MPTP, MAO-B is about 85% oxidised during serotonin turnover.

In summary, the spectrum of MAO activity (how much of each substrate is turned over) expressed at a given time in a given location depends on the predominant substrate which determines the steady-state oxidation level, on the rates for product dissociation and substrate binding which determine the free enzyme available for reaction with oxygen, and on the affinity of each substrate in the mixture for the reduced and oxidised forms of the enzyme.

Inhibitor binding

Similar factors determine the binding of inhibitors to the enzyme, namely, affinity for E_o or E_R and the availability of these species. Irreversible inhibitors, such as deprenyl, are in fact suicide substrates. Deprenyl, like a substrate, stimulates the reoxidation of reduced MAO-B (Sablin and Ramsay, unpublished) but then reduces and covalently modifies the flavin resulting in permanently inactivated enzyme. With the competitive inhibitor, amphetamine, Pearce and Roth (1985) showed elegantly that during turnover of benzylamine by MAO-B, the level of free reduced enzyme was too low to see significant binding to E_R. This experiment demonstrated, in the steady-state, the presence of a ternary complex during turnover. The same was shown for MAO A oxidising kynuramine and we established by stopped-flow spectrophotometry that the E_R-amphetamine complex was not oxidised (Ramsay, 1991).

Competitive inhibitors, like substrates, may bind differently to the oxidised and reduced forms of MAO. If an inhibitor binds more tightly to E_R, then E_R availability will influence the apparent K_i value — a good argument for determining true kinetic constants. If the inhibitor also dissociates slowly, anomolous kinetics may be observed. The slow development of steady-state inhibition of MAO-A by the β-carbolines may be an example of this (Kim et al., 1997). The new generations of good competitive MAO inhibitors for potential clinical use should be fully characterised to detect such differences because it will influence their efficacy against particular amine substrates. An inhibitor which binds better to the reduced form will be more effective against a substrate which results in higher E_R levels than against one for which E_o predominates in turnover.

3. Thermodynamics

With pure MAO-A and B now available, the redox potential of the cysteinyl-FAD in these enzymes was measured (Ramsay et al., 1995). It is a long standing problem that the redox potential for most primary amines is about $+1\,V$ whereas that for free flavin is around $-0.2\,V$, so that electron transfer is highly unfavourable. The two electron redox potential for the cysteinyl-FAD

Table 3. Redox potentials (2-electron) for MAO-A and MAO-B in the presence of substrates[a]

Substrate	Redox potential mV ± s.d. (n = 3)	
	MAO A	MAO B
None	−210	−221
5-Hydroxytryptamine	n.d.[b]	+194 ± 9
α-Methylbenzylamine	−116 ± 4	+281 ± 12
Benzylamine	+263 ± 15	n.d.[b]

[a]The enzyme and a reference dye were present at the beginning of the experiment and the reaction was initiated with substrate added anaerobically. Co-reduction of flavin and DCIP was monitored in the presence of catalytic amounts of methyl viologen. [b]n.d., not determined because the rate of reduction of this enzyme by this substrate was too fast to be certain of equilibrium between the enzyme and the dye

in MAO is also around −0.2 V but when substrate is bound the redox potential of the flavin is positively shifted by almost 0.5 V (Table 3). In contrast, inhibitors and products, which still bind well to the active site, do not induce a shift in potential. The large positive shift is the first indication of an explanation for the ability of the MAO flavin to oxidise amines. In addition, we can speculate that binding might reciprocally decrease the potential of the amine so that the barrier to the transfer of electrons from the more positive amine to the flavin would be lowered.

A positive effect of substrate and the lack of effect of product and inhibitor on the midpoint potential of MAO are seen not only for the redox properties of the flavin but also for the stimulation of the oxidative half-reaction. However, a midpoint potential of about +0.25 V for the complex of MAO-A with α-methylbenzylamine leaves only a small driving force for electron transport to oxygen, because the midpoint potential for the O_2/H_2O_2 couple is +0.3 V. The correlation between the potential shift and the increased rate of reoxidation of the flavin has not been studied.

4. Future investigation of how the substrate alters the molecular properties of MAO

Active site structure

The ideal starting point for rational drug design is a crystal structure, but membrane bound enzymes are notoriously difficult to crystallise. MAO conforms to this pattern and, despite the availability of good quantities of pure enzyme, no crystals have been obtained. Instead, an initial comparison of the dimensions of the binding sites of MAO-A and -B (Efange and Boudreau, 1991) could be refined using the power of new molecular modelling techniques. The earlier study was based on data for MPTP analogs but true kinetic constants are now available for a range of inhibitor structures for both MAO-A and -B.

Spectral studies

The effect of substrate on the covalently-bound flavin of MAO could be a direct interaction due to the proximity of the electron shells of the molecules or an indirect one mediated via the protein (i.e. a conformational change). To measure any direct interaction the effects of an inhibitor, a product, and a slow substrate on the spectrum of MAO-A were examined (Fig. 2). Clearly the substrate and inhibitor have very different (indeed inverse) effects on the spectrum. The direction of the alteration induced by product resembles an inhibitor effect, although the fine features are different. This initial experiment agrees with the trend of the kinetic experiments which show that substrate stimulates the reactivity of the flavin with oxygen but products and inhibitors decrease the reactivity. More data is currently being gathered to see if quantitative comparisons can be made.

Molecular biological approaches

The substrate site has proved elusive for many years but now mutagenesis has identified the cysteine at position 374 in MAO-A and 365 in MAO-B as the probable substrate site cysteine (Wu et al., 1993). Mutagenesis studies should eventually yield residues or sequences which when altered result in

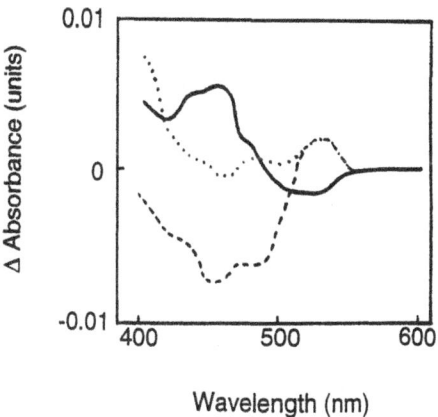

Fig. 2. Difference spectra for MAO-A. The spectrum for MAO-A (10 μM) alone was subtracted from the spectra obtained in the presence of 30 μM harmine (inhibitor, ————), 50 mM α-methylbenzylamine (slow substrate, ...), or 0.1 mM 1-methylstibazolium (product, --------)

MAO-A-like substrate stimulation in MAO-B (or vice-versa) without drastic alteration in the binding specificity. Ultimately this approach is the only way to identify the protein residues connecting the substrate and flavin binding regions.

Functional studies

Kinetic parameters and substrate competition experiments are needed to complete a pecking order for the wide range of amines likely to compete in the cell. On the whole body level, inhibitor influx studies combined with positron emission tomography (Fowler et al., 1996) could be used to look at how endogenous amines modify therapeutic inhibitor binding. The same method might also be extended to calculate true *in vivo* K_D values and, hence, to estimate the redox status of MAO in specific brain regions.

Conclusion

For a constitutive enzyme, effective MAO activity in a given cell is highly variable. (a) The mechanism for tissue- or cell-specific regulation of the level of expression of MAO-A and MAO-B is unknown. It is that which determines the range of activity available to the cell. (b) The selection of amines present in the cell influences the disposal rate of a particular amine by competition for both oxidised and reduced enzyme. Affinity of the amine for each redox form of each enzyme may differ. (c) The steady state redox level of MAO present is also determined by the amines being processed due to the amine-specific imbalance between the rate constants for the oxidative and reductive half-reactions.

However, the most distinctive influences of substrate amine on the catalytic processes of MAO are the increased redox potential and the increased reactivity with oxygen. The former explains how electrons can move from amine to flavin in the first place and the latter, reinforced by former, suggests that substrate binding induces a conformational change in the protein which alters the properties of the flavin. Spectral changes observed on ligand binding are consistent with this. It now remains to solve the molecular basis for these effects in order to progress with rational drug design in a way that does not neglect the influence of the varying amine effects on the enzyme in the cellular environment.

References

Efange SMN, Boudreau RJ (1991) Molecular determinants in the bioactivation of the dopaminergic neurotoxin, 1-methyl-4-phenyl-1,2,3,6-tetrahydropyridine(MPTP). J Computer-Aided Mol Des 5: 405–417

Fowler JS, Volkow ND, Wang G-J, Pappas N, Logan J, McGreror R, Alexoff D, Shea C, Schlyer D, Wolf AP, Warner D, Zezuiikova I, Cliento R (1996) Inhibition of MAO B in the brains of smokers. Nature 379: 733–736

Kim H, Sablin SO, Ramsay RR (1997) Inhibition of monoamine oxidase A by β-carboline derivatives. Arch Biochem Biophys 337: 137–142

McEwen CM Jr, Sasaki G, Lenz WR (1968) Human liver mitochondrial monoamine oxidase. I Kinetic studies of model interactions. J Biol Chem 213: 5217–5225

Pearce LB, Roth JA (1985) Human brain monoamine oxidase type B: mechanism of deamination probed by steady-state methods. Biochemistry 24: 1821–1826

Ramsay RR (1991) The kinetic mechanism of monoamine oxidase A. Biochemistry 30: 4624–4629

Ramsay RR, Koerber SC, Singer TP (1987) Stopped-flow studies on the mecchanism of oxidation of 1-methyl-4-phenyltetrahydropyridine by bovine liver monoamine oxidase B. Biochemistry 26: 3045–3050

Ramsay RR, Sablin SO, Bachurin SO, Singer TP (1993) Oxidation of tetrahydrostilbazole by monoamine oxidase A demonstrates the effect of alternate pathways in the kinetic mechanism. Biochemistry 32: 9025–9030

Silverman RB (1992) Electron transfer chemistry of monoamine oxidase. Adv Electron Transfer Chem 2: 177–213

Singer TP (1991) Monoamine oxidases. In: Muller F (ed) Chemistry and biochemistry of flavoenzymes, vol II. CRC Press, Boca Raton, pp 437–470

Tan AK, Ramsay RR (1993) Substrate-specific enhancement of the oxidative half-reaction of monoamine oxidases. Biochemistry 32: 2137–2143

Weyler W, Hsu Y-PP, Breakefield XO (1990) Biochemistry and genetics of monoamine oxidase. Pharmacol Ther 47: 391–417

Wu H-F, Chen K, Shih JC (1993) Site-directed mutagenesis of monoamine oxidase A and B: role of cysteines. Mol Pharmacol 43: 888–893

Author's address: Dr. R. R. Ramsay, School of Biomedical Sciences, University of St. Andrews, Irvine Building, North Street, St. Andrews KY16 9AL, United Kingdom

Monoamine oxidases and related amine oxidases as phase I enzymes in the metabolism of xenobiotics

M. Strolin Benedetti[1] and **K. F. Tipton**[2]

[1] Department of Preclinical Development and Human Pharmacology, Zambon Group, Bresso, Italy
[2] Department of Biochemistry, Trinity College, Dublin, Ireland

Summary. To date most of the interest in oxidative metabolism of xenobiotics has been devoted to the role of the microsomal cytochrome P-450 system and to establish the basis for classifying and naming P450 enzymes. The contribution of amine oxidases to the metabolism of xenobiotics has been largely neglected, with the exception of the contribution of monoamine oxidases (MAOs) to the metabolism of exogenous tyramine and the studies of the "cheese effect" produced as the result of ingestion of large amounts of tyramine-containing foods under particular conditions. A review of the involvement of the mitochondrial MAOs in drug metabolism was published in 1988. Since that time, considerable additional evidence has appeared in the literature to support the contribution of MAOs to drug metabolism. In addition, the involvement of other amine oxidases in the metabolism of foreign compounds has been established. A second review on the contribution of amine oxidases to the metabolism of xenobiotics was therefore published in 1994. On an arbitrary basis, the heterogeneous class of amine oxidases can be divided into two types according to their prosthetic group: the flavine-adenine dinucleotide (FAD)-dependent amine oxidases (Monoamine Oxidase and Polyamine Oxidase) and the amine oxidases not containing FAD (Semicarbazide-sensitive amine oxidases). In this overview, the contributions of these two types in xenobiotic metabolism are considered separately.

1. Introduction

The metabolic reactions of foreign compounds may be classified very broadly into four types occurring in two distinct phases. The first phase of metabolism involves either oxidation, reduction or hydrolysis. This is followed by the second phase in which the original compound or metabolite is linked with an endogenous molecule in a synthetic or conjugation reaction.

To date most of the interest in oxidative metabolism of xenobiotics has been devoted to the role of the microsomal cytochrome P-450 system. Increasing attention, however, is being given to the involvement of the cytosolic and mitochondrial enzymes known as the molybdenum hydroxylases (Beedham, 1985) and to the flavin-containing microsomal oxidases (Ziegler, 1988) in the metabolism of xenobiotics. In contrast, the contribution of amine oxidases to the metabolism of xenobiotics has been largely neglected, with the exception of the contribution of monoamine oxidases (MAOs) to the metabolism of exogenous tyramine and the studies of the "cheese effect" produced as the result of ingestion of large amounts of tyramine-containing foods under particular conditions. A review of the involvement of the mitochondrial MAOs in drug metabolism was published in 1988 (Strolin Benedetti et al., 1988). Since that time, considerable additional evidence has appeared in the literature to support the contribution of MAOs to drug metabolism. In addition, the involvement of other amine oxidases in metabolism has been established. A second review on the contribution of amine oxidases to the metabolism of xenobiotics was therefore published in 1994 (Strolin Benedetti and Dostert, 1994).

The oxidative deamination of xenobiotics by the amine oxidases affords the corresponding aldehydes. These compounds are generally further metabolized either to alcohols by aldehyde reductases or alcohol dehydrogenases, or to acids by aldehyde dehydrogenases or aldehyde oxidases. Although the final products, rather than the intermediate aldehydes, are often characterized in in vitro and in vivo metabolism studies involving amine oxidases, aspects relative to the contribution of aldehyde-metabolizing enzymes are not discussed in this article. On an arbitrary basis, the heterogeneous class of amine oxidases can be divided into two types according to their prosthetic group: the flavine-adenine dinucleotide (FAD)-dependent amine oxidases and the amine oxidases not containing FAD. The contributions of these two types in xenobiotic metabolism are considered separately.

2. Involvement of FAD-dependent amine oxidases in metabolism

A. Monoamine oxidase

Monoamine oxidase (EC 1.4.3.4: MAO) is often considered an endobiotic-metabolizing enzyme since it catalyses the oxidative deamination of biogenic amines, according to the general reaction shown in Fig. 1.

MAO is primarily a mitochondrial enzyme, although some MAO activity has also been reported in the microsomal fraction. In mammals, both MAO-A and MAO-B are present in most tissues examined. Very few tissues express only one form of the enzyme; examples in humans include the placenta, which mainly expresses MAO activity of the A type; and blood platelets, lymphocytes, and chromaffin cells in which only the B subtype has been demonstrated. In considering the contribution of MAO to the metabolism of xenobiotics such as environmental contaminants, food components, drugs and prodrugs, which are mainly inhaled or ingested by mouth, it is important to examine the activity of these enzymes in tissues such as the intestine, liver and lung. Data on the distribution of the two isoenzymes in peripheral organs of

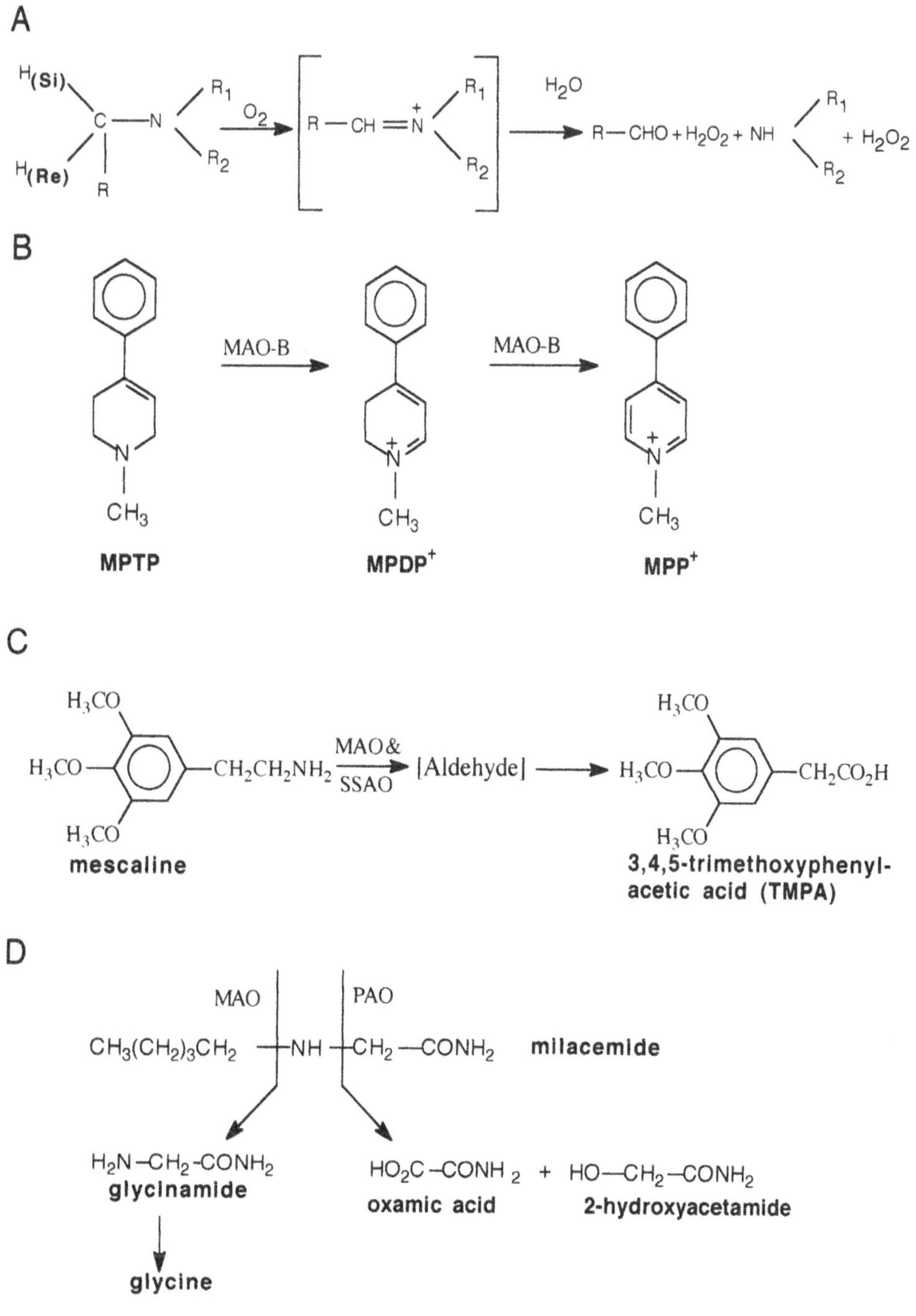

A

B

MPTP   MPDP⁺   MPP⁺

C

mescaline   3,4,5-trimethoxyphenyl-acetic acid (TMPA)

D

milacemide

glycinamide   oxamic acid   2-hydroxyacetamide

glycine

Fig. 1. Reactions catalysed by monoamine oxidase (MAO). **A** General mechanism of the oxidative deamination of substrates. **B** The oxidation of MPTP. **C** The oxidation of mescaline by MAO and the semicarbazide-sensitive amine oxidase (SSAO). **D** The oxidative cleavage of milacemide by MAO and polyamine oxidase (PAO)

rats, mice and humans, obtained with the same methodology, have recently been published (Saura et al., 1992, 1994, submitted): in contrast to tissues such as heart or pancreas, for which important species differences have been reported, MAO concentrations in the tissues directly exposed to the xenobiotics, and therefore mainly involved in their metabolism, do not differ markedly between species. Only the levels of MAO-A in mouse liver (0.9 pmol/mg protein) are particularly low compared to the corresponding values in rats and humans (7.3–7.5 and 7.5 pmol/mg protein, respectively). MAO is able to oxidize primary, secondary, and tertiary amines. The oxidation of amines by MAO is thought to occur by a radical mechanism, according to the general equation depicted in Fig. 1.

Whilst cytochrome P-450 and flavin-containing microsomal oxidases transfer one atom from molecular oxygen O_2 into their substrates (oxygenase reaction), the MAOs oxidize their substrates. Thus MAOs transform amines into the corresponding imines which are then hydrolyzed to the end product aldehydes with oxygen taken from water, not from molecular oxygen. Abstraction of an α-hydrogen by MAO has been proposed as an early step in the oxidation of amine substrates. It has been demonstrated that it is the pro-R-hydrogen which is removed by both MAO-A and MAO-B. It is worth noting that, since the molybdenum hydroxylases are known to produce the superoxide radical and hydrogen peroxide and the cytochrome P-450 enzymes to generate the superoxide radical under conditions of oxy-complex instability, all these amine oxidase reactions give rise to the formation of hydrogen peroxide. However, the superoxide radical itself is a poorly-reactive radical species and H_2O_2 is not a very reactive oxidizing agent. Therefore, the damaging effect of these products of phase I oxidative enzymes will depend on the availability of catalytic iron ions allowing the formation of the hydroxyl radical ˙OH (Halliwell, 1989).

Alterations of MAO activity occurring in diseases should be taken into account when considering the contribution of these enzymes to xenobiotic metabolism. However, alterations of MAO activity that have been studied so far have concentrated on psychiatric disorders and the activity determinations have been done mainly in tissues such as brain or platelets (for review see Strolin Benedetti and Dostert, 1992). It would thus be important to extend such a knowledge to diseases mainly involving alterations of the hepatic, pulmonary or gastrointestinal systems, measuring the enzyme activity in those tissues which are directly exposed to the xenobiotics, and therefore mainly involved in their metabolism.

Toxins

The ability of MAO to oxidize tertiary amines has been more recently documented.

The tertiary amine of exogenous origin, *1-methyl-4-phenyl-1,2,3,6-tetrahydro-pyridine (MPTP)*, has been shown to cause a condition resembling idiopathic Parkinson's disease (for review see Tetrud and Langston, 1989). MPTP is not toxic itself but is converted to the active toxin MPP^+ (the 1-methyl-4-phenylpyridinium ion) by the action of MAO. This process occurs by two successive MAO-catalysed oxidations with the dihydropyridinium species ($MPDP^+$) being formed as an intermediate (Fig. 1). Although MPTP is oxidised by both forms of MAO, it is a better substrate for MAO-B. As well as being a substrate for MAO-B, MPTP is also a mechanism-based inhibitor of that enzyme (Kinemuchi et al., 1987; Dostert and Strolin Benedetti, 1988). A number of analogues of MPTP have been studied as substrates of MAO and it has been shown that the 2'-ethyl derivative is a good substrate for both the MAOs whereas the 2'-n-propyl and 2'-isopropyl derivatives are preferred substrates for MAO-A (for review see Markey and Schmuff, 1986; Maret et al., 1990; Tipton and Singer, 1993).

Dimethylnitrosamine (DMN) is a stable compound which requires metabolic activation for its immunotoxicity, as well as carcinogenicity and mutagenicity. It has generally been accepted that the α-hydroxylation of DMN, and subsequent formaldehyde loss, induced by the P-450 monoxygenase system, the so-called DMN demethylase which has been characterized as alcohol-inducible P-450 2E1, is the major metabolic pathway of DMN (Thomas et al., 1987). On the other hand, it has also been reported that DMN can be metabolized by demethylases which are not P-450 proteins. In particular, Lake et al. (1978) demonstrated that DMN could be demethylated by an amine oxidase. They showed that inhibitors of MAO, including pargyline, can inhibit the metabolism of DMN without affecting P-450 activity (Phillips et al., 1982). However, it appears from a recent paper (Jeong et al., 1994) that pargyline inhibits several forms of cytochrome P-450 system in addition to MAO. Therefore, although the possible involvement of MAO in DMN activation cannot be completely ruled out, the cytochrome P-450 isoenzyme 2E1 plays the principle role in DMN activation.

Propranolol and related compounds

There is increasing evidence that MAOs participate in the metabolism of a number of amine derivatives used as drugs. In that respect, propranolol has received particular attention. Propranolol is N-dealkylated to N-desisopropylpropranolol by microsomal P-450 enzymes. From recent studies it appears that this reaction is mediated mainly by the P-450 1A2 (Yoshimoto et al., 1995) rather than by the 2C subfamily, as previously described (Strolin Benedetti and Dostert, 1994). Formation of the aldehyde corresponding to the oxidative deaminated metabolite of N-desisopropylpropranolol by MAO has been shown to occur upon incubation with rat liver mitochondria. This aldehyde derivative decomposes chemically to 1-naphthol or is further transformed to the glycol and the acid derivatives. The involvement of MAO in the oxidative deamination of N-desisopropylpropranolol has been ascertained using MAO inhibitors. Direct formation of the aldehyde from propranolol with concomitant generation of isopropylamine has also been shown to take place during incubation of propranolol with rat liver homogenates, and to contribute to a very minor degree to the oxidative deamination process. Inhibition of N-desisopropylpropranolol metabolism by clorgyline suggests that MAO-A plays the major role in the deamination process (Steiner et al., 1992). Furthermore, the observation that propranolol inhibits 5-HT oxidation in a reversible and competitive manner much more effectively than PEA oxidation (Teramoto et al., 1987) is consistent with N-desisopropylpropranolol being a preferential substrate for MAO-A. In man and various animal species, the presence in urine of the acid and glycol derivatives generated by the action of MAO on N-desisopropylpropranolol has been established. In a recent publication (Tateishi et al., 1995) the influence of aging on the oxidative metabolism of propranolol has been studied in patients with essential hypertension. An age-related reduction in both the ring-oxidation and the side-chain oxidation was found. However, as the production of naphthoxylactic acid is dependent on both the cytochrome P-450 and the MAO activities, it is impossible to conclude which enzyme system is mainly affected by age.

The involvement of MAO in the metabolism of other β-blocking agents related to propranolol, such as *alprenolol*, *oxprenolol* and *timolol*, has been discussed in a previous review (Strolin Benedetti et al., 1988).

β-hydroxyarylethylamine derivatives

The β-hydroxyarylethylamine moiety is present in norepinephrine and in a number of α- or β-adrenoceptor agonists and antagonists. There is no documented evidence for *phenylephrine* being a substrate for MAO in vitro. However, the acid and alcohol derivatives, which are likely to result from the oxidative deamination of phenylephrine by MAO, have been identified in rat and human urine (for review see Strolin Benedetti et al., 1988).

MAO does not appear to be involved in the in vivo metabolism of *isoprenaline*, which is mainly eliminated in urine as the sulfoconjugate or O-methyl derivative, depending on the route of administration.

Conversely, *pronethalol*, a β-adrenoreceptor antagonist which has the same side chain as isoprenaline, is primarily excreted in urine as the glycolic acid derivative, presumably following oxidative deamination of the corresponding primary amine, as has been shown for propranolol. In contrast to isoprenaline, the aromatic moiety of pronethalol, a naphthalene ring, is not substituted by hydroxy groups. The lack of formation of conjugates during absorption of pronethalol may explain its ability to serve as a substrate for MAO.

Terbutaline was shown not to be a substrate for MAO. The presence of a tert-butyl residue on the amine group is likely to protect terbutaline and *salbutamol* from N-dealkylation by cytochrome P-450 enzymes, and also from degradation by MAO. However, experimental evidence to support the suggestion of salbutamol being resistant to the action of MAO remains to be provided.

Arylethylamine derivatives

The arylethylamine moiety is present in dobutamine as well as in ibopamine (a prodrug of epinine), in docarpamine (a prodrug of dopamine) and in mescaline.

Dobutamine (3,4-dihydroxy-*N*-[3-(4-hydroxyphenyl)-1-Methylpropyl]-β-phenylethylamine) is a synthetic catecholamine, structurally related to dopamine but which differs pharmacologically in several aspects. The most significant of these being that dobutamine elicits its inotropic effect by acting directly on beta$_1$-adrenergic receptors whereas dopamine acts in part through the release of endogenous norepinephrine. The disposition of dobutamine has been examined in the dog after i.v. administration of the ^{14}C-labelled compound (Murphy et al., 1976). Dobutamine disappears from the plasma very rapidly. The circulating radioactivity consists mainly of the glucuronide conjugate of 3-O-methyldobutamine. During a 48-hour time period after administration of ^{14}C-dobutamine, 67% of the radiolabel was excreted in the urine and 20% in the feces. The major urinary metabolites were the glucuronide conjugates of dobutamine and 3-O-methyldobutamine. Therefore, dobutamine is a substrate of COMT that appears to be as good as dopamine, a natural substrate. However, the involvement of MAO in dobutamine metabolism in vivo, if any, is certainly minor. No mention is made by the authors of the presence of phenylacetic acid derivatives, free or conjugated, among the metabolites of dobutamine.

Ibopamine is a cardiovascular agent used in congestive heart failure. Ibopamine, the diisobutyrylester of epinine, is rapidly hydrolyzed by blood and hepatic esterases to yield the active moiety, epinine (N-methyl dopamine) (Fig. 2). Various animal and human studies (for a review see Spencer et al., 1993) have shown there to be a significant amount of presystemic and systemic metabolism of epinine after ibopamine administration. Epinine can be metabolized by conjugation with inorganic sulfate or glucuronic acid. Epinine can also be metabolized by MAO and COMT to yield homovanillic acid (HVA) and dihydroxyphenylacetic acid (DOPAC) and their respective sulfate and glucuronide conjugates. According to recent reports, epinine is a substrate for both MAO A and B in rat liver homogenates (Strolin Benedetti et al., this volume) and rat liver mitochondria (Kenny et al., 1996).

Docarpamine is an orally active dopamine prodrug that has been launched in Japan for acute cardiac insufficiency (Nishiyama, 1992). Docarpamine is rapidly metabolized to de-ethoxycarbonylated docarpamine, which is then converted to dopamine and consequently submitted to the classical dopamine metabolism (Yoshikawa et al., 1990) (Fig. 2). However, the acetyl methionyl group of docarpamine protects the amine moiety from rapid MAO metabolism, so that the free dopamine levels in plasma after oral docarpamine are much higher than the levels of free dopamine after oral dopamine (Murata et al., 1988).

The involvement of MAOs in the metabolism of *mescaline*, at least in the mouse, has been demonstrated. Pretreatment of mice with iproniazid or tranylcypromine (non selective MAO inhibitors) considerably diminished the levels of 3,4,5-trimethoxyphenylacetic acid (TMPA) and elevated those of N-acetylmescaline in the livers, plasma and urine of mice that had been treated i.p. with ^{14}C-mescaline (Fig. 1; Shah and Himwich, 1971).

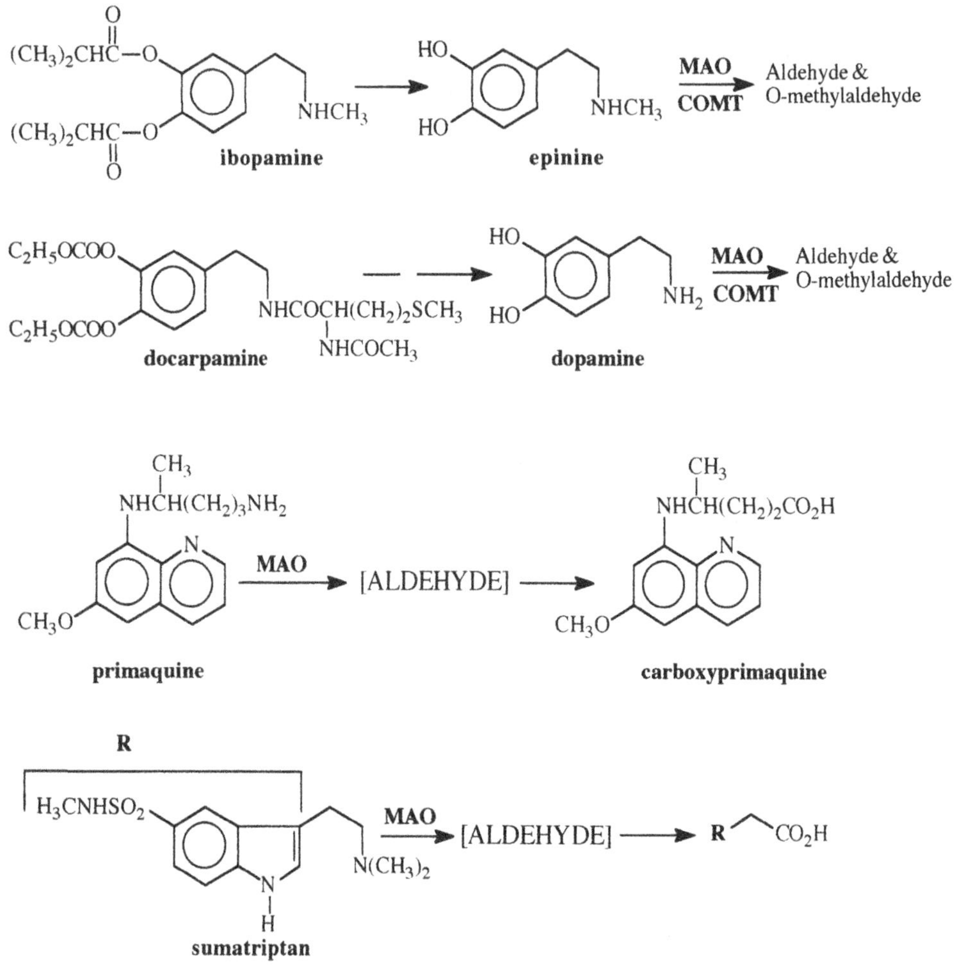

Fig. 2. The possible involvement of MAO and catechol-O-methyltransferase (COMT) in the metabolism of some drugs

MAO inhibitors and related compounds

MAO inhibitors such as *phenelzine*, a non selective irreversible MAO inhibitor (Fig. 3), behave as MAO substrates, giving rise to the formation of metabolites excreted in urine. Exposure of 1-[^{14}C]phenelzine to rat liver mitochondria was found to result in the formation of 1-[^{14}C]phenylacetic acid, the major urinary metabolite in rats given 1-[^{14}C]phenelzine. Yu and Tipton (1989) showed by using 1,1-dideuterated phenelzine that the mechanism of phenelzine oxidation by MAO appears to involve the hydrogen atoms of both the hydrazine group and carbon-1. The irreversible inhibition of MAO may result from the formation of a covalent adduct between the enzyme and a phenylethyldiazene intermediate, whereas the formation of phenylacetaldehyde, and subsequently of phenylacetic acid, appears to occur through the formation of a phenylethylidene hydrazine intermediate and is sensitive to deuterium substitution.

MAO is involved also in the metabolism of 1,2-disubstituted hydrazines, such as *procarbazine and 1,2-dimethylhydrazine* (Coomes and Prough, 1983).

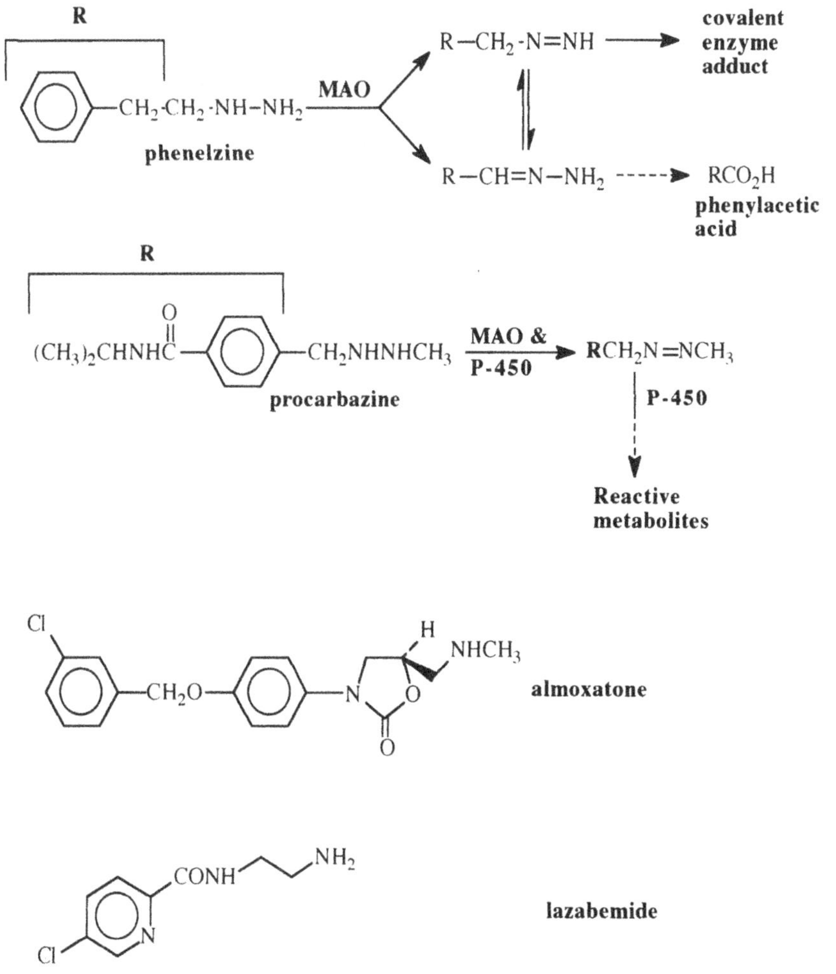

Fig. 3. Inhibitors that are substrates for MAO

Procarbazine is an antitumor agent that can cause several toxic effects. Procarbazine is biotransformed in a series of oxidations to reactive metabolites that irreversibly bind to tissue macromolecules. The first step in the oxidation sequence is formation of azoprocarbazine (Fig. 3), akin to diazene formation with monosubstituted derivatives like phenelzine. This oxidation can be mediated by both MAO and cytochrome P-450 (which in this case acts as an oxidase). Based on the specific activity of the NADPH-dependent cytochrome P-450 and NADPH-independent MAO activity, it appears that MAO is responsible for at least 40% of the total rate of procarbazine oxidation in the hepatocyte of untreated rats and 25% in the hepatocyte of phenobarbital-pretreated rats. Procarbazine not only is a substrate of MAO, but causes an "ex vivo" potentiation of MAO A activity in some rat tissues (Holt and Callingham, 1994).

Another MAO inhibitor giving rise to the formation of MAO-catalyzed metabolites is the (R)-5-aminomethyl-2-oxazolidinone derivative *almoxatone* (Fig. 3). This compound was shown to be a substrate for both forms of MAO, preponderantly of the A form, and to behave as a selective, time-dependent inhibitor of MAO-B in vitro. There are inconsistencies among studies as to the reversibility of MAO-B inhibition by almoxatone in vitro. Under ex vivo conditions, inhibition of MAO-B activity in brain and liver was found to be fully reversible. The urinary metabolite profile of almoxatone was studied in rat, dog, and man after administration of the ^{14}-C-labeled compound (Strolin Benedetti et al., 1984). The carboxylic acid derivative was the main metabolite in urine of all three species. The corresponding alcohol derivative was found only in trace amounts. Although debenzyl derivatives were also found in urine, indicating some contribution of the cytochrome P-450 system to the metabolism of almoxatone in the three species, MAO plays the major role in the metabolism of this compound. In vivo, the formation of almoxatone metabolites appears to result mainly from the action of MAO-A, and to a smaller degree, from MAO-B.

MAO-dependent metabolism is also required for enzyme inhibition by the selective MAO-B inhibitor *lazabemide* (Fig. 3). Lazabemide was shown to be a substrate for MAO-B, resulting in the formation of a high-affinity intermediate (Cesura et al., 1988). However, in contrast to what has been reported for almoxatone, the MAO-B inhibition induced by lazabemide in vitro is rapidly reversed by dialysis. The main metabolite of lazabemide in rat, dog, and human urine was found to be the acid derivative of the intermediate aldehyde formed by the action of MAO.

Miscellaneous drugs

Primaquine. Based on chemical structure and the formation of acid metabolites, it has been suggested that MAO may be involved in the metabolism of a variety of drugs, including primaquine (Fig. 2). Using rat liver microsomes or mitochondria, it has recently been established that the oxidative deamination of primaquine is mainly brought about by MAO (Ni et al., 1992).

Both enantiomers of primaquine also have weak MAO inhibitory properties. Carboxyprimaquine [8-(3-carboxy-1-methylpropylamino)-6-methoxy-quinoline], the acid derivative formed by the action of MAO, was found to be the main metabolite of primaquine in rat and human plasma (Mihaly et al., 1984; Clark et al., 1984). Brain MAO activity was reported to be enhanced in mice infected with Plasmodium yoelii. However, no change in the area under the curve for plasma carboxyprimaquine was observed in patients with falciparum malaria before and after recovery from infection (Edwards et al., 1993).

Haloperidol. The possible involvement of MAO in the metabolism of the antipsychotic drug haloperidol was suggested on the basis of the structural analogy between haloperidol and the tetrahydropyridine derivative MPTP, and the presence of 4-fluorobenzoylpropionic acid in urine of rats and humans treated with haloperidol. In contrast to MPTP, where biotransformation into the neurotoxic species MPP is mediated by MAO type B, no formation of the pyridinium derivative from haloperidol was found to occur on incubation with purified beef liver MAO-B or with crude mitochondrial preparation of MAO from albino mice (Subramanyam et al., 1990; Fang et al., 1993; Fang and Gorrod, 1991). However, the pyridinium derivative was demonstrated in rat brain and urine (Subramanyam et al., 1990), and in the urine of schizophrenic patients (Subramanyam et al., 1991) after administration of haloperidol, probably resulting from the action of cytochrome P-450 enzyme(s). Although the involvement of MAO in the formation of the 4-fluorobenzoylpropionic acid cannot be ruled out a priori, the role of cytochrome P-450 enzymes in this biotransformation appears more likely, particularly as tertiary amines with bulky substituents are known to be resistant to the action of MAO.

Sumatriptan. Sumatriptan is a highly selective agonist of the serotoninergic 5-HT$_1$ receptor subtype and is effective in the acute treatment of migraine headache. Following oral administration of ^{14}C-labeled sumatriptan to humans, 43% of radioactivity was recovered in urine as the indoleacetic acid derivative, mostly in the free form (Fowler et al., 1991) (Fig. 2). In a recent paper (Dixon et al., 1994) MAO A was characterized as the enzyme mainly responsible for the metabolism of sumatriptan in human liver, in spite of the fact that this compound is a tertiary amine.

Flurazepam. Flurazepam metabolism is interesting, because this is perhaps the first example of a relatively high molecular weight aldehyde being detected "in vivo" in human plasma from the deamination of a drug (Schwartz and Postma, 1970; Garland et al., 1983). Flurazepam is a tertiary amine, which seems to be metabolized by successive N-dealkylation to yield the monodeethyl and the dideethyl derivatives, both of which are excreted in human and dog urine. The principal human urinary metabolite is the alcohol, in the conjugated form, whereas in dog it is the carboxylic acid.

The possible involvement of MAO in the metabolism of drugs, such as the calcium antagonist *diltiazem* or the non-steroidal antiestrogens *tamoxifen* and *toremifene* cannot be excluded when the structures of some of their metabolites are examined.

Milacemide and related compounds

The anticonvulsant activity of milacemide [2-(pentylamino)-acetamide] (Fig. 1) has been suggested to be related, at least in part, to the formation of glycine. By analogy with n-pentylamine, a known MAO-B substrate, the formation of glycine from milacemide has been shown to result from the action of MAO-B (Janssen de Varebeke et al., 1988). Although the efficacy of milacemide as an anticonvulsant has been questioned, as has the role of glycine in its mechanism of action (O'Brien et al., 1991), these findings supported the hypothesis that MAO-B may act as a trigger enzyme to deliver amino acids or other active agents into the brain. Such an approach presents some similarities with the design of γ-L-glutamyl-L-dopa, where γ-glutamyltransferase acts as a trigger enzyme to deliver another aminoacid, L-dopa, into the kidney; in this tissue L-dopa is further transformed to dopamine, the active drug (Worth et al., 1985), which is then metabolized, at least partially, by renal MAO-B to DOPAC (Freestone et al., 1992). In line with the milacemide concept, Yu and Davis showed that the branched aliphatic amine derivatives *2-[(2-propyl)pentylamino]acetamide* and *2-propyl-1-aminopentane* are metabolized by MAO-B, resulting in the formation of valproic acid (Yu and Davis, 1991a,b). Although detectable levels of valproic acid were found in the brain after intraperitoneal administration of either molecule to mice, no anticonvulsant activity, but rather an intense tremor syndrome, was noted. The reason for this proconvulsant effect remains to be elucidated.

The interactions of a series of analogues of milacemide with MAO have been studied (O'Brien et al., 1994). In these the aminoacetamide portion of milacemide was retained but the pentyl moiety was replaced with aromatic residues similar to those present in a number of MAO substrates and inhibitors. These compounds were found to differ markedly in their abilities to act as substrates for ox liver MAO-B although none was oxidised by MAO-A to any detectable extent. Like milacemide itself, most were found to be time-dependent and irreversible inhibitors as well as substrates for MAO-B. However, *2-(4-(3-chlorobenzyloxy)phenethylamino)acetamide* showed no time-dependent inhibition of the enzyme but was a potent reversible inhibitor of both MAO-A- and B (O'Brien et al., 1994). Comparisons of the behaviour of MAO-B from rat, mouse, human and ox tissues revealed marked species differences in their interactions with milacemide and its analogues, both substrates and inhibitors (O'Brien et al., 1995). The alaninamide derivative, *(S)-2-(4-(3-fluorobenzyloxy)benzylamino)-propionamide* (FCE 26473) has been shown to be a selective MAO-B inhibitor (Strolin Benedetti et al., 1994).

B. Polyamine oxidase

In addition to several copper-containing amine oxidases which are capable of oxidatively deaminating the terminal amino function of diamines and polyamines, a FAD-dependent polyamine oxidase (EC 1.5.3: PAO) has been shown to be present in virtually all rat and human tissues including the brain (Seiler et al., 1980). This enzyme is involved in the interconversion of polyamines, its role being to oxidize a secondary amino group in substrates such as monoacetylspermine and -spermidine to form spermidine and putrescine, respectively (Fig. 4). Using N^1, N^{12}-diacetylspermine as substrate, Seiler et al. (1980) found high PAO activities in rat tissues, with the highest activities being measured in pancreas and liver. With N^1-monoacetylspermine as substrate, PAO activity was detected in all human tissues tested (Suzuki et al., 1984), with the highest activity being found in the liver, followed by the testis and the kidney. Activity was found to be much higher in rat tissues, but this may be confounded by instability of the enzyme in the human postmortem samples.

PAO activity has been shown to be increased in human blood during pregnancy. Although PAO is not localized in mitochondria (Seiler, 1987), this enzyme resembles MAO in many ways. Both enzymes require molecular oxygen as electron acceptor, both produce hydrogen peroxide, and both have

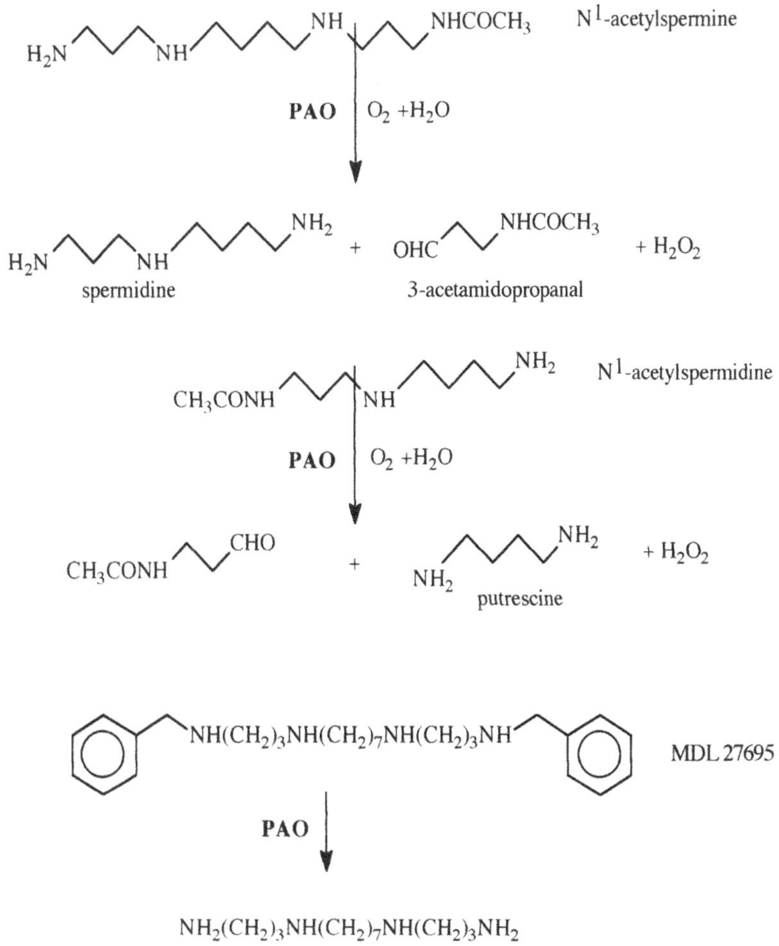

Fig. 4. Some reactions catalysed by polyamine oxidase (PAO)

the same cofactor. The N,N'-bis (2,3-butadienyl) derivative of putrescine (MDL 72527) was found to be a potent, selective, time-dependent, irreversible inhibitor of PAO (Bey et al., 1985).

Using this inhibitor, it was recently shown that the *N-N'-bis(benzyl)polyamine derivative, MDL 27695* (Fig. 4), a potent antimalarial agent with rat hepatoma (HTC) cell growth inhibitory activity, is metabolized by PAO to its debenzyl analogue (Bitonti et al., 1990). *Dopexamine* hydrochloride is a novel synthetic catecholamine, structurally related to dopamine. It has marked intrinsic agonist activity at β_2-adrenoreceptors and lesser agonist activity at dopamine DA_1 and DA_2-receptors and β_1-adrenoceptors. An inhibitory action on neuronal catecholamine uptake was also demonstrated (for review, see Fitton and Benfield, 1990). Dopexamine hydrochloride is extensively metabolized in animals and humans via O-methylation and O-sulphation of the catechol hydroxyl groups to yield two major, pharmacologically inactive products. The compounds which might be produced by PAO have not been reported among the metabolites of dopexamine, so that it is not known whether they are absent or whether their presence has not yet been adequately investigated.

Milacemide (see Fig. 1) represents the first example of the involvement of PAO in the metabolism of a xenobiotic, the chemical structure of which is not related to polyamines. Milacemide is cleaved by MAO on one side of the secondary amino group, resulting in the formation of glycinamide and by PAO on the other side and causing the formation of oxamic acid and 2-hydroxyacetamide (Strolin Benedetti et al., 1992). The latter pathway is inhibited when rats are pretreated with MDL 72527.

3. Involvement of semicarbazide-sensitive amine oxidases in metabolism

The class of amine oxidases not containing FAD can be referred to as semicarbazide-sensitive amine oxidase (SSAO) enzymes. All the SSAO enzymes are classified as EC 1.4.3.6 with the systematic name "amine: oxygen oxidoreductase (deaminating) (copper containing)" and the trivial name "copper containing amine oxidases". However, in view of the sensitivity of this class of enzymes to carbonyl reagents typified by semicarbazide, and owing to the fact that not all have been shown to contain copper, the abbreviation "SSAO" has now gained a reasonable degree of acceptance. SSAO has also sometimes been called benzylamine oxidase, although this is inappropriate as benzylamine is a preferred substrate in most but not all cases, and MAO-B also deaminates benzylamine. SSAO enzymes catalyze the oxidative conversion of primary amines only, whether the amino group is present in mono-, di-, or polyamines. Therefore, the enzymes which carry out the terminal oxidation of diamines or polyamines belong to this class. Recent controversy has surrounded the identity of the organic cofactor of SSAO. Pyridoxal phosphate was long thought to be the most likely candidate, until pyrroloquinoline quinone was proposed (for review, see Duine, 1991). However, it now appears that the cofactor of SSAO is an enzyme-bound quinone of trihydroxyphenylalanine (topaquinone) (McIntire, 1992). Similarly to MAO, SSAO enzymes also produce aldehydes, ammonia, and hydrogen peroxide. However, in contrast to MAO, all possible modes of proton abstraction at C-1 (pro-R, pro-S, and an apparent nonstereospecificity) have been reported for different SSAO enzymes (for review see Dostert et al., 1989; Scaman and Palcic, 1992).

The SSAO enzymes have been divided into plasma and tissue enzymes (for review see Callingham et al., 1990), although it is worth noting that the arterial wall has been suggested to be the site of synthesis of the circulating plasma SSAO. A recent review paper deals with xenobiotic aromatic and aliphatic amines as potential SSAO substrates (Lyles, 1994).

A. Plasma amine oxidases

A group of SSAO enzymes, whose substrate specificities are distinct from those of diamine oxidase, has been found circulating freely in the blood plasma of many species, including man. In most, but not all, cases benzyalmine is a preferred substrate. This enzyme, in addition to its ability to oxidize some primary monoamines, is also capable of deaminating exclusively the primary amino groups of the aminopropyl moieties present in the polyamines spermine and spermidine. Putrescine is not a substrate.

Large differences in the activity of polyamine-oxidizing enzymes have been reported in serum from a variety of species, with extremely high activities being found in ruminants and very low activities in some other species, for example, rodents.

Older publications (McEwen and Harrison, 1965; Nilsson et al., 1968) reported an increased activity of "serum monoamine oxidase" in congestive heart failure, and the authors suggested that levels of serum monoamine oxidase are a measure of the severity of congestive heart failure. Benzylamine was used as substrate and, since MAO is not present in plasma, it is highly probable that what the authors called "serum monoamine oxidase" is in reality serum SSAO. Also the high plasma levels of benzylamine oxidase observed in hepatic cirrhosis (McEwen and Castell, 1967; Lewinsohn, 1984) probably refer to levels of circulating SSAO. For a review on benzylamine oxidase in human tissues, in health and disease, see Lewinsohn (1984).

Plasma SSAO has been suggested to play a role in the metabolism of the calcium channel blocker amlodipine (Fig. 5). Identification of the urinary metabolites of amlodipine has shown that the oxidative deamination of the 2-aminoethoxymethyl side-chain is a major metabolic pathway in dogs and man but not in rats (Beresdorf et al., 1988a,b). The loss of the amino group upon incubation of amlodipine at 37°C in dog, but not in rat, plasma is in agreement with the low activities of plasma SSAO in rodents.

B. Tissue amine oxidases

The SSAO enzymes which preferentially oxidize aliphatic diamines — such as putrescine and cadaverine, and histamine — are termed diamine oxidases (DAO, histaminase). Although DAO was first considered to be a cytoplasmic enzyme, there is recent evidence showing that DAO is bound to the membrane of the endoplasmic reticulum in rabbit kidney cortex. DAO activities are present in many mammalian tissues, with the highest activity being found in placenta and the small intestine, but its activity in the brain is very low. Although DAO is considered a tissue enzyme, low but measurable activity is also found in plasma. Examples of drugs metabolized by DAO have, to our knowledge, not yet been reported. However, since *histamine* is not only an endogenous amine but is also an amine present in high amounts in some foods, DAO might be regarded as playing a role in xenobiotic metabolism.

Apart from those containing DAO activity, tissues where activity of SSAO enzymes is important and/or has been extensively studied are those of the cardiovascular system, such as heart and blood vessels, as well as brown adipose tissue, small intestine, lung, anococcygeus muscle and dental pulp. The SSAO activity in lung might play an important role in the metabolism of

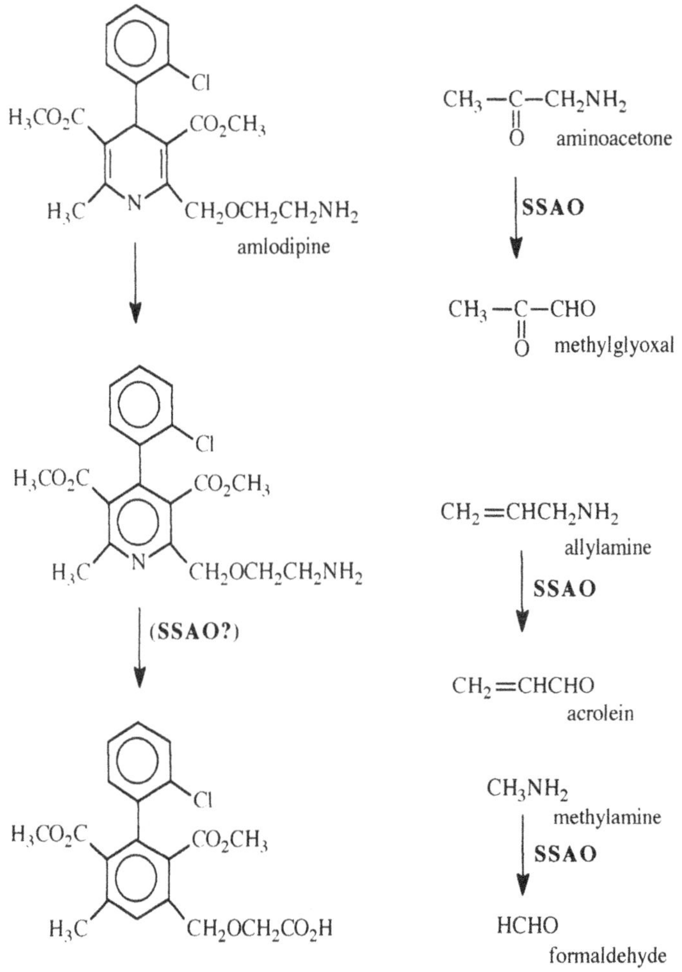

Fig. 5. Some reactions catalysed by the semicarbazide-sensitive amine oxidase (SSAO) enzymes. Each reaction involves the stoichiometric formation of ammonia and hydrogen peroxide

inhaled volatile amines (Liczano et al., 1994). Although a "soluble" oxidizing activity was reported in some studies, most tissue SSAO activity has been shown to be membrane bound and appears to be a constituent of plasma-lemmal and/or microsomal membranes. The physiological roles of the tissue SSAO enzymes have not yet been established.

Aminoacetone has been shown to be deaminated to methylglyoxal by human umbilical artery homogenates (Fig. 5) (Lyles and Chalmers, 1992) and the K_m value for metabolism of this compound by SSAO was far lower than that of the other aliphatic and aromatic biogenic amines examined previously as potential physiological substrates for the human vascular enzyme. However, whether metabolism of aminoacetone by blood vessel SSAO represents a significant pathway for methylglyoxal formation in vivo remains to be investigated.

Another amine which has been considered as a possible physiological substrate for tissue SSAO in vivo is *methylamine*. Methylamine has been shown to be metabolized by SSAO from both rat aorta and human umbilical artery, with the enzyme of the human tissue being particularly active (Precious et al., 1988). Furthermore, formaldehyde production from methylamine (see Fig. 5) by rat and porcine aortic preparations has been recently demonstrated. Methylamine may arise from endogenous metabolic degradation of sarcosine and creatinine, and is known to be a product of the intracellular deamination of the secondary amine epinephrine by MAO. Methylamine is elevated in the serum and urine in several physiological (e.g. pregnancy) and pathological (e.g. uremia, diabetes mellitus) situations. Methylamine is present in some foods and can also be produced and absorbed after gut bacterial degradation of dietary creatinine, choline, and lecithin. In addition, methylamine is a major product in nicotine metabolism and in the hydrolysis of certain insecticides and herbicides. Enhanced urinary excretion of methylamine was found in rats treated with semicarbazide or hydralazine, two irreversible inhibitors of SSAO.

Apart from any physiological function, it is now well documented that tissue SSAO is involved in the metabolism of allylamine, which is a relatively specific cardiovascular toxin. Chronic exposure to *allylamine* was shown to result in lesions which mimic acute myocardial necrosis and atherosclerosis. It has been established that allylamine is metabolized by SSAO in cardiovascular tissues in both animals and humans, and that this produces acrolein (Fig. 5) (Nelson and Boor, 1982). The SSAO inhibitors semicarbazide and phenelzine afford protection from allylamine injury both in vitro and in vivo, suggesting that acrolein, formed by the action of SSAO, contributes toward allylamine toxicity.

2-phenylallylamine was also found to be a substrate and a reversible inhibitor of rat aorta SSAO (Lyles et al., 1987).

In the oxazolidinone series, the *N-demethyl derivative of almoxatone* (see Fig. 3), MD 220661, was found to be an inhibitor and a substrate for SSAO in rat aorta and also to be a substrate for SSAO in vivo in the rat.

The enzyme system responsible for *mescaline* metabolism to TMPA in rabbit lung, liver and kidney homogenates seems to be SSAO (Fig. 1).

Finally, pretreatment of rats with benserazide, an inhibitor of SSAO, significantly enhanced the central nervous system depressant action of systemically administered *kojic amine*, suggesting that this enzyme system may be important for the metabolism of this compound (Ferkany et al., 1981).

4. Conclusions

It is now well-established that monoamine oxidases and related amine oxidases participate in the metabolism of xenobiotics, affording protection against exogenous agents or, in some cases, producing toxic metabolites (Kopin, 1993). However, little information is available on the extent to which factors, such as nutrients and other dietary factors, fasting, environmental contaminants, cigarette smoke, alcohol and age, can modify the capacity of these enzyme systems to metabolise xenobiotics. In contrast, the extent to which these factors can modify the activity of other phase I oxidative enzymes, the cytochromes P-450 system in particular, has been extensively studied (for a recent review see Guengerich, 1995). Frequently it is even known which cytochrome P-450 isoenzyme is affected by a given factor or in which tissue this occurs. For instance, high-protein diet enhances the rates of oxidations catalyzed by P-450 1A2 and 3A4, whereas high-carbohydrate diets decrease P-450 activities. Smoking induces not only hepatic P-450 1A2 but also P-450 1A1, an essentially extrahepatic P-450. Chronic ethanol administration induces P-450 2E1, in liver as well as in other tissues (Roberts et al., 1994), and an attenuation in the responsiveness to cytochrome P-450 induction by several agents has been observed with age (Lee and Werlin, 1995).

Information is available on the effect of smoke, alcohol and age on the activities of the monoamine oxidases, but again almost exclusively in platelets and in brain (Fowler et al., 1996; see also Strolin Benedetti and Dostert, 1992, for review), and not in the tissues mainly involved in xenobiotic metabolism, such as intestine, liver and lung. As such determinations in vivo are not always feasible, an alternative approach might be investigation of the effect of these agents on MAOs and related amine oxidases by measuring the plasma and/or urinary pharmacokinetic parameters of xenobiotics metabolized by these enzymes.

Alterations of MAO A and B activities with age in such tissues that are major sites of exposure have been reported in a few studies. In the rat both enzyme activities decrease in the lung, MAO-B increases in the liver but there were no effects of age on either A or B activities in the duodenum (Cao Danh et al., 1984). In the rat heart there are large increases of MAO-A activity with age. This has been shown to reflect an increase in the total amount of MAO-A protein and to reflect a decreased rate of protein degradation (Della Corte and Callingham, 1977). Corresponding information in humans is, to our knowledge, not available.

It would be particularly interesting to investigate whether monoamine oxidases and/or related amine oxidases can be induced by exogenous agents (Mageed et al., 1993), as is the case for most of the cytochrome P-450 isoforms. Enzyme induction is defined in the strict sense as an increased rate of biosynthesis of the enzyme or, in a looser sense, as an increase in the amount of the enzyme protein irrespective of the underlying mechanism (increased rate of synthesis or decreased rate of degradation). MAO may be inducible, by endogenous as well as by exogenous agents, at least in some species and in some tissues. Typical inducers of hepatic P-450 enzymes, such as cigarette smoke and alcohol, decrease rather than increase MAO activity, although the enzyme activity determinations were performed only in platelets and brain. Furthermore, if exogenous agents are able to induce MAOs and/or related amine oxidases, there is no reason to assume that these will be the same as those inducing cytochrome P-450 enzymes.

Further studies are clearly needed to define the physiological role(s) of the SSAO enzymes. In the case of the tissue-bound enzymes, the different sites of product formation (plasma membrane/endoplasmic reticulum for SSAO, mitochondrial outer-membrane for MAO) might be an important factor in the physiological and or toxicological behaviour of the aldehyde, peroxide and ammonia formed.

Acknowledgements

The authors wish to thank Prof. F. Oesch for helpful advice and comments and Miss D. Riboldi for skilful secretarial assistance. K. F. Tipton is grateful to the EC Human Capital and Mobility Programme (Contract No. ERBCHRXCT 93 0256) for financial assistance.

References

Beedham C (1985) Molybdenum hydroxylases as drug-metabolizing enzymes. Drug Metab Rev 16: 119–156

Beresdorf AP, Macrae PV, Stopher DA (1988a) Metabolism of amlodipine in the rat and the dog: a species difference. Xenobiotica 18: 169–182

Beresdorf AP, McGibney D, Humphrey MJ, Macrae PV, Stopher DA (1988b) Metabolism and kinetics of amlodipine in man. Xenobiotica 18: 245–254

Bey P, Bolkenius FN, Seiler N, Casara P (1985) N-2,3-butadienyl-1,4-butanediamine derivatives: potent irreversible inactivators of mammalian polyamine oxidase. J Med Chem 28: 1–2

Bitonti AJ, Dumont JA, Bush TL, Stemerick DM, Edwards ML, McCann PP (1990) Bis(benzyl)polyamine analogs as novel substrates for polyamine oxidase. J Biol Chem 265: 382–388

Callingham BA, Holt A, Elliot J (1990) Some aspects of the pharmacology of semicarbazide-sensitive amine oxidases. J Neural Transm [Suppl] 32: 279–290

Cao Danh H, Strolin Benedetti M, Dostert P (1984) In: Tipton KF, Dostert P, Strolin Benedetti M (ed) Monoamine oxidase and disease: prospects for therapy with reversible inhibitors. Academic Press, London, p 301

Cesura AM, Imhof R, Galva MD, Kettler R, Da Prada M (1988) Interactions of the novel inhibitors of MAO-B Ro 19-6327 and Ro 16-6491 with the active site of the enzyme. Pharmacol Res Commun 20 [Suppl IV]: 51–61

Clark AM, Baker JK, McChesney JD (1984) Excretion, distribution, and metabolism of primaquine in rats. J Pharm Sci 73: 502–506

Coomes MW, Prough RA (1983) The mitochondrial metabolism of 1,2-disubstituted hydrazines, procarbazine and 1,2-dimethylhydrazine. Drug Metab Dispos 11(6): 550–555

Della Corte L, Callingham BA (1977) The influence of age and adrenalectomy on rat heart monoamine oxidase. Biochem Pharmacol 26: 407–415

Dixon CM, Parkt GR, Parbit MH (1994) Characterization of the enzyme responsible for the metabolism of sumatriptan in human liver. Biochem Pharmacol 42: 1253–1257

Dostert P, Strolin Benedetti M (1988) Les bases de la neurotoxicité du MPTP. L'Encéphale 14: 399–412

Dostert P, Strolin Benedetti M, Tipton KF (1989) Interaction of monoamine oxidase with substrates and inhibitors. Med Res Rev 9: 45–89

Duine JA (1991) Quinoproteins: enzymes containing the quinonoid cofactor pyrroloquinoline quinone, topaquinone or tryptophan-tryptophan quinone. Eur J Biochem 200: 271–284

Edwards G, McGrath CS, Ward SA, Suparanond W, Pukrittayakamee S, Davis TME, White NJ (1993) Interactions among primaquine, malaria infection and other antimalarials in Thai subjects. Br J Clin Pharmacol 35: 193–198

Fang J, Gorrod JW (1991) Dehydration is the first step in the bioactivation of haloperidol to its pyridinium metabolite. Toxicol Lett 59: 117–123

Fang J, Rosen A, Gorrod JW (1993) Investigation of heterocyclic amines as potential substrates and inhibitors of monoamine oxidase. Pharm Sci Comm 4: 75–79

Ferkany JW, Andree TH, Clarke DE, Enna SJ (1981) Neurochemical effects of kojic amine, a gabamimetic, and its interaction with benzylamine oxidase. Neuropharmacology 20: 1177–1182

Fitton A, Benfield P (1990) Dopexamine hydrochloride: a review of its pharmacodynamic and pharmacokinetic properties and therapeutic potential in acute cardiac insufficiency. Drugs 39(2): 308–330

Fowler JS, Volkow ND, Wang GJ, Pappas N, Logan J, MacGregor R, Alexoff D, Shea C, Schlyer D, Wolf AP, Warner D, Zezulkova I, Cilento R (1996) Inhibition of monoamine oxidase B in the brains of smokers. Nature 379: 733–736

Fowler PA, Lacey LF, Thomas M, Keene ON, Tanner RJ, Baber NS (1991) The clinical pharmacology, pharmacokinetics and metabolism of sumatriptan. Eur Neurol 31: 291–294

Freestone S, Li Kam Wa TC, Lee MR (1992) The effect of the monoamine oxidase-B inhibitor selegiline on the metabolism of y-L-glutamyl-L-dopa (gludopa) in man. Proceedings of the BPS, September 9–11, 1992, p 74

Garland WA, Miwa BJ, Dairman W, Kappell B, Chiueh MC, Divoll M, Greenblatt DJ (1983) Identification of 7-chloro-5-(2'-fluorophenyl)-2,3-dihydro-2-oxo-1H-1, 4-benzodiazepine-1-acetaldehyde, a new metabolite of flurazepam in man. Drug Metab Dispos 11: 70–72

Guengerich FP (1995) Influence of nutrients and other dietary materials on cytochrome P-450 enzymes. Am J Clin Nutr 61 [Suppl]: 651S–658S

Halliwell B (1989) Oxidants and the central nervous system: some fundamental questions. Acta Neurol Scand 126: 23–33

Holt A, Callingham BA (1994) The ex vivo effects of procarbazine and methylhydrazine on some rat amine oxidase activities. J Neural Transm 41: 439–443

Janssen de Varebeke P, Cavalier R, David-Remacle M, Youdim MBH (1988) Formation of the neurotransmitter glycine from the anticonvulsant milacemide is mediated by brain monoamine oxidase B. J Neurochem 50: 1011–1016

Jeong TC, Yang KH, Holsapple MP (1994) Recovery of dimethylnitrosamine-induced immunosuppression by pargyline in the mixed cultures of murine hepatocytes and splenocytes. Life Sci 54(9): 605–613

Kenny P, Tipton KF, Strolin Benedetti M (1996) 6th Amine Rappaport Symposium and 7th International Amine Oxidase Workshop, Haifa, June 10–13, 1996. Book of Abstracts, p 40

Kinemuchi H, Fowler CJ, Tipton KF (1987) The neurotoxicity of 1-methyl-4-phenyl-1,2,3,6-tetrahydropyridine (MPTP) and its relevance to Parkinson's disease. Neurochem Int 11: 359–373

Kopin IJ (1993) Monoamine oxidase (MAO): relationships to foods, poisons and medicines. Biogen Amines 9: 355–365

Lake BG, Phillips JC, Cottrell RC, Gangolli SD (1978) Biological oxidation of nitrogen. Elsevier/North Holland, New York, p 131

Lee PC, Werlin SL (1995) The induction of hepatic cytochrome P450 3A in rats: effects of age. P.S.E.B.M. 210: 134–139

Lewinsohn R (1984) Mammalian monoamine-oxidizing enzymes, with special reference to benzylamine oxidase in human tissues. Braz J Med Biol Res 17: 223–256

Liczano JM, Fernández de Arriba A, Lyles GA, Unzeta M (1994) Several aspects of the amine oxidation by semicarbazide-sensitive amine oxidase (SSAO) from bovine lung. J Neural Transm [Suppl] 41: 415–420

Lyles GA (1994) Properties of mammalian tissue-bound semicarbazide-sensitive amine oxidase: possible clues to its physiological function? J Neural Transm [Suppl] 41: 387–396

Lyles GA, Chalmers J (1992) The metabolism of aminoacetone to methylglyoxal by semicarbazide-sensitive amine oxidase in human umbilical artery. Biochem Pharmacol 43: 1409–1414

Lyles GA, Marshall CM, McDonald IA, Bey P, Palfreyman MG (1987) Inhibition of rat aorta semicarbazide-sensitive amine oxidase by 2-phenyl-3-haloallylamines and related compounds. Biochem Pharmacol 36: 2847–2853

Mageed AA, Williams D, Faray B, Messina C, Ragab AH (1993) The effect of prednisone therapy on platelet monoamine oxidase and plasma catecholamine levels in children with acute lymphocytic leukemia. Cancer Res Ther Control 3: 79–85

Maret G, Testa B, Jenner P, el Tayar N, Carrupt PA (1990) The MPTP story: MAO activates tetrahydropyridine derivatives to toxins causing parkinsonism. Drug Metab Rev 22: 291–332

Markey SP, Schmuff NR (1986) The pharmacology of the parkinsonian syndrome producing neurotoxin MPTP (1-methyl-4-phenyl-1,2,3,6-tetrahydropyridine) and structurally related compounds. Med Res Rev 6: 389–429

McEwen CM, Harrison DC (1965) Abnormalities of serum monoamine oxidase in chronic congestive heart failure. J Lab Clin Med 65(4): 546–559

McEwen CM Jr, Castell DO (1967) Abnormalities of serum monoamine oxidase in chronic liver disease. J Lab Clin Med 70: 36–47

McIntire WS (1992) Wither PQQ. Essays Biochem 27: 119–134

Mihaly GW, Ward SA, Edwards G, L'E Orme M, Breckenridge AM (1984) Pharmacokinetics of primaquine in man: identification of the carboxylic acid derivative as a major plasma metabolite. Br J Clin Pharmacol 17: 441–446

Murata K, Noda K, Kohno K, Samejima M (1988) Bioavailability and pharmacokinetics of oral dopamine in dogs. J Pharm Sci 77(7): 565–568

Murphy PJ, Williams TL, Kau DLK (1976) Disposition of dobutamine in the dog. J Pharmacol Exp Ther 199(2): 423–431

Nelson TJ, Boor PJ (1982) Allylamine cardiotoxicity-IV metabolism to acrolein by cardiovascular tissues. Biochem Pharmacol 31: 509–514

Ni Y, Xu Y, Wang M (1992) In: Proceedings of the Fourth North American ISSX Meeting, Bal Harbour, FL, November 2–6, 1992, p 70

Nilsson SE, Tryding N, Tufvesson G (1968) Serum monoamine oxidase (MAO) in diabetes mellitus and some other internal diseases. Acta Med Scand 184: 105–108

Nishiyama S, Yoshikawa M, Yamaguchi I (1992) An orally effective peripheral dopamine prodrug: docarpamine (TA-870). Cardiovasc Drug Rev 10: 101–116

O'Brien EM, Tipton KF, Strolin Benedetti M, Bonsignori A, Marrari P, Dostert P (1991) Is the oxidation of milacemide by monoamine oxidase a major factor in its anticonvulsant actions? Biochem Pharmacol 41: 1731–1737

O'Brien EM, Dostert P, Pevarello P, Tipton KF (1994) Interactions of some analogues of the anticonvulsant milacemide with monoamine oxidase. Biochem Pharmacol 48: 905–914

O'Brien EM, Dostert P, Tipton KF (1995) Species differences in the interactions of the anticonvulsant milacemide and some analogues with monoamine oxidase-B. Biochem Pharmacol 50: 317–324

Phillips JC, Bex C, Lake BG, Cottrell RC, Gangolli SD (1982) Inhibition of dimethylnitrosamine metabolism by some heterocyclic compounds and by substrates and inhibitors of monoamine oxidase in the rat. Cancer Res 42: 3761–3765

Precious E, Gunn CE, Lyles GA (1988) Deamination of methylamine by semicarbazide-sensitive amine oxidase in human umbilical artery and rat aorta. Biochem Pharmacol 37: 707–713

Roberts BJ, Shoaf SE, Jeong KS, Song BJ (1994) Induction of CYP2E1 in liver, kidney, brain and intestine during chronic ethanol administration and withdrawal: evidence that CYP2E1 possesses a rapid phase half-life of 6 hours or less. Biochem Biophys Res Commun 205: 1064–1071

Saura J, Kettler R, Da Prada M, Richards JG (1992) Quantitative enzyme radioautography with ^{3}H-Ro 41-1049 and ^{3}H-Ro 19-6327 in vitro: localization and abundance of MAO-A and MAO-B in rat CNS, peripheral organs, and human brain. J Neurosci 12: 1977–1999

Saura J, Nadal E, van den Berg B, Vila' M, Richards JG, Bombi JA, Mahy N (1994) Localization of monoamine oxidases in human peripheral tissues. 6th Amine Oxidase Workshop and 5th Trace Amine Conference, Saskatoon, 31 July–3 August 1994. Book of Abstracts No A 03

Saura J, Richards JG, Mahy N (1994) Differential age-related changes of MAO-A and MAO-B in mouse brain and peripheral organs. Neurobiol Aging 15: 399–408

Scaman CH, Palcic MM (1992) Stereochemical course of tyramine oxidation by semicarbazide-sensitive amine oxidase. Biochemistry 31: 6829–6841

Schwartz MA, Postma E (1970) Metabolism of flurazepam, a benzodiazepine, in man and dog. J Pharm Sci 59: 1800–1806

Seiler N (1987) In: McCann PP, Pegg AE, Sjoerdsma A (eds) Inhibition of polyamine metabolism. Academic Press, London, p 49

Seiler N, Bolkenius FN, Knodgen B, Mamont P (1980) Polyamine oxidase in rat tissues. Biochim Biophys Acta 615: 480–488

Shah NS, Himwich HE (1971) Study with mescaline-8-C^{14} in mice: effect of amine oxidase inhibitors on metabolism. Neuropharmacology 10: 547–556

Spencer C, Faulds D, Fitton A (1993) Ibopamine: a review of its pharmacodynamic and pharmacokinetic properties, and therapeutic use in congestive heart failure. Drugs & Aging 3(6): 556–584

Steiner A, Walle UK, Walle T (1992) Induction of propranolol metabolism in the Hep G2 human hepatoma cell line. J Pharm Pharmacol 44: 611–614

Strolin Benedetti M, Dostert P (1992) Monoamine oxidase: from physiology and pathophysiology to the design and clinical application of reversible inhibitors. Drug Res 23: 65–125

Strolin Benedetti M, Dostert P (1994) Contribution of amine oxidases to the metabolism of xenobiotics. Drug Metab Rev 26: 507–535

Strolin Benedetti M, Rovei V, Thiola A, Donath A (1984) In: Tipton KF, Dostert P, Strolin Benedetti M (eds) Monoamine oxidase and disease: prospects for therapy with reversible inhibitors. Academic Press, London, p 203

Strolin Benedetti M, Dostert P, Tipton KF (1988) Contributions of monoamine oxidase to the metabolism of xenobiotics. Progr Drug Metab 11: 149–174

Strolin Benedetti M, Allievi C, Cocchiara G, Pevarello P, Dostert P (1992) Involvement of FAD-dependent polyamine oxidase in the metabolism of milacemide in the rat. Xenobiotica 22(2): 191–197

Strolin Benedetti M, Marrari P, Colombo M, Castelli MG, Arand M, Oesch F, Dostert P (1994) The anticonvulsant FCE 26743 is a selective and short-acting MAO-B inhibitor devoid of inducing properties towards cytochrome P450-dependent testosterone hydroxylation in mice and rats. J Pharm Pharmacol 46: 814–819

Strolin Benedetti M, Sanson G, Bona L, Gallina M, Persiani S, Tipton KF (1996) 6th Amine Rappaport Symposium and 7th International Amine Oxidase Workshop, Haifa, June 10–13, 1996. Book of Abstracts, p 14

Subramanyam B, Rollema H, Woolf T, Castagnoli N Jr (1990) Identification of a potentially neurotoxic pyridinium metabolite of haloperidol in rats. Biochem Biophys Res Commun 166: 238–244

Subramanyam B, Pond SM, Eyles DW, Whiteford HA, Fouda HG, Castagnoli N Jr (1991) Identification of potentially neurotoxic pyridinium metabolite in the urine of schizophrenic patients treated with haloperidol. Biochem Biophys Res Commun 181: 573–578

Suzuki O, Matsumoto T, Katsumata Y (1984) Determination of polyamine oxidase activities in human tissues. Experientia 40: 838–839

Tateishi T, Fujimura A, Shiga T, Ohashi K, Ebihara A (1995) Influence of aging on the oxidative and conjugative metabolism of propranolol. Int J Clin Pharm Res XV(3): 95–101

Teramoto T, Takahashi N, Yamada F, Fukazawa I, Oguchi K, Yasuhara H (1987) Inhibitory effects of antiarrhythmic drugs on rat heart and liver monoamine oxidase. Asia Pac J Pharmacol 2: 275–279

Tetrud JW, Langston JW (1989) MPTP-induced parkinsonism as a model for Parkinson's disease. Acta Neurol Scand [Suppl] 126: 35–40

Thomas PE, Bandiera S, Maines SL, Ryan DE, Levin W (1987) Regulation of cytochrome P-450j, a high-affinity N-nitrosodimethylamine demethylase, in rat hepatic microsomes. Biochemistry 26: 2280–2289

Tipton KF, Singer TP (1993) Advances in our understanding of the mechanisms of the neurotoxicity of MPTP and related compounds. J Neurochem 61: 1191–1206

Worth DP, Harvey JN, Brown J, Lee MR (1985) y-L-Glutamyl-L-dopa is a dopamine prodrug, relatively specific for the kidney in normal subjects. Clin Sci 69: 207–214

Yoshikawa M, Endo H, Komatsu K, Fujihara M, Takaiti O, Kagoshima T, Umehara M, Ishikawa H (1990) Disposition of a new orally active dopamine prodrug, N-(N-acetyl-L-methionyl)-O,O-bis (ethoxycarbonyl) dopamine (TA-870) in humans. Drug Metab Dispos 18(2): 212–217

Yoshimoto K, Echizen H, Chiba K, Tani M, Ishizaki T (1995) Identification of human CYP isoforms involved in the metabolism of propranolol enantiomers-N-desisopropylation is mediated mainly by CYP1A2. Br J Clin Pharmacol 39: 421–431

Yu PH, Davis BA (1991a) 2-propyl-1-aminopentane, its deamination by monoamine oxidase and semicarbazide-sensitive amine oxidase, conversion to valproic acid and behavioral effects. Neuropharmacology 30: 507–515

Yu PH, Davis BA (1991b) Simultaneous delivery of valproic acid and glycine to the brain. Deamination of 2-propylpentylglycinamide by monoamine oxidase B. Mol Chem Neuropathol 15: 37–49

Yu PH, Tipton KF (1989) Deuterium isotope effect of phenelzine on the inhibition of rat liver mitochondrial monoamine oxidase activity. Biochem Pharmacol 38(23): 4245–4251

Ziegler DM (1988) Flavin-containing monooxygenases: catalytic mechanism and substrate specificities. Drug Metab Rev 19: 1–32

Authors' address: M. Strolin Benedetti, Preclinical Development and Human Pharmacology, Zambon Group S.p.A., Via Lillo del Duca, 10, I-20091 Bresso (MI), Italy

Monoamine oxidases: from brain maps to physiology and transgenics to pathophysiology*

J. G. Richards[1], J. Saura[1], J. M. Luque[1], A. M. Cesura[1], J. Gottowik[1], P. Malherbe[1], E. Borroni[1], and J. Gray[2]

[1] Pharmaceuticals Division, Preclinical CNS Research, and [2] Pharma Business Development & Strategic Marketing, F. Hoffmann-La Roche Ltd, Basel, Switzerland

Summary. The present report reviews recent advances in mapping the cellular sites of synthesis and catalytic activity, as well as age- and disease-related changes of monoamine oxidases A and B in the brain. A transgenic model of oxidative stress is also described. The relevance of these findings for the physiological and pathophysiological roles of monoamine oxidases is briefly discussed.

Introduction

Monoamine oxidases (MAO, EC 1.4.3.4) are integral proteins of outer mitochondrial membranes. Isoenzymes (MAO-A and MAO-B) occur in various cells (both neuronal and non-neuronal in the CNS and peripheral organs) where they oxidatively deaminate biogenic and xenobiotic amines (Cesura and Pletscher, 1992; Waldmeier, 1987; Strolin-Benedetti and Dostert, 1992). In the CNS they play not only a physiological role in the metabolic inactivation of released monoamine transmitters (catecholamines, serotonin) and in the detoxification of xenobiotic amines but perhaps also a pathophysiological role by indirectly generating cytotoxic free radicals during aging and in neurodegenerative diseases (Cohen, 1986; Olanow, 1993). Hence, the therapeutic potential of selective reversible MAO inhibitors lies not only in their ability to increase the biological half-life of monoamine transmitters (*symptomatic effects*) but also to slow down the process of neurodegeneration (*neuroprotective effects*).

The isoenzymes were first identified by their inhibitor sensitivity and substrate selectivity and later by differences in their primary structure (Abell et al., 1994). In human brain, serotonin and noradrenaline are presumed to be preferentially metabolized by MAO-A, the trace amine phenethylamine and

---

*This review, of recent advances in our knowledge of the molecular neuroanatomy of monoamine oxidases, describes the fruits of research initiated by our late friend and colleague Mosé Da Prada to whom this is dedicated

tele-methylhistamine by MAO-B, and tyramine, octopamine as well as dopamine by both enzymes. Reversible inhibitors of MAO-A (RIMAs; proto-type: moclobemide) are effective antidepressants with unusually low side-effect profiles (Da Prada et al., 1994a; Haefely et al., 1992; Scatton et al., this volume; Youdim, 1995). *Lazabemide*, a selective reversible inhibitor of MAO-B (Da Prada et al., 1994b; Parkinson Study Group, 1993), has thera-peutic potential for the treatment of Parkinson's disease (as add-on therapy with L-dopa) as well as Alzheimer's disease.

Knowledge of the brain distribution of these enzymes is important since their tissue compartmentation and cellular localization determine to a large extent which substrate has access to which isoenzyme. Moreover, disease-related changes in enzyme distribution could provide a rationale for drug therapy. Previous studies (Chan-Palay et al., 1993; Konradi et al., 1988, 1989; Moll et al., 1990; Nakamura et al., 1990; Westlund et al., 1985, 1988; Zetsche and Chan-Palay, 1992) have used enzyme histochemical and immunohis-tochemical techniques to map the distribution of MAOs in rodent, monkey and human brain. The recent development of two high resolution assays- *in situ hybridization histochemistry* (using enzyme-selective oligonucleotide probes) to map the sites of enzyme synthesis and *quantitative enzyme radio-autography* (using enzyme-selective reversible inhibitors, Ro 41-1049 and lazabemide (Ro 19-6327) for MAO-A and -B, respectively) to map the abun-dance of catalytic sites and their de novo synthesis-, has enabled more exten-sive cellular and quantitative analyses to be carried out (Luque et al., 1995, 1996; Saura et al., 1992, 1994a,b, 1996, 1997). Moreover, in order to investigate the role of MAO-B in oxidative stress in an animal model, we and others (Anderson et al., 1994) have generated transgenic mice over-expressing MAO-B (up to 4-fold) under the regulation of the neuron-specific enolase promoter (Gottowik et al., submitted).

Results and discussion

I. Mapping sites of synthesis and catalytic activity in rat and human CNS

In situ hybridization histochemistry. ^{35}S-labelled oligonucleotide probes, selec-tively complementary to MAO-A or MAO-B mRNA sequences have been recently used for transcript mapping of the CNS at a cellular level (Luque et al., 1995, 1996; Saura et al., 1996). These studies have revealed the extremely high levels of expression of MAO-A mRNA in noradrenergic neurons of the locus coeruleus in rat, monkey and human brain (Fig. 1a–c). In contrast, MAO-A was only moderately expressed in raphe neurons and weakly but significantly in neurons of the substantia nigra (evidence from studies of rat brain) (Fig. 2a–c). MAO-A was also heterologously expressed in neurons of the hippocampal formation (CA1, CA3 and dentate gyrus granule cells) and spinal cord motoneurons. In other words, these non-aminergic neurons ap-pear to have the potential to synthesize this isoform of the enzyme, although evidence for the protein is lacking to date. MAO-B mRNA was highly ex-

pressed in serotoninergic neurons of the raphe (Fig. 3a,b) as well as in histaminergic neurons of the posterior hypothalamus in rat, monkey and human brain. High levels of MAO-B transcripts were also found in circumventricular organs, including subfornical organ, subcommissural organ, area postrema and ventricular ependyma, and in Bergmann glia. A moderate hybridization signal was, furthermore, observed in putative astroglia throughout the CNS.
Quantitative enzyme radioautography. The use of radiolabelled selective enzyme inhibitors which bind to their corresponding enzyme, has, in analogy to receptor radioautography, allowed high resolution measurements of catalytic sites in CNS and peripheral tissues (Saura et al., 1992, 1994a,b, 1996, 1997) (Fig. 3c,d).

Several observations lend support to the utility of the binding assays to measure the respective enzymes selectively.

The binding of both inhibitors is of a high affinity, saturable, reversible, tissue-specific and pharmacologically selective. Moreover, the binding assays can be used to successfully determine the rates of de novo enzyme synthesis after their irreversible inhibition, as well as age- and disease-related enzyme changes (see below). In a recent study (Ordway et al., 1996), it was found that [^{3}H]Ro 41-1049 binding (to MAO-A) in the locus coeruleus in 10 drug-free subjects with major depression is unaltered (vs. age-matched controls), although there was a significant correlation between the number of noradrenergic cells per section and the specific binding (further supporting the utility of the binding assay).

The binding ratio of A:B (at K_D) in rat CNS is maximal (~4) in locus coeruleus, internal granular layer of the olfactory bulb, nucleus of the olfactory tract, deep mesencephalic nucleus and C1 cells, for example, and minimal (~0.2) in the inner part of the olfactory nerve layer, circumventricular organs and posterior pituitary. The present investigations revealed a correlative, but not necessarily identical, distribution to the corresponding transcripts. In human brain, in contrast, B:A ratios were between 1 and 8. For example, the B_{max} of lazabemide binding (to MAO-B) in the substantia nigra was ~3-fold that of Ro 41-1049 binding (to MAO-A), although MAO-A was more concentrated in the pars compacta than in the pars reticulata, the reverse being true for MAO-B (Fig. 4a,b).

Mapping de novo synthesis in rat CNS and peripheral organs

The half-life of the isoenzymes in rat brain and peripheral organs have been determined with high resolution by measuring the recovery of binding of the tritiated inhibitors in different brain regions after irreversible inhibition of their respective enzymes. Animals were allowed to recover from treatment with clorgyline or l-deprenyl (5 µmol/kg s.c.), irreversible inhibitors of MAO-A and MAO-B respectively, for various times (from 2 hours to 28 days), then tissue sections were pre-incubated to remove excess clorgyline or l-deprenyl and subsequently incubated with the respective radiolabelled inhibitor to determine the binding levels (= catalytic activity). The binding data was

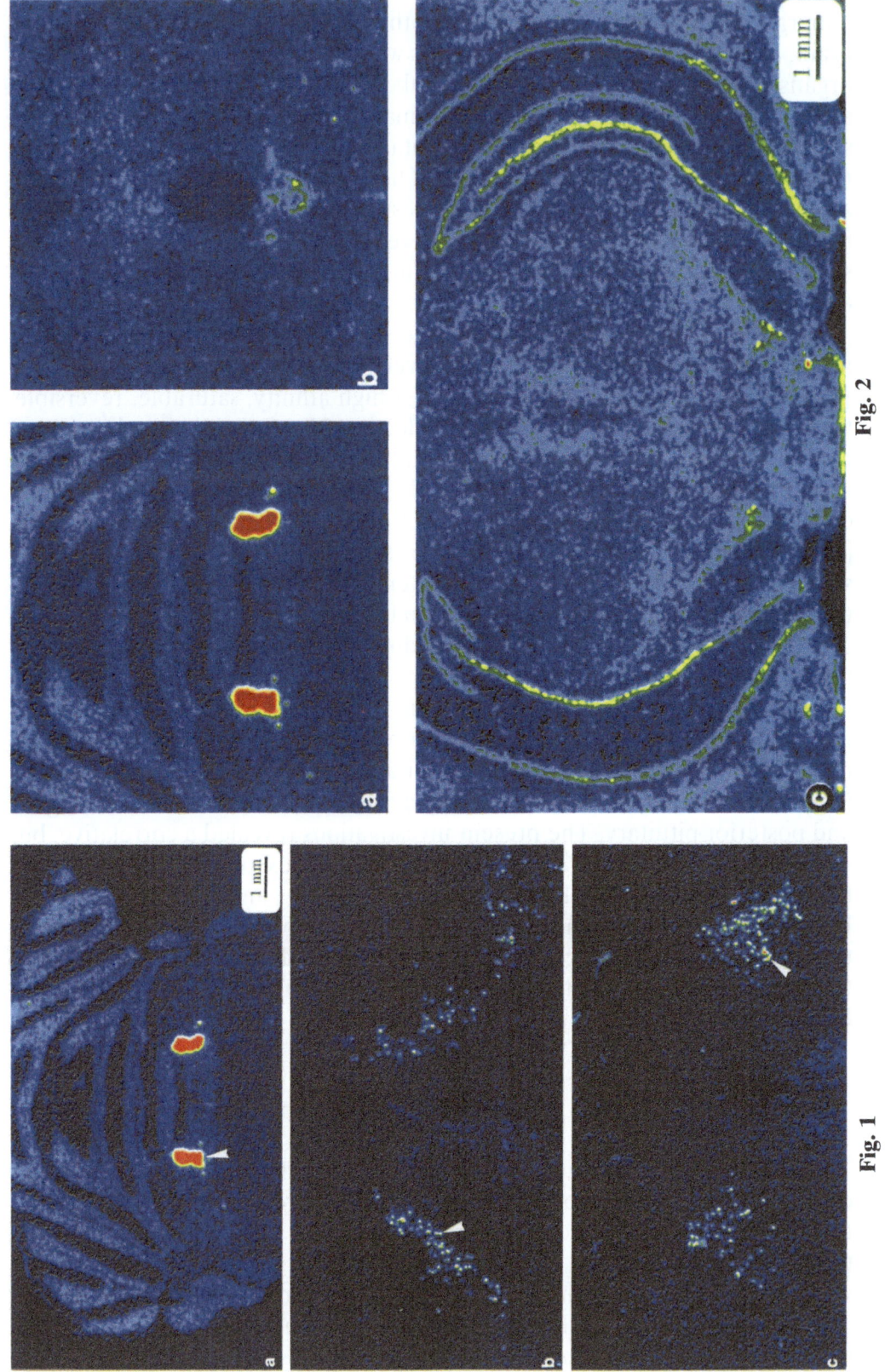

Fig. 2

Fig. 1

analyzed by non-linear regression and the $t_{1/2}$ for each isoenzyme in various tissues was estimated.

Two hours after enzyme blockade with supramaximal doses of the respective inhibitors, no specific binding was detectable with either radioligand as determined by quantitative radioautography. However, after increasing survival times differentially radiolabelled tissues appeared. De novo synthesis of both MAO-A and MAO-B in peripheral organs (average $t_{1/2}$ = 4.5 days) was found to be much faster than in brain regions (13 days) (Fig. 5). For both enzymes, markedly different turnover rates among brain regions were observed. For example, $t_{1/2}$ estimates for MAO-A ranged from 7.7 days (hippocampal CA3 oriens layer) to 20.5 days (molecular layer of cerebellum) and for MAO-B from 4.8 days (ventricular ependyma) to 29.9 days (white matter of cerebellum).

Previous estimates by Goridis and Neff (1971) also pointed out the faster turnover rate in the periphery vs. brain. Other (low resolution) studies (Arnett et al., 1987; Goridis and Neff, 1973; Nelson et al., 1979; Oreland et al., 1990) have produced values within the range described here for both brain and peripheral organs. The present study opens the way for high resolution investigations of drugs influencing the turnover of MAOs in discrete brain regions.

Comments on the physiological roles of MAOs

In accordance with the substrate-selectivity of MAO isoforms, the locus coeruleus and posterior hypothalamus are sites of synthesis of MAO-A and MAO-B, respectively. To a lesser degree (because of the observed moderate to low levels of transcripts) this is also true of the raphe and substantia nigra pars compacta which are possible sites of synthesis of MAO-A. The potential of raphe neurons to synthesize both isoforms, however, is unexpected, since only MAO-B protein has been detected to date. It might indicate a unique role of MAO-A in these neurons. Aggressive behaviour and altered amounts of brain serotonin and norepinephrine in MAO-A knockout mice have been recently reported by Cases et al. (1995). The observed abnormal levels of serotonin in catecholaminergic neurons implies that serotonin would normally accumulate in these neurons as a false transmitter were it not for the

Fig. 1. Distribution of MAO-A mRNAs in rat (**a**), monkey (**b**) and human (**c**) locus coeruleus revealed by in situ hybridization histochemistry. Pseudocolour computer images (warm colours, e.g. red, yellow, indicate high levels of hybridization signal and cool colours, e.g. blue, green, low levels). Note the abundance of transcripts in single neurons [arrowheads, obscured in (a)], revealed by film radioautography

Fig. 2. Distribution of MAO-A mRNAs in the locus coeruleus (**a**), dorsal raphe nucleus (**b**) and substantia nigra pars compacta (**c**) of rat brain revealed by in situ hybridization histochemistry. Pseudocolour computer images. Note the relative decrease in abundance of transcripts (a > b > c) revealed by film radioautography

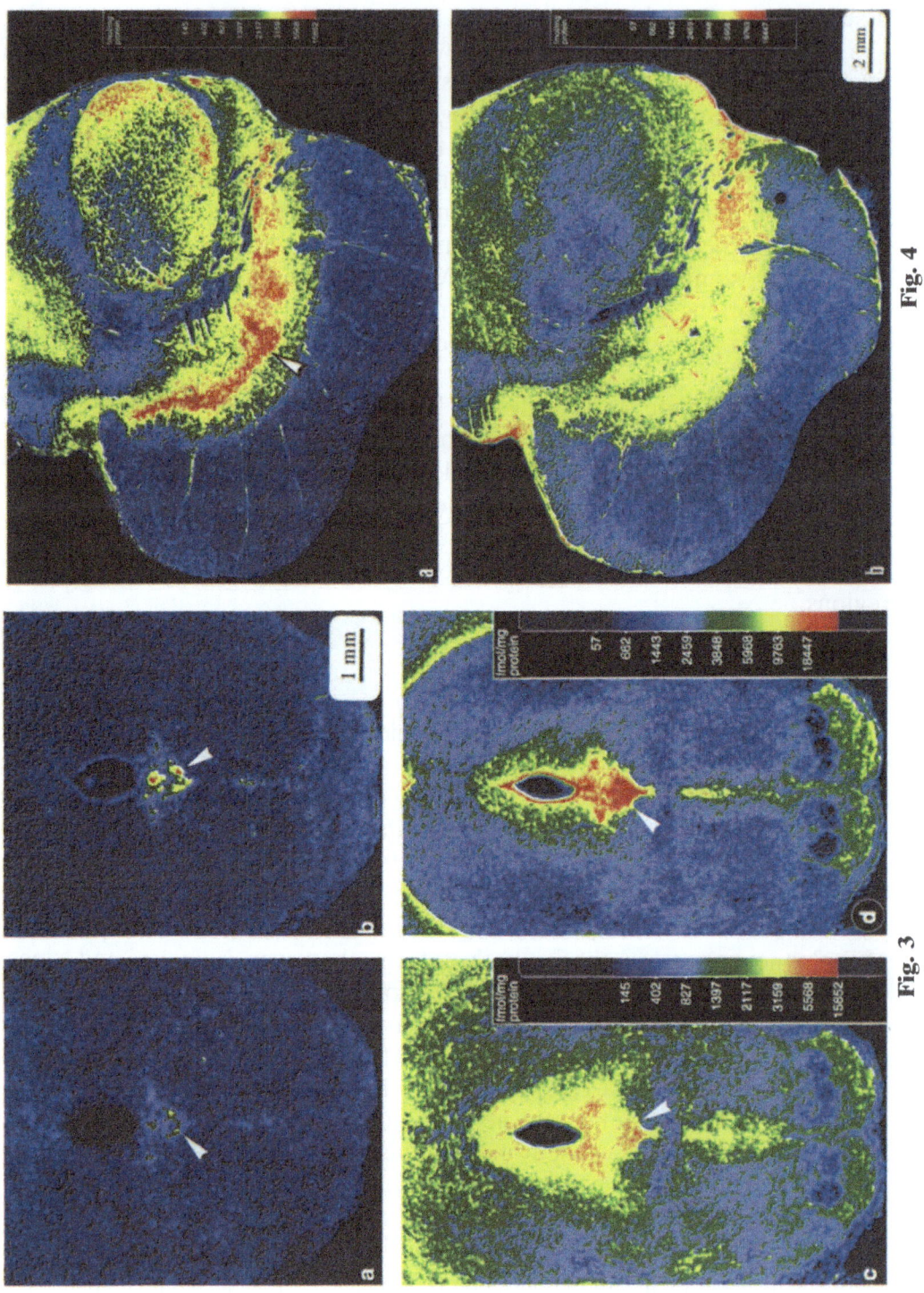

Fig. 4

Fig. 3

presence of MAO-A. The weak to moderate expression of MAO-B mRNA in putative astrocytes throughout the CNS indicates a slow turnover of the enzyme despite histochemical evidence for a high protein content. The functional meaning of the observed heterologous expression of transcripts (see also Chan-Palay et al., 1993; Nakamura et al., 1993, 1995) is not known, although it might suggest a completely different (hitherto unknown) substrate for the respective enzyme in these neurons.

Targeted mutations or knockouts of one or both isoenzymes might provide further clues for their physiological roles. Moreover, further detailed examination of the neurochemical and clinical phenotypes of individuals with genetic deficiencies of MAO-A and/or -B, due to different X-chromosomal microdeletions or a point mutation (see Murphy et al., Lenders et al., Brunner et al., this volume), may also shed new light on the possible functions of these enzymes and provide a rationale for drug therapy. Severe to borderline mental retardation has been observed in Dutch kindred with either combined MAO-A and -B deficiencies or selective MAO-A deficiency, respectively. In contrast, those with selective MAO-B deficiency appear to be mentally normal. The authors concluded i.a. that MAO-A is considerably more important than MAO-B in the metabolism of biogenic amines and that MAO-B has a limited capacity to deaminate catecholamines.

II. Excessive MAO activity as a source of free radicals?

The products of the oxidative metabolism of monoamines are, under physiological conditions, disposed of through several pathways. However, during aging and in certain neurodegenerative diseases abnormally increased enzyme activity (particularly MAO-B) may lead to the failure to adapt and the consequent increased production of potentially cytotoxic free radicals (Fig. 6). A causative role for MAO-B in Parkinson's disease has been implied by several findings, namely the observed increase in platelet MAO-B (Stevenson et al., 1989) and the association of a MAO-B allele with a genetic predisposition for the disease (Girmen et al., 1992; Kurth et al., 1993) but has been questioned by others (Jarman et al., 1993).

Quantitative enzyme radioautography has been used to demonstrate, with high resolution, both age- and disease-related changes in enzyme activity in rat and human brain (Saura et al., 1994a,b, 1997). Thus, in brains of aging BL/C57 mice, whereas MAO-A activity decreased in the first 2 months of life,

---

Fig. 3. Distribution of MAO-A and MAO-B mRNAs and proteins in the dorsal raphé nucleus (arrowhead) of rat brain revealed by in situ hybridization histochemistry (**a** and **b**, respectively) and quantitative enzyme radioautography (**c** and **d**, respectively). Pseudocolour computer images. Note the greater abundance of MAO-B transcripts and protein
Fig. 4. Distribution of MAO-A (**a**) and MAO-B (**b**) catalytic sites in human substantia nigra revealed by quantitative enzyme radioautoradiography. MAO-A is highly concentrated in the pars compacta enriched in melanin-containing neurons (arrowhead). Pseudocolour computer images

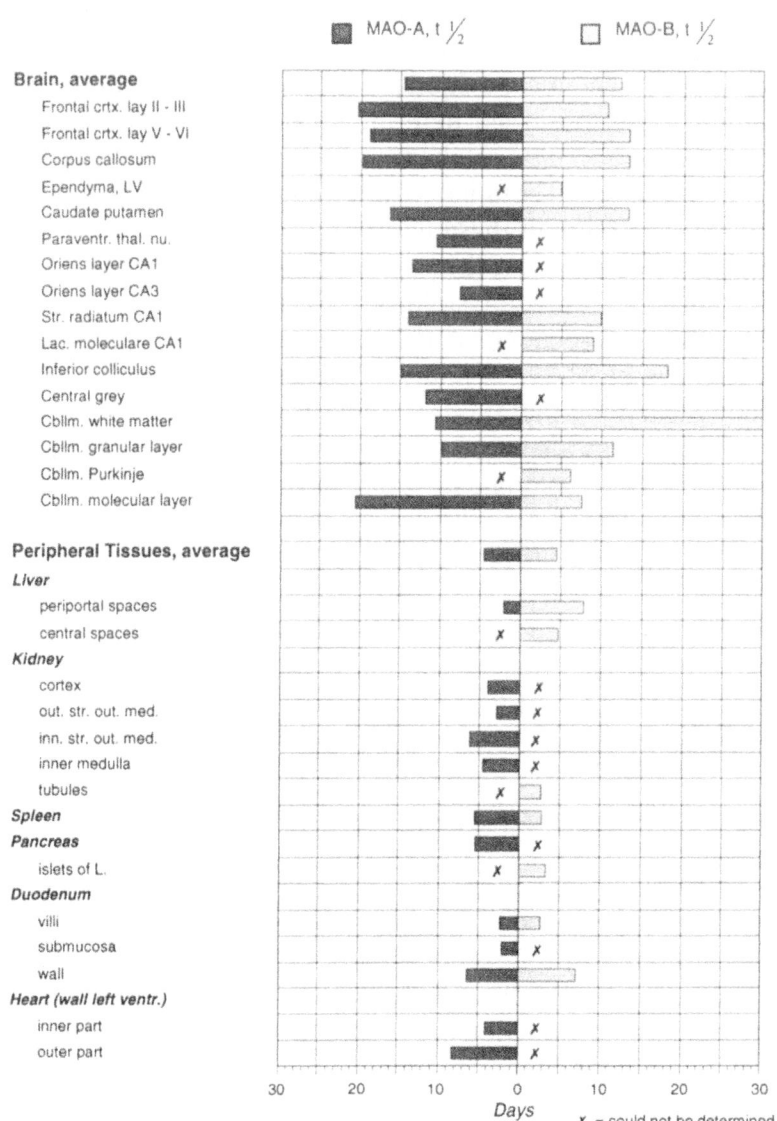

Fig. 5. De novo synthesis of monoamine oxidases in different rat brain regions and peripheral organs revealed by quantitative enzyme radioautography

MAO-B showed a clearcut age-related increase in activity, e.g. at 19 and 25 months. The levels of the isoenzymes in 27 human subjects (age range: 17–93 years) has been recently studied by quantitative enzyme radioautography (Saura et al., 1997). Eighteen brain structures of temporal cortex, precentral gyrus, hippocampal formation, striatum, cerebellum and brainstem were analysed. A marked age-related increase in MAO-B (not MAO-A) was observed in most structures (notably not in substantia nigra; see also Damier et al., 1996), beginning at the age of 50–60 years. The post-mortem delay or duration of tissue storage did not seem to affect enzyme levels. These findings confirm and extend previous biochemical investigations on aging rodent and

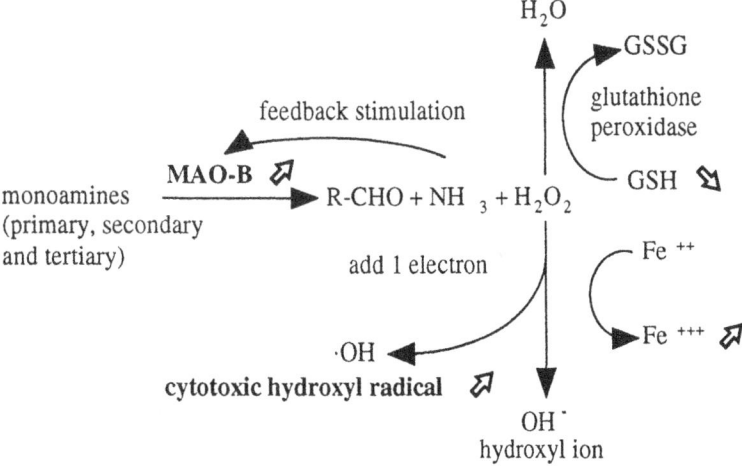

Fig. 6. The role of MAO-B in the pathomechanism of Parkinson's disease (and other neurodegenerative diseases?) (modified after Gerlach et al., 1994). The increased catalytic activity of MAO-B (under positive feedback stimulation) is stoichiometrically linked to the overproduction of hydrogen peroxide. Normally, detoxifying mechanisms are in place for the clearance of peroxide (glutathione peroxidase) and excess Fe^{3+} (sequestered by ferritin). In Parkinson's disease, lower levels of reduced glutathione and increased iron in the substantia nigra, suggest decreased protection against hydroxyl radical formation

human brain (Adolfsson et al., 1980; Fowler et al., 1980; Gottfries et al., 1983; Oreland et al., 1986; Reinikainen et al., 1988; Robinson et al., 1975; Sparks et al., 1991).

Alzheimer's disease

In a recent study (Saura et al., 1994) of Alzheimer brains, a selective increase of MAO-B in plaque-associated astrocytes was observed (Fig. 7a,b) with no change in MAO-A. This finding confirms and extends an earlier immunocytochemical study (Nakamura et al., 1990). Although it has been suggested that astrocytes (via catalase activity) protect neurons from hydrogen peroxide toxicity (Desagher et al., 1996), the local increased activity of MAO-B in astroglioses surrounding senile plaques might exacerbate the hypothesized ongoing cytotoxic effects of amyloid β-protein in Alzheimer brains (see Yankner, 1996). This being the case, selective MAO-B inhibitors, such as lazabemide, might offer a means of slowing down the rate of cognitive decline. Indeed, initial clinical findings with lazabemide indicate promising efficacy and good tolerance in early Alzheimer patients.

MAO-B transgenics

To gain insight into the possible role of MAO-B in Parkinson's and Alzheimer's diseases, we have generated lines of transgenic mice in which neurons overexpress human MAO-B protein under the control of the neuron-specific enolase promoter (Gottowik et al., submitted). Heterozygous transgenic mice (5–8 weeks old) were analysed biochemically, histologically, by quantitative enzyme radioautography and in situ hybridization histochemistry.

The transgenic mice express 4–6 fold higher MAO-B activity in brain as compared to non-transgenic littermates (Fig. 8a–d), whereas the activity of MAO-B in the liver was equal to that of control animals. Analysis of brain monoamines showed a 30–40% increase in the basal levels of the dopamine metabolites DOPAC and HVA, whereas no changes were observed in the basal levels of dopamine (DA). Further analyses of brain regions showed an increase in the levels of these metabolites in the substantia nigra and striatum of transgenic mice, whereas no changes in neurotransmitter levels were observed in the locus coeruleus and frontal cortex. The transgenic mice metabolize ~20% more DA than control animals under basal conditions. The higher rate of DA metabolism observed in these mice is likely to result in an increased oxidant stress since the conversion of DA to DOPAC, catalysed by MAO, is stoichiometrically linked to the production of hydrogen peroxide (as evidenced by enzyme histochemistry). However, no histological or astroglial changes were observed in l-dopa-treated or -untreated heterozygotes up to 3 months old. Interestingly, however, L-ferritin mRNA was markedly increased in l-dopa-treated and -untreated heterozygotes (Fig. 9), suggesting that a protective adaptive response to the formation of reactive oxygen species, due to increased enzyme activity, had indeed occurred. No changes in other markers of anti-oxidant activity were observed. Behavioural changes are currently under investigation. A selective atrophy of catecholaminergic neurons in MAO-B transgenic mice has been described by Anderson et al. (1994), further supporting this mouse as a model for studying the effects of free-radical damage during aging and in neurodegenerative disease.

Comments on pathophysiological roles of MAO

Further evidence for the potential cytotoxic role of increased glial MAO-B occuring in aging and Alzheimer's disease is required before it can be un-

Fig. 7. Increased MAO-B protein in plaque-associated astrocytes in cortices of Alzheimer individuals revealed by quantitative enzyme radioautography with [³H]lazabemide. **a** Pseudocolour computer image of patches of increased enzyme activity (red, arrowhead) in the temporal cortex of an Alzheimer individual. **b** Quantitative analysis of [³H]lazabemide binding in controls vs. Alzheimer individuals

Fig. 8. Distribution of MAO-B protein (**a,b**) and mRNA (**c,d**) in adjacent parasagittal sections of transgenic (a,c) vs. wild-type (b,d) mouse brains revealed by quantitative enzyme radioautography and in situ hybridization histochemistry, respectively. Pseudocolour computer images. Note the discrete localization of protein and mRNA in the raphe nucleus (arrowhead) of wild-type brain in contrast to their much broader distribution in the transgenic brain

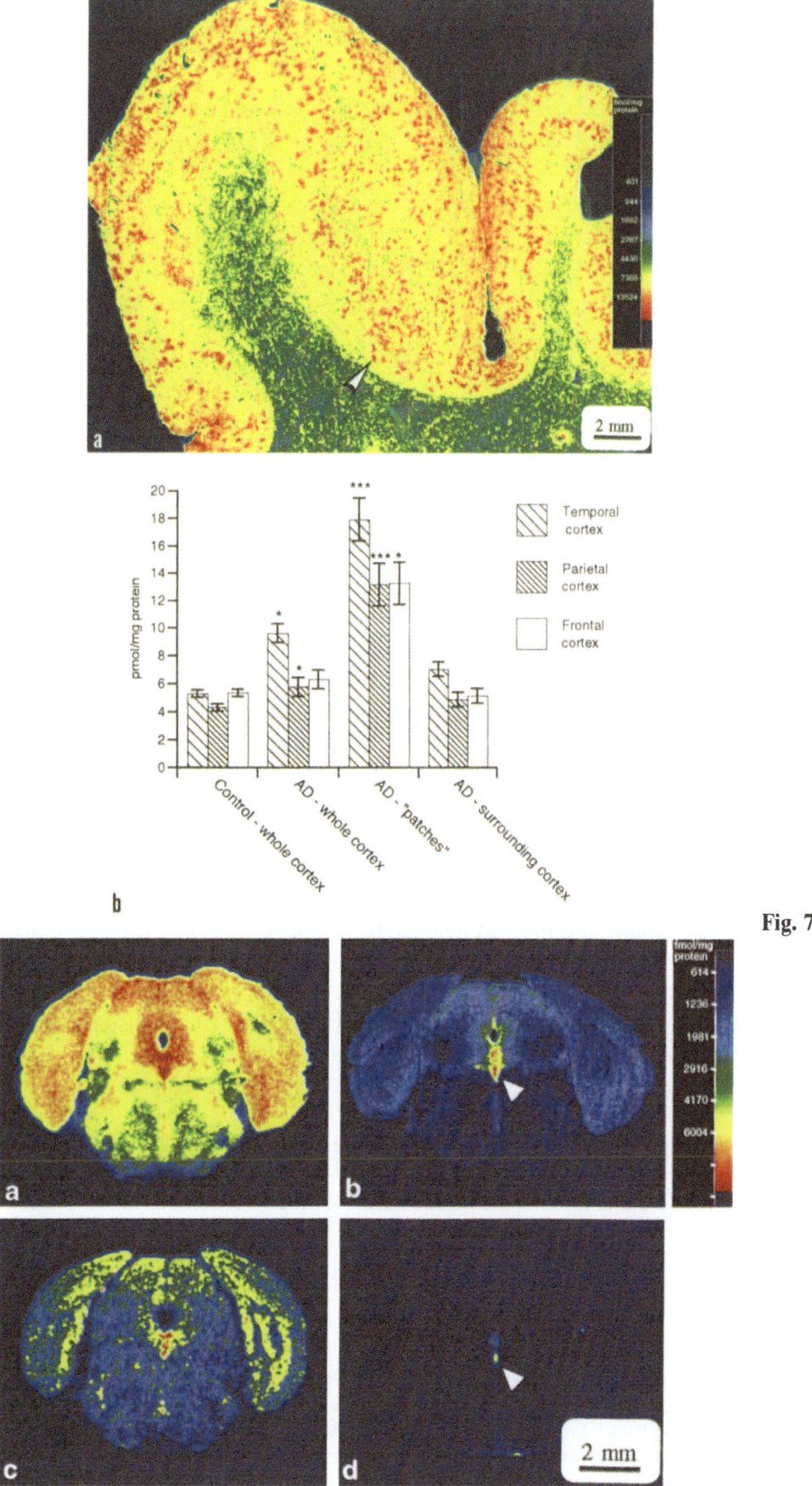

Fig. 7

Fig. 8

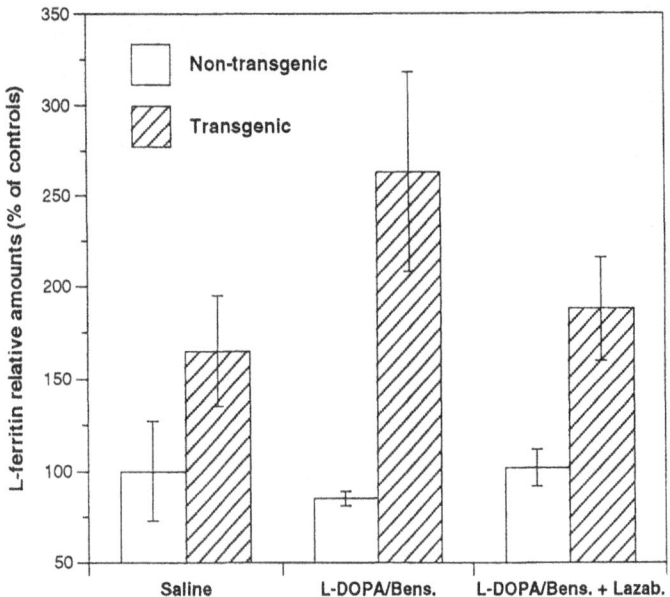

Fig. 9. Upregulation of L-ferritin mRNA in the brains of treated and untreated transgenic vs. non-transgenic mice

equivocally identified as a drug target for the treatment of neurodegenerative diseases. Nevertheless, the clinical findings to-date with the selective MAO-B inhibitor lazabemide are, indeed, promising. The efficacies of selective MAO-B inhibitors (whether symptomatic or neuroprotective) in the therapy of Parkinson's disease has yet to be demonstrated. Recently, inhibition of MAO-B in the brains of smokers was demonstrated in a PET study (Fowler et al., 1996). Their findings may explain the resistance of cigarette smokers to Parkinson's disease, and may contribute to the performance-enhancing properties of smoking.

It remains to be seen whether selective MAO inhibitors can be used diagnostically to determine (by the use of PET or SPECT) disease-related enzymatic changes clinically in Parkinson's or Alzheimer's diseases.

Acknowledgements

The authors are grateful to J. Messer and Z. Bleuel for their excellent technical assistance and to Prof. C. Köhler and Dr. J. Kemp for critical evaluation of the manuscript.

References

Abell CW, Stewart RM, Andrews PJ, Kwan S-W (1994) Molecular and functional properties of the monoamine oxidases. Heterocycles 39: 933–955

Adolfsson R, Gottfries CG, Oreland L, Wiberg A, Winblad B (1980) Increased activity of brain and platelet monoamine oxidase in dementia of Alzheimer type. Life Sci 27: 1029–1034

Anderson JK, Frim DM, Isacson O, Breakfield XO (1994) Catecholaminergic cell atrophy in a transgenic mouse aberrantly overexpressing MAO-B in neurons. Neurodegen 3: 97–109

Arnett CD, Fowler JS, MacGregor RR, Schlyer DJ, Wolf AP, Langstrom B, Halldin C (1987) Turnover of brain monoamine oxidase measured in vivo by positron emmission tomography using L-[11C]deprenyl. J Neurochem 49: 522–527

Cases O, Seif I, Grimsby J, Gaspar P, Chen K, Pournin S, Muller U, Aguet M, Babinet C, Shih JC, De Maeyer E (1995) Aggressive behavior and altered amounts of serotonin and norepinephrine in mice lacking MAO-A. Science 268: 1763–1766

Cesura AM, Pletscher A (1992) The new generation of monoamine oxidase inhibitors. Prog Drug Res 38: 171–297

Chan-Palay V, Hochli M, Savaskan E, Hungerecker G (1993) Calbindin D-28k and monoamine oxidase A immunoreactive neurons in the nucleus basalis of Meynert in senile dementia of the Alzheimer type and Parkinson's disease. Dementia 4: 1–15

Cohen G (1986) Monoamine oxidase, hydrogen peroxide and Parkinson's disease. Adv Neurol 45: 119–125

Damier P, Kastner A, Agid Y, Hirsch EC (1996) Does monoamine oxidase type B play a role in dopaminergic nerve cell death in Parkinson's disease? Neurology 46: 1262–1269

Da Prada M, Pieri L, Cesura AM, Kettler R (1994a) The pharmacology of moclobemide. Revs Contemp Pharmacol 5: 1–18

Da Prada M, Zurcher G, Kettler R, Dingemanse K, Jorga K, Dubuis R (1994b) Remodelling the kinetics and dynamics of levodopa therapy in Parkinson's disease by inhibiting MAO-B with lazabemide and COMT with tolcapone. In: Poewe W, Lees AJ (eds) Levodopa — the first 25 years. Proceedings of the Symposium "20 Years of Madopar^R, New Avenues", Berlin, October 1993. Editiones Roche, Basel, pp 99–117

Desagher S, Glowinski J, Premont J (1996) Astrocytes protect neurons from hydrogen peroxide toxicity. J Neurosci 16: 2553–2562

Fowler CJ, Wiberg A, Oreland L, Marcusson J, Winblad B (1980) The effect of age on the activity and molecular properties of human brain monoamine oxidase. J Neural Transm 49: 1–20

Fowler JS, Volkow ND, Wang G-J, Pappas N, Logan J, MacGregor R, Alexoff D, Shea C, Schyler D, Wolf AP, Warner D, Zezulkova I, Cilento R (1996) Inhibition of monoamine oxidase B in the brains of smokers. Nature 379: 733–736

Gerlach M, Ben-Shachar D, Riederer P, Youdim MBH (1994) Altered brain metabolism of iron as a cause of neurodegenerative diseases? J Neurochem 63: 793–807

Girman AS, Baenziger J, Hotamisligil GS, Konradi C, Shalish C, Sullivan JL, Breakfield XO (1992) Relationship between platelet monoamine oxidase B activity and alleles at the MAO-B locus. J Neurochem 59: 2063–2066

Goridis C, Neff NH (1971) Monoamine oxidase: an approximation of turnover rates. J Neurochem 18: 1673–1682

Goridis C, Neff NH (1973) Neuronal and hormonal influences on the turnover of monoamine oxidase in salivary gland. Biochem Pharmacol 22: 2501–2510

Gottfries CG, Adolfsson R, Aquilonius S-M, Carlsson A, Eckernas S-A, Nordberg A, Oreland L, Svennerholm L, Wiberg A, Winblad B (1983) Biochemical changes in dementia disorders of Alzheimer type (AD/SDAT). Neurobiol Aging 4: 261–271

Haefely W, Burkard WP, Cesura AM, Kettler R, Lorez HP, Martin JR, Richards JG, Scherschlicht R, Da Prada M (1992) Biochemistry and pharmacology of moclobemide, a prototype RIMA. Psychopharmacology 106: S6–S14

Jarman J, Glover V, Sandler M, Turjanski N, Stern G (1993) Platelet monoamine oxidase B activity in Parkinson's disease: a re-evaluation. J Neural Transm [P-D Sect] 5: 1–4

Konradi C, Svoma E, Jellinger K, Riederer P, Denney P, Thibault J (1988) Topographical immunocytochemical mapping of MAO-A, MAO-B and tyrosine hydroxylase in human post-mortem brainstem. Neuroscience 26: 791–802

Konradi C, Kornhuber J, Froelich L, Fritze J, Heinsen H, Beckmann H, Schulz E, Riederer P (1989) Demonstration of monoamine oxidase -A and -B in the human brainstem by a histochemical technique. Neuroscience 33: 383–400

Kurth JH, Kurth MC, Poduslo SE, Schwankhaus JD (1993) Association of a monoamine oxidase B alle with Parkinson's disease. Ann Neurol 33: 368–372

Luque JM, Kwan S-W, Abell CW, Da Prada M, Richards JG (1995) Cellular expression of mRNAs encoding monoamine oxidases A and B in the rat central nervous system. J Comp Neurol 363: 665–680

Luque JM, Bleuel Z, Hendrickson A, Richards JG (1996) Detection of MAO-A and MAO-B mRNAs in monkey brainstem by cross-hybridization with human oligonucleotide probes. Mol Brain Res 36: 357–360

Moll G, Moll R, Riederer P, Gsell W, Heinsen H, Denney RM (1990) Immunofluorescence cytochemistry on thin frozen sections of human substantia nigra for staining of monoamine oxidase A and monoamine oxidase B, a pilot study. J Neural Transm [Suppl] 32: 67–77

Nakamura S, Kawamata T, Akiguchi I, Kameyama M, Nakamura N, Kimura H (1990) Expression of monoamine oxidase B activity in astrocytes of senile plaques. Acta Neuropathol 80: 319–325

Nakamura S, Akigachi I, Kimura J (1993) A subpopulation of mouse striatal cholinergic neurons show monoamine oxidase activity. Neurosci Lett 161: 141–144

Nakamura S, Akigachi I, Kimura J (1995) Topographic distribution of monoamine oxidase-B -containing neurons in the mouse striatum. Neurosci Lett 184: 29–31

Nelson DL, Herbet A, Glowinski J, Hamon M (1979) [3H]Harmaline as a specific ligand of MAO-A. II. Measurement of the turnover rates of MAO A during ontogenesis in rat brain. J Neurochem 32: 1829–1836

Olanow CW (1993) A radical hypothesis for neurodegeneration. Trends Neurosci 16: 439–444

Ordway GA, Klimek V, Richards JG, Overholser JC, Meltzer HY, Dilley G, Stockmeier CA (1996) [^{3}H]Ro 41-1049 binding to monoamine oxidase-A in the locus coeruleus is not altered in major depression. Soc Neurosci Abstr 368.2

Oreland L, Gottfries CG (1986) Brain monoamine oxidase in aging and in dementia of Alzheimer type. Prog Neuropsychopharmacol Biol Psychiatry 10: 533–540

Oreland L, Jossan SS, Hartvig P, Aquilonius SM, Langstrom B (1990) Turnover of monoamine oxidase B (MAO-B) in pig brain by positron emission tomography using 11C-L-deprenyl. J Neural Transm [Suppl] 32: 55–59

Parkinson Study Group (1993) A controlled trial of lazabenmide (Ro 19-6327) in untreated Parkinsons disease. Ann Neurol 33: 350–356

Reinikainen KJ, Paljarvi L, Halonen T, Malminen O, Kosma V-M, Laakso M, Riekkinen PJ (1988) Dopaminergic system and monoamine oxidase-B activity in Alzheimer's disease. Neurobiol Aging 9: 245–252

Robinson DS (1975) Changes in monoamine oxidase and monoamines with human development and aging. Fed Proc 34: 103–107

Saura J, Kettler R, Da Prada M, Richards JG (1992) Quantitative enzyme radioautography with ^{3}H-Ro 41-1049 and ^{3}H-Ro 19-6327 in vitro: localization and abundance of MAO-A and MAO-B in rat CNS, peripheral organs, and human brain. J Neurosci 12: 1977–1999

Saura J, Luque JM, Cesura AM, Da Prada M, Chan-Palay V, Huber G, Loffler J, Richards JG (1994a) Increased monoamine oxidase -B activity in plaque-associated astrocytes of Alzheimer brains revealed by quantitative enzyme radioautography. Neuroscience 62: 15–30

Saura J, Richards JG, Mahy N (1994b) Differential age-related changes of MAO-A and MAO-B in mouse brain and peripheral organs. Neurobiol Aging 15: 399–408

Saura J, Bleuel Z, Ulrich J, Mendelowitsch A, Chen K, Shih JC, Malherbe P, Da Prada M, Richards JG (1996) Molecular neuroanatomy of human monoamine oxidases A and B revealed by quantitative enzyme radioautography and in situ hybridization histochemistry. Neuroscience 70: 755–774

Saura J, Andres N, Andrade C, Ojuel J, Eriksson K, Mahy N (1997) Biphasic and region-specific MAO-B response to aging in control human brain. Neurobiol Aging (in press)

Sparks DL, Woeltz VM, Markesbery WR (1991) Alterations in brain monoamine oxidase activity in aging, Alzheimer's disease, and Pick's disease. Arch Neurol 48: 718–721

Stevenson GB, Sturman SG, Heafield MTE (1989) Platelet monoamine oxidase B activity in Parkinson's disease. J Neural Transm 1: 255–261

Strolin-Benedetti M, Dostert P (1992) Monoamine oxidase: from physiology and pathophysiology to the design and clinical applications of reversible inhibitors. Adv Drug Res 23: 65–125

Waldmeier PC (1987) Amine oxidases and their endogenous substrates (with special reference to monoamine oxidase and the brain). J Neurotransm [Suppl] 23: 55–72

Westlund KN, Denney RM, Kochersperger LM, Rose RM, Abell CW (1985) Distinct monoamine oxidase A and B populations in primate brain. Science 230: 181–183

Westlund KN, Denney RM, Rose RM, Abell CW (1988) Localization of distinct monoamine oxidasde A and monoamine oxidase B cell populations in human brainstem. Neuroscience 25: 439–456

Yankner BA (1996) Mechanisms of neuronal degeneration in Alzheimer's disease. Neuron 16: 921–932

Youdim MBH (1995) The advent of selective monoamine oxidase A inhibitor antidepressants devoid of the cheese reaction. Acta Psychiatr Scand 91 [Suppl 386]: 5–7

Zetsche T, Chan-Palay V (1992) MAO-A and MAO-B immunoreactivity in the hippocampus, temporal cortex and cerebellum of normal controls and of patients with senile dementia of the Alzheimer type. Dementia 3: 270–281

Authors' address: Dr. J. G. Richards, Pharmaceuticals Division, Preclinical CNS Research, Building 69, Room 234, F. Hoffmann-La Roche Ltd, CH-4070 Basel, Switzerland

Structure-function relationships of mitochondrial monoamine oxidase A and B: chimaeric enzymes and site-directed mutagenesis studies

A. M. Cesura, J. Gottowik, G. Lang, P. Malherbe, and **M. Da Prada**[†]

Pharma Division, Preclinical Research, Nervous System Diseases,
F. Hoffmann-La Roche Ltd, Basel, Switzerland

Summary. To gain insight into the structure of monoamine oxidases (MAO) A and B, we investigated the properties of various chimaeric enzymes, engineered by moving progressively the junction between the NH_2- and the COOH-termini of each MAO form. Whereas exchange of the ADP-binding sequence did not modify the catalytic properties of either MAO isoforms, chimaeras with increasing length of the NH_2-terminus of MAO-A (up to position 256) showed a marked decrease in affinity towards substrates and inhibitors. Two sequences, spanning position 62 to 103 and 146 to 220, appeared of particular importance in putatively constituting the binding site of MAO-B. Conversely, the catalytic properties and specificity of MAO-A were insensitive to substitution of both the NH_2- (up to position 112) and COOH-termini (from residue 395), but further modification of the central sequence of MAO-A was not compatible with activity. None of the engineered chimaeras showed a shift in substrate and inhibitor specificity. Investigation on MAO-B by site-directed mutagenesis revealed that His382 and Thr158 may represent residues relevant for MAO-B catalytic mechanism.

Introduction

Mitochondrial, flavin-containing monoamine oxidases (MAO; EC 1.4.3.4), i.e. MAO-A and MAO-B, are encoded by very similar but separate genes (Grimsby et al., 1991). Human MAO-A and MAO-B consist of 527 and 520 amino acids respectively, showing a high degree of sequence similarity (~85%) and identity (~72%) (Bach et al., 1988). In both MAO forms, the co-factor binding site appears to be constituted by at least two regions: i) the NH_2-terminal domain (residues Asp15 to Glu43 and Asp6 to Glu34 for MAO-A and -B, respectively) constituting the non-covalent FAD co-factor binding site which contains a sequence fingerprint characteristic of dinucleotide-binding site $\beta\alpha\beta$ supersecondary structure (Wierenga et al., 1986), and a

---

[†] Mosé Da Prada died on April 25, 1995

highly conserved tyrosine residue found in several dinucleotide-dependent enzymes (Zhou et al., 1995), and ii) a region containing a cysteine residue (Cys406 and Cys397 in MAO-A and B, respectively) which covalently anchors FAD through a 8α-methyl-S-cysteinyl bond (Kearney et al., 1971). In both isoenzymes, the mitochondrial targeting peptide appears to be located at the COOH-terminus (Mitoma and Ito, 1992; Weyler, 1994), although the precise mode of association with the outer mitochondrial membrane remains to be clarified.

Investigation on the topology of the active site of MAOs and on residues involved in the catalytic mechanism has greatly relied on the use of substrates and inhibitors (see e.g. Dostert and Strolin-Benedetti, 1991; Silverman, 1991; Cesura and Pletscher, 1992; Efange et al., 1993). However, data on the structural features of MAO-A and MAO-B proteins are still limited due to the fact that no information on the three-dimensional structure of MAOs or other FAD-containing amine oxidases is available so far, and, except for the ADP-binding site, MAOs share no sequence similarity with other proteins of known structure. Therefore, attempts to construct molecular models for MAO structure are at present of limited value. In lieu of three-dimensional structure analysis, indirect approaches can be used to determine the structure-function relationships of enzymes and a number of recent studies using mutant and chimaeric forms of MAOs are providing useful information for unraveling the structural features of the two enzymes forms (Gottowik et al., 1993, 1995; Wu et al., 1993; Kwan et al., 1995; Zhou et al., 1995a; Cesura et al., 1996) as well as the mechanism of incorporation of the co-factor into MAOs (Zhou et al., 1995b). In the present paper, we summarise some of our findings on the structure-function relationships of chimaeric and mutant forms of MAOs.

Materials and methods

Engineering of MAO-A/MAO-B chimaeras and of MAO-B mutations

A 1,670 bp MAO-B EcoRI/HindIII fragment and a 1,680 bp MAO-A fragment were separately subcloned into the vector pCMV (Bertocci et al., 1991). Construction of the chimaera was performed by recombinant polymerase chain reaction (PCR) using overlapping primers according to the procedure described by Horton et al. (1989). Fragments from the MAO-A and -B genes which were to be recombined were generated in separate PCRs. The primers were designed to be complementary at the joining regions and were chosen at regions of the MAO-A and -B gene coding for the same amino acids. The PCR products were separated from excess primers by agarose gel electrophoreses, followed by phenol-chloroform extraction, and subsequently mixed, denatured, and eventually reannealed. The strands having the matching sequences at their 3'-ends overlapped and acted as primers for each other. Primer extension and amplification by PCR of this overlap with Pfu DNA polymerase (Stratagene) therefore produced a molecule in which the original sequences are joined together to form chimaeric cDNAs. For convenient subcloning, EcoRI and XbaI restriction sites were added by PCR at the fragment 5'- and 3'-ends, respectively. For PCR, the DNA was denatured at 95°C for 30 s, annealed at 42–60°C (depending on the primer used) for 30 s and amplified at 75°C for 1.5 min for 20 cycles. A final extension cycle of 10 min was done at 75°C. The recombinant DNAs were digested with EcoRI/XbaI according to the restriction sites introduced by PCR, gel purified and finally subcloned into the vector pCMV for expression in the mammalian cells and for DNA sequencing (Bertocci et al., 1991).

Site-directed mutagenesis of MAO-B was performed by PCR with overlapping mismatch primers according to the procedure used by Higuchi et al. (1988) (see also Gottowik et al., 1993; Cesura et al., 1996).

The sequence of the chimaeric and mutant constructs was confirmed by single-stranded dideoxy-DNA sequence analysis (Sanger et al., 1980) of the entire PCR fragment.

Expression of constructs in HEK-293 cells

For transfection, HEK-293 cells (human embryonic kidney cells, ATCC CRL 1573) were seeded at a concentration of 2.5×10^5 cells/ml and grown for 24 h at 37°C under 5% CO_2, in MEM+ (Gibco-BRL) with the addition of 20 mM HEPES, 10% fetal calf serum, 2 mM L-glutamine, 100 IU/ml penicillin, 100 µg/ml streptomycin. The subconfluent cells were then washed with DMEM (Gibco-BRL) and incubated in the same medium with a mixture of 1 µg of expression vector and 5 µg of Transfectam® (IBF Biotechnics) per ml, for 2 h at 37°C under 5% CO_2 (Löffler and Feltz, 1990; Gottowik et al., 1993). The cells were grown for another 48 h in MEM+ before collection. Transfection was performed with the wild-type MAO-A and B constructs, both in sense and antisense orientation (mock-transfection) to the promoter, and with the chimaeric and mutant constructs described above. For assay of enzyme activity, the transfected HEK-293 cells were homogenized in 20 mM Tris HCl buffer, pH 8.0, containing 0.5 mM EGTA and 0.5 mM phenylmethansulfonyl fluoride. Enzymatic activity was determined by a radiochemical method (Da Prada et al., 1989) by incubating aliquots of the homogenized cells (5–50 µl) in the presence of the radiolabeled substrates. For kinetic analysis, the cell homogenates were incubated in the presence of at least six different substrate concentrations. Kinetic constants were calculated using the Enzyme Kinetics (Trinity Software) or Ultrafit (Biosoft) computer programs. The data obtained by non-linear regression analysis of the experimental data were used.

Turnover numbers (k_{cat}) were calculated from the obtained V_{max} values divided by the concentrations of wild-type or mutant MAO forms expressed in the corresponding batch of transfected cells. The amounts of wild-type MAOs and mutant forms transiently expressed in HEK-293 cells were either directly measured by [^{3}H]RO 41-1049 (MAO-A) and [^{3}H]lazabemide (MAO-B) binding (Cesura et al., 1989, 1990) and/or by immuno blot analysis using monoclonal antibodies specific against MAO-A and MAO-B as previously described (Gottowik et al., 1993, 1995).

Protein content was determined by the Pierce BCA bicinchoninic acid method.

Results

Enzymatic activity of the MAO chimaeric proteins

HEK-293 cells were chosen for transient transfection because a relatively high level of protein expression can be achieved in these cells and they contain low levels of endogenous MAO activity which is exclusively of the A-type (>0.5 pmol mg protein^{-1}, see Gottowik et al., 1993). Mock-transfection with MAO-A and -B antisense constructs did not alter the enzyme activity expressed by these cells.

For assessing which chimaeric protein showed significant enzymatic activity over background activity (due to the presence of endogenous MAO-A),

Table 1. Summary of enzymatic activity displayed by the various constructed chimaeric MAO forms towards the substrates used

MAO form	Substrate		
	5-HT	PEA	TYR
MAO-B WT	−	+	+
MAO(A45)-B	−	+	+
MAO(A70)-B	−	+	+
MAO(A112)-B	−	+	−
MAO(A154)-B	−	+	−
MAO(A229)-B	−	+	+
MAO(A256)-B	−	+	+
MAO(A316)-B	−	−	−
MAO(A394)-B	+	−	+
MAO(A440)-B	+	−	+
MAO-A WT	+	+	+
MAO(B36)-A	+	+	+
MAO(B61)-A	+	+	+
MAO(B103)-A	+	+	+
MAO(B145)-A	−	−	−
MAO(B220)-A	−	−	−
MAO(B247)-A	−	−	−
MAO(B307)-A	−	−	−
MAO(B385)-A	−	−	−
MAO(B431)-A	−	−	−

5-HT 5-hydroxy-tryptamine, *PEA* phenylethylamine, *TYR* tyramine

and was therefore suitable for reliable analysis of enzyme kinetics, the MAO activity of the transfected cells was first determined using a single concentration of the substrates, i.e 5-HT, phenylethylamine (preferential substrates for MAO-A and MAO-B, respectively) and tyramine (nonselective). The enzymatic activity pattern displayed by the various contructed chimaeric MAO forms is schematically summarised in Table 1.

Kinetic properties of MAO chimaeric forms

Table 2 shows the kinetic constants of the wild-type and of the active chimaeric forms of MAO using phenylethylamine as substrate. Among the enzyme chimaeras with the NH_2-terminus of MAO-A and the COOH-terminal sequence of MAO-B, MAO(A45)-B showed virtually identical enzymatic properties as the wild-type MAO-B, whereas MAO(A70)-B showed a ~2-fold increase in the K_m value. Conversely, a marked and progressive drop in phenylethylamine affinity was observed for the other chimaeras of this series, with increases in K_m values ranging from ~25-fold for MAO(A112)-B and MAO(A154)-B, to ~40-fold for MAO(A229)-B and MAO(A256)-B

Table 2. Kinetic constants of phenylethylamine deamination and inhibition by lazabemide of wild-type and chimaeric forms of MAO

MAO form	K_m (μM)	k_{cat} (s^{-1})	Lazabemide IC_{50} (M)
MAO-B wild-type	2.4 ± 0.3	2.57 ± 0.26	8.9 ± 2.2 × 10^{-9}
MAO(A45)-B	1.6 ± 0.6	2.78 ± 0.27	6.7 ± 1.0 × 10^{-9}
MAO(A70)-B	4.5 ± 0.4	2.45 ± 0.16	2.7 ± 0.4 × 10^{-8}
MAO(A112)-B	56.2 ± 10.0	2.93 ± 0.61	8.0 ± 2.0 × 10^{-7}
MAO(A154)-B	36.1 ± 4.8	2.20 ± 0.22	1.4 ± 0.2 × 10^{-6}
MAO(A229)-B	74.0 ± 6.5	5.09 ± 0.50	5.2 ± 0.4 × 10^{-5}
MAO(A256)-B	96.2 ± 13.4	5.09 ± 1.38	6.8 ± 0.8 × 10^{-5}
MAO-A wild-type	154.4 ± 12.3	3.07 ± 0.37	1.2 ± 0.4 × 10^{-4}
MAO(B61)-A	182.0 ± 38.2	3.28 ± 0.33	1.1 ± 0.2 × 10^{-4}
MAO(B103)-A	127.6 ± 8.3	3.91 ± 0.20	7.1 ± 1.0 × 10^{-5}

Results are the mean ± SEM of 3–5 experiments performed with cells obtained from three separate transfections

(Table 2). No major differences in the turnover number (k_{cat} values) were observed, with MAO(A229)-B and MAO(A256)-B showing only a ~2-fold increase in k_{cat} when compared to wild-type MAO-B. As inferred from the ratios between the apparent second-order rate constants (k_{cat}/K_m) of the chimaeras and that of wild-type MAO-B, the decrease in catalytic efficiency of these mutant enzymes, was mainly due to a decreased affinity of the substrate for the chimaeric enzymes. Regarding the enzyme forms with the NH_2-terminus of MAO-B and COOH-terminus of MAO-A, the kinetic constants of MAO(B61)-A and MAO(B103-A) towards phenylethylamine were very similar to that observed for wild-type MAO-A (Table 2).

The IC_{50} values of the selective MAO-B inhibitor lazabemide (Cesura et al., 1989, 1996; Da Prada et al., 1990) towards the chimaeric enzymes using phenylethylamine as substrate are shown in Table 2. Among the chimaeras with the COOH-terminus of MAO-B, the inhibitory potency of lazabemide towards MAO(A45)-B and MAO(A70)-B was similar to that observed for wild-type MAO-B, with only a ≈2-fold increase in IC_{50} value for the latter chimaera. As observed with the substrate phenylethylamine (see Table 1), further substitution of the NH_2-terminus of MAO-B with the corresponding sequence of MAO-A caused a marked and progressive drop in the lazabemide inhibitory potency, until, in MAO(A229)-B and MAO(A256)-B, the IC_{50} values approached that displayed by lazabemide for inhibiting wild-type MAO-A. When the selective MAO-A inhibitor Ro 41-1049 (Da Prada et al., 1990; Cesura et al., 1990) was used, the observed IC_{50} values for all these chimaeras were in the mid μM range, and did not differ from the IC_{50} value of Ro 41-1049 for wild-type MAO-B (~50μM). The potency of lazabemide in inhibiting MAO(B61)-A and MAO(B103)-A was identical to that observed for wild-type MAO-A (Table 2).

Compared to wild-type MAO-A, the four constructed chimaeras which displayed activity using 5-HT as substrate revealed minor, if any, differences

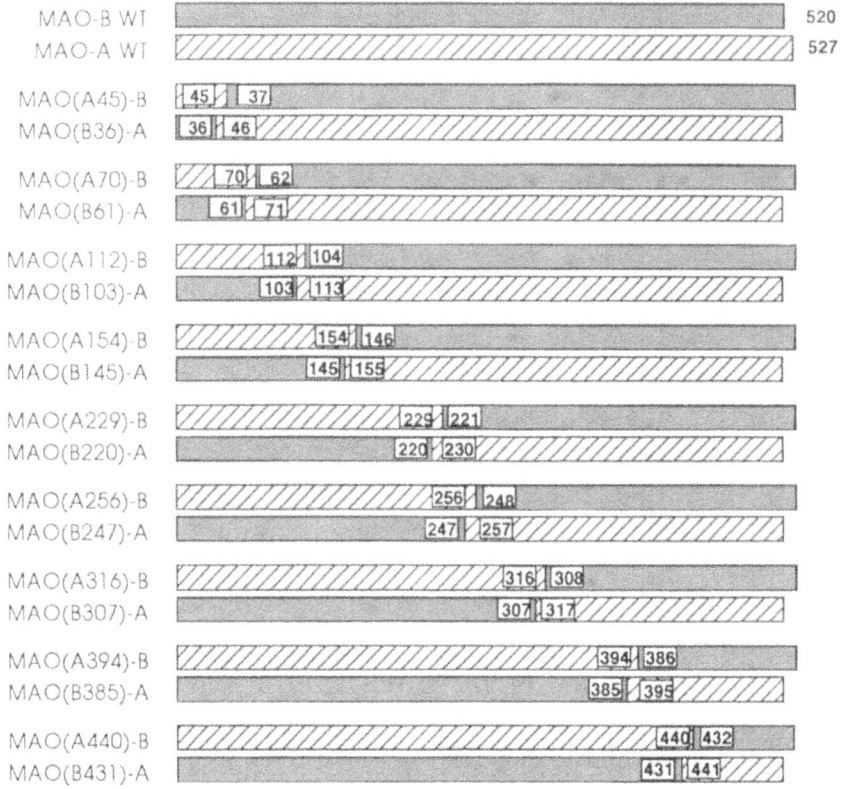

Fig. 1. Diagram of MAO-A and MAO-B and of their engineered chimaeras. The wild-type and the chimaeric enzymes are schematically summarized: solid and hatched bars correspond to sequences of MAO-A and MAO-B, respectively. The numbers shown in the bars are the amino acid positions where the sequences of the isoenzymes were joined together

in their catalytic properties towards 5-HT (Table 3). Regarding the inhibition of 5-HT deamination by Ro 41-1049, no significant differences in comparison to wild-type MAO-A were observed in the IC_{50} values for these chimaeric enzymes (Table 3). The same was true when lazabemide was used, which displayed inhibitory potencies in the high μM range (IC_{50}, ~100 μM).

Site-directed mutagenesis of MAO-B residues

A number of mutant MAO-B forms carrying single amino acid mutation were generated and their enzymatic properties determined after transient transfection of the corresponding cDNA into HEK-293 cells. All the constructed mutants were found to be expressed to a similar level in the range of 5–7 pmol mg protein^{-1}.

Substitution of Cys-397 of MAO-B, the residue covalently anchoring FAD through a thioether bridge, with a neutral amino acid, i.e. Ala ([Cys397 → Ala]MAO-B), resulted in the expression of inactive MAO-B (Table 4). We

Table 3. Kinetic constants of 5-HT deamination and inhibition by Ro 41-1049 of wild-type and chimeric forms of MAO

MAO form	K_m (μM)	k_{cat} (s^{-1})	Ro 41-1049 IC_{50} (M)
MAO-A wild-type	108.0 ± 23.8	11.6 ± 1.0	$3.3 \pm 0.3 \times 10^{-8}$
MAO(A394)-B	151.0 ± 7.6	8.1 ± 0.3	$3.7 \pm 0.6 \times 10^{-8}$
MAO(A440)-B	161.6 ± 19.4	10.1 ± 0.7	$3.5 \pm 0.7 \times 10^{-8}$
MAO(B36)-A	80.3 ± 10.9	11.9 ± 1.1	$3.3 \pm 0.3 \times 10^{-8}$
MAO(B61)-A	165.6 ± 32.6	9.7 ± 1.4	$2.6 \pm 0.8 \times 10^{-8}$
MAO(B103)-A	147.0 ± 3.2	5.3 ± 0.8	$4.8 \pm 1.0 \times 10^{-8}$

Results are the mean $\pm$ SEM of 3–4 experiments performed in cells from three separate transfections

Table 4. Kinetic parameters of phenylethylamine deamination by wild-type and mutant forms of MAO-B

Mutant	K_m (μM)	k_{cat} (sec^{-1})
MAO-B wild-type	2.45 ± 0.11	4.06 ± 0.46
[Cys397 $\rightarrow$ Ala]MAO-B	No activity	
[Cys397 $\rightarrow$ His]MAO-B	No activity	
[Cys365 $\rightarrow$ Ala]MAO-B	2.94 ± 0.45	0.73 ± 0.25
[His382 $\rightarrow$ Arg]MAO-B	2.49 ± 0.08	0.06 ± 0.02
[Lys386 $\rightarrow$ Met]MAO-B	2.47 ± 0.10	5.76 ± 1.14
[Cys389 $\rightarrow$ Ala]MAO-B	No activity	
[Ser394 $\rightarrow$ Ala]MAO-B	1.99 ± 0.20	1.68 ± 0.52
[Thr158 $\rightarrow$ Ala]MAO-B	2.76 ± 0.08	0.14 ± 0.04

Results are the mean $\pm$ SEM of 3–4 experiments performed in cells from three separate transfections

also investigated the effect of substituting the flavin-modified cysteine of MAO-B with histidine ([Cys397 $\rightarrow$ His]MAO-B). Also in the case of this mutant the expressed protein was catalytically inactive. These results suggests that covalent FAD linkage with the apoprotein is required for enzymatic activity (see also Zhou et al., 1995b). In addition, the strict requirement for a cysteine is stressed by the fact that this residue cannot be replaced by a histidine, the residue to which FAD is found to be covalently linked in other flavoproteins.

We recently found that the pseudosubstrate MAO-B inhibitor lazabemide is incoproprated, after $NaBH_3CN$ reduction of the inhibitor-enzyme complex (Cesura et al., 1989), into an amino acid of the enzyme located in the peptide region containing the Cys residue covalently anchoring FAD (Cesura et al., 1996). Although the precise incorporation site of lazabemide could not be determined, it can be postulated that this region contains, in addition to the FAD-modified Cys, at least another amino acid relevant for catalysis, i.e. the one incorporating lazabemide. We therefore analysed the function of putative reactive amino acids (His382, Lys386, Cys389, Ser394) present in this peptide by site-directed mutagenesis. The enzymatic activity of the MAO-B variant carrying an Arg residue instead of His at position 382 ([His382 → Arg]MAO-B) was much lower than that determined for the wild-type enzyme. Since the imidazole side chain of this amino acid is known to have properties that make it a very effective nucleophile, it conceivable that His382 may play a relevant role in MAO-B catalytic mechanism (Cesura et al., 1996). Interestingly, whereas substitution of Cys389 with a Ser has been reported to have a minor effect on the enzyme activity (Wu et al., 1993), the MAO-B mutant with an Ala at this position ([Cys389 → Ala]MAO-B) was found to be completely inactive. In accordance to what previously reported (Wu et al., 1993), substitution of Cys365 with an Ala residue ([Cys365 → Ala]MAO-B) greatly reduced the enzymatic activity. On the other hand, no major changes in catalytic activity were observed for the other mutant forms produced, i.e. [Lys386 → Met]MAO-B and [Ser394 → Ala]MAO-B.

During construction of the various MAO-B mutants, the random mutation Thr158 to Ala was also generated, likely as a result of a PCR artifact. Subsequent expression of the [Thr158 → Ala]MAO-B mutant showed that the corresponding protein displayed very low enzymatic activity (Table 4). This serendipitous finding appears to be of interest. In fact this residue is located within a sequence whose relevance for MAO-B activity is suggested by the presence of the essential Cys156 (Wu et al., 1993). Although this Cys residue is conserved in both MAO isoenzymes, it appears to be essential for enzymatic activity exclusively in the case of MAO-B, but not of MAO-A (Wu et al., 1993). In addition, the studies on the structure-function relationships of MAO chimaeric forms suggest that the MAO-B sequence containing these residues could be part of the binding site of the enzyme (see above and Gottowik et al., 1995). Although the precise role of Thr158 deserves further investigation, studies on the function of the corresponding threonine in MAO-A may be instrumental in possibly identifying a protein moiety which may play a role in the different catalytic functions of the two MAO isoforms.

With the exception of the enzyme mutated in the FAD-anchoring residue and of [Cys389 → Ala]MAO-B, for which no activity could be detected, the K_m values for phenylethylamine of the other MAO-B mutants produced appeared to be similar to that of wild-type MAO-B. This suggests that the mutations introduced did not affect the binding of the substrate.

Discussion

The purpose of the reported experiments was to identify amino acid sequences and residues of MAO-A and MAO-B which could belong to the recognition/active site of the two enzymes. The approach used was the construction of several MAO chimaeric forms produced by progressively moving the junction between the NH_2-terminus of one MAO isoform with COOH-terminal sequence of the other enzyme subtype, as well as the introduction of single amino acid mutations.

The results obtained with the chimaeric MAO forms indicate that the various substitutions in amino acid sequences introduced in MAO-A and MAO-B differentially affected the enzymatic activity of the two proteins and that none of the functionally expressed chimaeras showed a clear shift in their pharmacological specificity from B- to A-like MAO activity or vice versa. Whereas exchange of the NH_2-terminal region comprising the noncovalent FAD binding domain (ADP-binding site) of MAO-B with that of MAO-A did not result in any significant changes in catalytic properties in comparison to the wild-type activity (Gottowik et al., 1993), further substitution of the NH_2-terminal part of MAO-B with the corresponding MAO-A sequence caused a progressive and marked decrease in the affinity of MAO-B selective substrates and inhibitors. Despite the large drop in substrate K_m values, all functional chimaeras in this series maintained a substrate turnover number similar to that of wild-type MAO-B, suggesting that the alteration introduced in these artificial proteins affected mainly the structural features of the substrate binding site, rather than those determining its catalytic mechanism. The dramatic decrease in affinity for phenylethylamine and for the inhibitor lazabemide observed for chimaera MAO(A112)-B in comparison to MAO(A70)-B (see Table 2) suggests that the MAO-B amino acid stretch spanning residues 62 to 103 may contain or be part of a domain conferring MAO-B specificity. The fact that the chimaeras MAO(A112)-B and MAO(A154)-B did not show any significant activity towards tyramine (see Table 1), a nonselective substrate with considerably lower affinity for MAO-B than phenylethylamine, further supports the involvement of the above mentioned region in contributing to the substrate recognition domain of MAO-B.

We have also obtained evidence for the existence of another MAO-B sequence, encompassing amino acids 146 to 220, which may be more relevant for interaction with the inhibitors used than with substrates. In fact, substitution of this region had a greater effect in decreasing inhibitor potency rather than substrate affinity. It can be therefore speculated that this sequence may constitute a second region to which compounds with high affinity for MAO-B may bind, thereby strengthening their interaction with the enzyme. The progressive drop in the affinity of phenylethylamine and lazabemide for this chimaera series was not accompanied by a shift towards A-like activity and these chimaeras (up to MAO(A256)-B) can, therefore, be viewed as MAO forms with lower catalytic efficiency, containing a binding/active site lacking the structural features necessary for binding with high affinity MAO-B substrate and inhibitors.

A quite different picture emerges from the chimaeras constructed by progressively substituting the NH_2-terminus of MAO-A. In fact, only two of them appeared to be functionally expressed, i.e. MAO(B61)-A and MAO(B103)-A, and their catalytic behaviour was very similar to that of wild-type MAO-A. Therefore, MAO-A appears to be relatively insensitive to replacement of a substantial portion of its sequence both at its NH_2- (up to position 112) and at its COOH-terminal sequence (from residue 395). In contrast, further modification in the central 283 amino acid sequence of

MAO-A did not appear to be compatible with the expression of functional proteins, therefore precluding the assignment of MAO-A region(s) conferring affinity towards substrate and inhibitors.

Accordingly to previous biochemical studies (for review see Weyler et al., 1991), the data presented here support the view that, despite their high degree of amino acid similarity, MAO-A and MAO-B may have quite distinct structural features. Investigation on the secondary structure of human MAO-A and bovine MAO-B by infrared spectroscopy (Fourier transform attenuated total reflection spectroscopy) have indicated that the two proteins may have a different folding pattern (Wouters et al., 1995). In addition, by applying the results obtained with this technique to secondary structure prediction using the statistical GOR method, it has been proposed that MAO-A and MAO-B fold differently in the region spanning residues 100 to 200 (Wouters et al., 1995). Although secondary structure predictions have to be cautiously considered, these results would, nevertheless, add weight to our finding on the relevance of the MAO-B sequences 62–103 and 146–220 as part of binding/active site of the enzyme. Mutation of selected amino acids in these sequences may provide a deeper insight in the structural requirements of this form of MAO. Yet, precise elucidation of the structure of MAOs will ultimately rely on the availability of enzyme crystals suitable for X-ray crystallography, a difficult task for membrane proteins. Alternatively, knowledge on the three-dimensional structure of homologous amine oxidases should help in attempting reliable molecular modeling of MAOs.

References

Bach AJW, Lan NC, Johnson DL, Abell CW, Bembenek ME, Kwan SW, Seeburg P, Shih JC (1988) cDNA cloning of human liver monoamine oxidase A and B: molecular basis of differences in enzymatic properties. Proc Natl Acad Sci USA 85: 4934–4938

Bertocci B, Miggiano V, Da Prada M, Denbic Z, Lahm H-W, Malherbe P (1991) Human catechol-O-methyltransferase: cloning and expression of the membrane-associated form. Proc Natl Acad Sci USA 88: 1416–1420

Cesura AM, Pletscher A (1992) The new generation of monoamine oxidase inhibitors. Prog Drug Res 38: 171–297

Cesura AM, Galva MD, Imhof R, Kyburz E, Picotti GB, Da Prada M (1989) [^{3}H]Ro 19-6327: a reversible ligand and affinity labelling probe for monoamine oxidase-A. Eur J Pharmacol 162: 457–465

Cesura AM, Bös M, Galva MD, Imhof R, Da Prada M (1990) Characterisation of the binding of [^{3}H]Ro 41-1049 to the active site of human monoamine oxidase-A. Mol Pharmacol 37: 358–366

Cesura AM, Gottowik J, Lahm H-W, Lang G, Imhof R, Malherbe P, Röthlisberger U, Da Prada M (1996) Investigation on the structure of the active site of monoamine oxidase-B by affinity labeling with the selective inhibitor lazabemide and by site-directed mutagenesis. Eur J Biochem 236: 996–1002

Da Prada M, Kettler R, Keller HH, Burkard WP, Muggli-Maniglio D, Haefely WE (1989) Neurochemical profile of moclobemide, a short-acting and reversible inhibitor of monoamine oxidase type A. J Pharmacol Exp Ther 248: 400–414

Da Prada M, Kettler R, Keller HH, Cesura AM, Richards JG, Saura Marti J, Muggli-Maniglio D, Wyss P-C, Kyburz E, Imhof R (1990) From moclobemide to Ro 19-6327 and Ro 41-1049: the development of a new class of reversible, selective MAO-A and MAO-B inhibitors. J Neural Transm [Suppl] 29: 279–292

Dostert P, Strolin-Benedetti M (1991) Structure-modulated recognition of substrates and inhibitors by monoamine oxidase A and B. Biochem Soc Trans 19: 207–211

Efange SMN, Michelson RH, Tan AK, Krueger MJ, Singer TP (1993) Molecular size and flexibility as determinants of selectivity in the oxydation of N-methyl-4-phenyl-1,2,3,6-tetrahydropyridine analogs by monoamine oxidase A and B. J Med Chem 36: 1278–1283

Gottowik J, Cesura AM, Malherbe P, Lang G, Da Prada M (1993) Characterisation of wild-type and mutant forms of human monoamine oxidase A and B expressed in a mammalian cell line. FEBS Lett 317: 152–156

Gottowik J, Malherbe P, Lang G, Da Prada M, Cesura AM (1995) Structure/function relationships of mitochondrial monoamine oxidase A and B chimeric forms. Eur J Biochem 230: 934–942

Grimsby J, Chen K, Wang L-J, Lan NC, Shih JC (1991) Human monoamine oxidase A and B genes exhibit identical exon-intron organisation. Proc Natl Acad Sci USA 88: 3637–3641

Higuchi R, Krummel B, Saiki RK (1988) A general method of in vitro preparation and specific mutagenesis of DNA fragments: study of protein and DNA interactions. Nucl Acids Res 16: 7351–7367

Horton R, Hunt HD, Ho SN, Pullen JK, Pease LR (1989) Engineering hybrid genes without the use of restriction enzymes: gene splicing by overlap extension. Gene 77: 61–68

Kearney EB, Salach JI, Walker WH, Seng RL, Kenney W, Zeszotek E, Singer TP (1971) The covalently bound flavin of hepatic monoamine oxidase. 1. Isolation and sequence of a flavin peptide and evidence for binding in the 8α position. Eur J Biochem 24: 321–327

Kwan S-K, Lewis DA, Zhou BP, Abell CW (1995) Characterization of a dinucleotide-binding site in monoamine oxidase B by site-directed mutagenesis. Arch Biochem Biophys 316: 385–391

Löffler J, Feltz A (1990) Lipoplyamine-mediated transfection allows gene expression studies in primary neuronal cells. J Neurochem 54: 1812–1815

Mitoma J, Ito A (1992) Mitochondrial targeting signal of rat liver monoamine oxidase B is located at its carboxy terminus. J Biochem 111: 20–24

Sanger F, Coulson A, Barrell B, Smith A, Roe B (1980) Cloning in single-stranded bacteriophage as an aid to rapid DNA sequencing. J Mol Biol 143: 161–178

Silverman RB (1991) The use of mechanism-based inactivators to probe the mechanism of monoamine oxidase. Biochem Soc Trans 19: 201–206

Weyler W (1994) Functional expression of C-terminally truncated human monoamine oxidase type A in Saccharomyces cerevisiae. J Neural Transm [Suppl] 41: 3–15

Weyler W, Hsu Y-PP, Breakefield XO (1991) Biochemistry and genetics of monoamine oxidase. Pharmacol Ther 47: 391–417

Wierenga RK, Terpstra P, Hol WGJ (1986) Prediction of the occurrence of the ADP-binding βαβ-fold in proteins, using an amino acid sequence fingerprint. J Mol Biol 187: 101–107

Wu H-F, Chen K, Shih JC (1993) Site-directed mutagenesis of monoamine oxidase A and B: role of cysteines. Mol Pharmacol 43: 888–893

Wouters J, Ramsay R, Goormaghtigh E, Ruysschaert J-M, Brasseur R, Duranr F (1995) Secondary structure of monoamine oxidase by FTIR spectroscopy. Biochem Biophys Res Commun 208: 773–778

Zhou BP, Lewis DA, Kwan S-W, Kirskey TJ, Abell CW (1995a) Mutagenesis of a highly conserved tyrosine residue in monoamine oxidase B affects FAD incorporation and catalytic activity. Biochemistry 34: 9526–9531

Zhou BP, Lewis DA, Kwan S-W, Abell CW (1995b) Flavynation of monoamine oxidase B. J Biol Chem 270: 23653–23660

Authors' address: Dr. A. M. Cesura, Pharma Division, Preclinical Research (PRPN, 70/304), F. Hoffmann-La Roche Ltd, CH-4070 Basel, Switzerland

Deamination of methylamine and angiopathy; toxicity of formaldehyde, oxidative stress and relevance to protein glycoxidation in diabetes

P. H. Yu

Neuropsychiatry Research Unit, College of Medicine, University of Saskatchewan, Saskatoon, Saskatchewan, Canada

Summary. Semicarbazide-sensitive amine oxidase (SSAO) is located in the vascular smooth muscles, retina, kidney and the cartilage tissues, and it circulates in the blood. The enzyme activity has been found to be significantly increased in blood and tissues in diabetic patients and animals. Methylamine and aminoacetone are endogenous substrates for SSAO. The deaminated products are formaldehyde and methylglyoxal respectively, as well as H_2O_2 and ammonia, which are all potentially cytotoxic. Formaldehyde and methylglyoxal are cytotoxic towards endothelial cells. Excessive SSAO-mediated deamination may directly initiate endothelial injury and plaque formation, increase oxidative stress, which can potentiate oxidative glycation, and/or LDL oxidation and damage vascular systems. Formaldehyde is also capable of exacerbating advanced glycation, and thus increase the complexity of protein cross-linking. Uncontrolled SSAO-mediated deamination may be involved in the acceleration of the clinical complications in diabetes.

Abbreviations

AGEs Advanced glycation end products, *MDL-72974A* (E)-4-Fluoro-β-fluoroethylenebenzenebutamine, *LDH* Lactate dehydrogenase, *LDL* Low density lipoprotein, *MAO* Monoamine oxidase, *SSAO* Semicarbazide-sensitive amine oxidase, *STZ* Streptozotocin

Introduction

Accelerated atherosclerosis and subsequent increase in vascular risk probably cause various diabetic complications, such as retinopathy (Rand, 1991), nephropathy (D'Elia et al., 1985) and neuropathy (Thomas, 1991). Much effort has been devoted to understanding diabetic microvascular disease and its impact on diabetic complications. A number of hypotheses have been advanced regarding atherogenesis in diabetes, such as via endothelial injury by oxidative stress, nitric oxide, dyslipidemia, lipoprotein modifications, advanced glycation or factors affecting hormones, growth factors, cytokines, smooth muscle cell proliferation etc. (see Bierman, 1992). Recently, excessive SSAO-mediated deamination was proposed to be related to diabetic angiopathy. The production of formaldehyde, methylglyoxal, and H_2O_2 via deamination of methylamine and aminoacetone respectively by SSAO seem to be consistent with several of the existing hypotheses for diabetic complications.

Oxidative modification of low-density lipoprotein (LDL) and diabetic atherosclerosis

The polyunsaturated fatty acid component of arterial wall LDL is subject to free radical attack and produces oxidized LDL, which has been considered to be a crucial step in the development of atherosclerosis (Regnstrom et al., 1987). Oxidative LDL becomes harmful towards endothelial cells (Morel et al., 1983) and it can attract adhesion of circulating monocytes, which may lead to plaque formation. Diabetes has been considered to be a state of increased oxidative stress (a disturbance of balance between oxidative stress factors and antioxidant factors in favor of the former) (see review by Baynes, 1991). Increased levels of circulating lipid peroxides (Sato et al., 1979), conjugated dienes (Jennings et al., 1987), malonaldehyde (Bucala et al., 1993), or hydroxyl free radicals (Ohkuwa et al., 1995) were found to be associated with diabetic atherosclerosis. It is not well established, however, how oxidative stress is increased in diabetes.

Advanced oxidative glycation and the pathogenesis of diabetic atherosclerosis

Glucose reacts chemically with amino groups of amino acids or nucleic acids and subsequently forms Amidori products (Brownlee et al., 1988), which in turn rearrange into advanced glycation end products (AGEs) (Brownlee, 1992). This leads to marked changes in the structure and function of various proteins. The formation of AGEs from the Amidori product is also known to be via an oxidative process (Fu et al., 1992). It is accelerated in the presence of transition metals and reduced by ascorbate (Baynes, 1991; Hunt et al., 1993), suggesting that the reaction is mediated by free radicals (Hunt et al., 1990). There is accumulating evidence that the formation of AGEs increases with age (Moonier et al., 1984) and is implicated in the pathogenesis of diabetic atherosclerosis (Brownlee, 1992). AGEs can cross link with collagen and other structural proteins (Brownlee et al., 1988). They increase the rigidity of collagen, decrease arterial compliance and enhance hypertension (Lo et al., 1986; Uusitupa et al., 1993). Advanced glycated collagen is capable of covalently trapping LDL (Brownlee et al., 1985). LDL, once trapped in the arterial wall, is susceptible to attack by free radicals and subject to oxidative modification (Mullarkey et al., 1990). AGEs may also affect monocyte migration and induce an inflammatory response (Brett et al., 1990; Vlassara et al., 1990).

Aminoguanidine has been found to react with Amidori fragmentation products and thus prevent glycoxidation and AGEs formation (Edelstein and Brownlee, 1992). Aminoguanidine has been shown to prevent experimental diabetic nephropathy (Soulis-Liparota et al., 1991) and to inhibit lipid peroxidation in vivo (Bucala et al., 1993). It also inhibits the oxidative modification of LDL and its subsequent uptake by macrophages in a euglycemic in vitro system (Picard et al., 1992). The drug reduces the development of atherosclerotic plaque without altering serum cholesterol levels in cholesterol-fed rabbits (O'Brien et al., 1992). However aminoguanidine also inhibits nitric oxide synthase, and was claimed to be related to prevention of diabetic vascular dysfunction (Tilton et al., 1993).

Oxidative deamination of methylamine and aminoacetone leads to production of bioactive toxic aldehydes and H_2O_2

As indicated below methylamine (Precious et al., 1988; Yu, 1990; Boor et al., 1992) and aminoacetone (Lyles, 1995; Kalapos et al., 1992; Lyles and Chalmers, 1992) are readily deaminated by semicarbazide-sensitive amine oxidase (EC 1.4.3.6, SSAO) in vitro and in vivo. Formaldehyde and methylglyoxal respectively, and hydrogen peroxide and ammonia are produced.

$$SSAO$$

$$CH_3NH_2 + O_2 + H_2O \rightarrow HCHO + H_2O_2 + NH_3$$

$$CH_3CO\ CH_2NH_2 + O_2 + H_2O \rightarrow CH_3COCHO + H_2O_2 + NH_3$$

A large quantity of methylamine is excreted in human urine (Blau, 1961; Yu et al., unpublished), and is present in blood (Baba et al., 1984; Asatoor and Kerr, 1961) and tissues (Lyles and McDougall, 1989; Smith and Jepson, 1967; Nixon, 1972). It can be derived from several metabolic reactions, such as the deamination of adrenaline (Schayer et al., 1952), sarcosine and creatinine (Dar et al., 1985) and it can be ingested from food and drink or inhaled from cigarette smoke (US Dept Health and Human Services, 1982; Zeisel et al., 1983, 1986). When rats are treated with semicarbazide, an SSAO inhibitor, their urinary excretion of methylamine is substantially increased (Lyles and McDougall, 1989). We confirmed this result using the selective SSAO inhibitor MDL-72974A (unpublished). Although methylamine is endogenously present and subject to deamination by SSAO, the product formaldehyde is not easily detected in vivo, because it rapidly interacts with cellular constituents or is quickly metabolized. Tracing residual radioactivity after the administration of ^{14}C-methylamine in the presence or absence of specific SSAO inhibitors confirmed the conversion of methylamine to formaldehyde, and the irreversible interaction of formaldehyde with tissue components in vivo (Yu and Zuo, 1995).

Cytotoxicity of endogenous formaldehyde, methylglyoxal and H_2O_2

Formaldehyde is an extremely reactive chemical. It interacts with different cellular components such as cysteine to form thiazolidine-4-carboxylate and with the free amino groups or amides to form a methylene bridge. It can produce irreversible adducts with proteins and single strand DNA, and it induces covalent cross linkage to produce protein-protein or protein-DNA molecule complexes (Bolt, 1987). It is extremely cytotoxic and has been considered to be potentially carcinogenic; it is the subject of major environmental concern (Grafstrom et al., 1985; Gibson, 1983; WHO, 1989). Formaldehyde is normally metabolized to formic acid and is detoxified intra-cellularly, e.g. by liver cytosol NAD-dependent dehydrogenase in the presence of reduced glutathione (GSH) (Cooper and Kini, 1962; Tsuboi et al., 1992). Interestingly, serum does not contain formaldehyde dehydrogenase (Helander and Tottmar, 1987). Formaldehyde in blood therefore cannot be metabolized unless it is first transported into erythrocytes (Malorny et al., 1965). This is a very interesting finding with respect to formaldehyde-induced toxicity to blood vessels.

H_2O_2 is a major reactive oxygen species, which is generated in a number of biochemical reactions. In the presence of transition metals H_2O_2 can be converted to toxic hydroxyl free radicals via the Fenton reaction ($H_2O_2 + Fe^{++} \rightarrow {}^*OH + OH^- + Fe^{+++}$). It has been implicated in several diseases (Sies, 1991). Interestingly, free radicals can also be generated from formaldehyde in the presence of hydrogen peroxide (2 HCHO + $H_2O_2 \rightarrow$ 2 H-*C=O + 2 H_2O) (Trézl et al., 1992 and personal communication). It is intriguing that both formaldehyde and H_2O_2 are simultaneously formed from deamination of methylamine. It is reasonable to suggest that oxidative stress can be induced by this reaction (due to increased serum SSAO activity) in diabetes, and thus be involved in the oxidation of LDL and glycoxidation. Methylglyoxal is also a bioactive aldehyde. Its cytotoxicity has been previously reported (Egyud and Szent-Gyorgyi, 1968).

SSAO-catalyzed deamination and potential cytotoxicity

SSAO has been shown to be involved in the bioactivation of a cardiovascular toxin, allylamine, which is an industrial chemical causing extensive and progressive vascular and myocardial lesions in several mammalian species (Boor and Hysmith, 1987). The vascular damage induced by allylamine exhibits features very similar to those seen in human atherosclerosis. Allylamine (CH_2=$CHCH_2NH_2$) is converted to the toxic aldehyde acrolein (CH_2=CHCHO) by vascular SSAO in vitro and in vivo (Nelson and Boor, 1982). Acrolein, therefore, acts as a distal toxin of allylamine (Boor et al., 1987) responsible for cellular damage. The SSAO inhibitor semicarbazide protects experimental animals against damage caused by allylamine (Boor and Nelson, 1980).

Unlike allyamine, which is an industrial toxin, methylamine is present endogenously (i.e. derived from adrenaline, creatinine, sarcosine, lecithin, nicotine, etc.). Both methylamine and SSAO are present in circulating blood. It is conceivable that the deamination of methylamine can occur in blood. The products, formaldehyde and H_2O_2, if not detoxified, could become harmful to the blood vessels. Methylamine in the presence of SSAO is indeed toxic to human endothelial cells (Yu and Zuo, 1993) and forms patch-like lesions (Yu et al., unpublished). Our recent results have further shown that chronic administration of methylamine alters urinary excretion of prorenin, suggesting that the kidney may be damaged (Yu et al., 1996).

Repeated or chronic endothelial damage is required to initiate atherosclerosis (More, 1981; Ross, 1986). Any form of arterial injury (i.e. mechanical, chemical, or immunological processes) that is sufficient to cause endothelial cell death or desquamation, can evoke a series of complicated "wound-healing" responses in the vascular wall. If the injury is a single event, the lesions may be reversible and regress. If, in contrast, the injury is continuous, the lesions may become progressive. Experimental models have shown that homocysteinemia (Harker et al., 1976), endotoxinemia (Reidy and Bowyer, 1978), dietary-induced hyperlipidemia (Ross and Harker, 1976), oxidative stress (Tesfamariam and Cohen, 1992) and nicotine (Lin et al., 1992) may induce endothelial injury with subsequent intimal hyperplasia. Formaldehyde (which cross-links proteins) and H_2O_2 (which enhances oxidative stress) derived from methylamine deamination can play such a role to induce typical chronic stress triggering such repeated endothelial damage (Yu and Zuo, 1993) and may cause atherosclerosis. It is interesting to note that exposure to exogenous formaldehyde does not necessarily cause endothelial damage, since because it is so reactive, it interacts with surface components in the respiratory tract before it enters the blood stream. Inhaled formaldehyde primarily causes epithelial damage in the respiratory tract. The compartment in which endogenous formaldehyde is produced is very important. If SSAO-catalyzed deamination of methylamine occurs in the blood, the formaldehyde produced cannot be immediately detoxified. It is also interesting to note that bound SSAO was found to be facing out of the plasma membrane of the vascular smooth muscle (Callingham et al., 1995), suggesting that not only the circulating SSAO but also the tissue bound SSAO is capable of deaminating the circulating amines. SSAO is selectively distributed in the retina, kidney, and cartilage tissues, where there are often vulnerable sites for diabetic complications.

Deamination of methylamine and aminoacetone by SSAO is related to diabetic angiopathy

Deamination of methylamine by serum SSAO is cytotoxic to human endothelial cells in vitro (Yu and Zuo, 1993). Serum amine oxidase (i.e. SSAO) activity has been found to be increased in diabetic patients (Nilsson et al.,

1968; Tryding et al., 1969; Yuen et al., 1987; and recently confirmed by Boomsma et al., 1995). SSAO activity has also been found to be increased in the blood and kidney of diabetic rats (STZ-treated) (Hayes and Clarke, 1990) and diabetic sheep (alloxan-treated) (Elliott et al., 1991). Methylamine has been shown in an earlier study to be increased in the urine of diabetics (Kapeller-Adler and Toda, 1932). Unfortunately, in this latter study only three patients were assessed and the analytical method used was rather primitive. Our preliminary results indicate that urinary methylamine in diabetic patients with nephropathy is, in fact, drastically decreased (unpublished), suggests that either methylamine is metabolized or the patients were unable to clear methylamine. Blood methylamine levels in uremia patients have been found to be 20-fold higher than in control populations (Baba et al., 1984). This is probably due to the accumulation of creatinine, which converts to methylamine. Nephropathy is a major type of complication in diabetes and cardiovascular disorder is a common problem for patients exhibiting renal failure.

Aminoacetone can be converted to toxic methylglyoxal by SSAO (Kalapos et al., 1992; Lyles and Chalmers, 1992). It has been suggested that aminoacetone may be increased in nutritional deprivation. Interestingly, methylglyoxal is increased 2–4 fold in the blood and kidney of diabetics (McLellan et al., 1992). Methylglyoxal was observed to form adducts with albumin and the reaction is also blocked by aminoguanidine (Selwood and Thornalley, 1993). This has been implicated in the pathogenicity of diabetes mellitus (Thornalley, 1994). It remains, however, to be clarified whether methylglyoxal is primarily derived from glyceraldehyde 3-phosphate (Thornalley, 1993) or is from aminoacetone catabolized by SSAO.

Properties and localization of SSAO

SSAO is an enzyme or group of enzymes, residing predominantly in the plasma membrane of vascular smooth muscle cells, such as blood vessels (Wibo et al., 1980; Lewinsohn, 1981; Lyles and Singh, 1985), retina and brain microvessels (Zuo and Yu, 1994) and cartilage (Lyles and Bertie, 1987). SSAO is inhibited by hydrazines, such as semicarbazide (Lyles, 1984) and aminoguanidine (Yu and Zuo, 1996). The properties of SSAO's from different species vary considerably (Yu et al., 1993). The enzyme has not been purified and its molecular properties are largely unknown. It is probably a copper enzyme with 6-hydroxydopa as a prosthetic group (Jane et al., 1990). The findings of high SSAO activities selectively in cardiovascular smooth muscles and cartilage tissues suggests that the enzyme might play some role in these tissues. It may be involved in the deamination of circulating biogenic amines (Elliott et al., 1989). Histochemical studies revealed that most SSAO is located in the tunica media of the rat aorta (Lyles and Singh, 1985) and human placental blood vessels (Ryder et al., 1979).

SSAO and pathophysiology

The increase of serum SSAO activity in diabetes could be a result of genetic up-regulation, or possibly a consequence of initial vascular damage of diabetes or may be due to uncontrolled SSAO leakage into the blood stream from SSAO-rich tissues, such as the vascular smooth muscle cells or kidney. Excessive deamination of methylamine (and/or aminoacetone) in the blood will increase toxic aldehyde levels, enhance oxidative stress and cause vascular injury. Damaged vascular or kidney tissues would cause more SSAO leakage and this may form a vicious cytotoxic cycle. Aminoguanidine prevents the formation of AGEs and reduces diabetic albuminuria and nephropathy (Soulis-Liparota et al., 1991). We have recently found that aminoguanidine is a potent SSAO inhibitor both in vitro and in vivo (Yu and Zuo, 1996). MDL-72974A (a selective SSAO inhibitor, but not a hydrazide) significantly reduces urinary LDH excretion (a nephropathy indicator) in the STZ-induced diabetic rats (Yu and Zuo, 1996). It suggests that SSAO-mediated deamination is perhaps also related to the advanced glycation. Clearly, more research is required in order to substantiate this hypothesis.

Endogenous formaldehyde and chronic pain

Pain is one of the most common complications of arthritis or diabetes (Courteix et al., 1993). It is possible that such pain may also be related to the increased production of formaldehyde via deamination of methylamine by SSAO, which is known to be rich in the cartilage tissues (Lyles and Bertie, 1987). Formaldehyde is well-known to cause severe pain and has been used to produce chronic pain in animal research. This method is now banned for ethical reasons.

The relationship of cigarette smoking and chronic stress to SSAO-mediated deamination of methylamine

Smoking has been established to be a major risk factor for coronary heart disease (Wilens and Plair, 1962; Sackett and Winkelstein, 1967) and atherosclerosis (Auerbach et al., 1965; Strong and Richards, 1976) and may exacerbate the degenerative complications associated with diabetes (Pirart, 1978). The effects of cigarette smoking on insulin action (Gerich, 1988) and glucoregulation (Epifano et al., 1992) have been proposed. Deamination of methylamine by SSAO may perhaps also play an important role in smoking induced vascular damage. Both formaldehyde and methylamine are major components of cigarette smoke (US Dept Health and Human Services, 1982). The inhaled formaldehyde, due to its reactivity, will probably not penetrate very far. Methylamine derived from cigarettes could be metabolized by serum SSAO and become cytotoxic in the blood vessels. Smoking is also capable of stimulating the sympatho-adrenal system (Cryer et al., 1976) and increase adrenaline release, which in turn could be metabolized to methylamine by MAO (Yu, 1986). Methylamine is, in fact, a major metabolic end product of nicotine (McKennis et al., 1962). Long lasting residual radioactivity was detected in the tissues of rats following injection of N-methyl-^{3}H-nicotine. Such adduct formation can also be blocked by specific SSAO inhibitors (Yu et al., in preparation). These observations are consistent with the hypothesis, namely, increased deamination of methylamine could be a stress factor for endothelial damage and atherosclerosis. It explains why cigarette smoking worsens diabetic complications (Pirart, 1978).

Cardiovascular and cerebrovascular disorders are well known to be associated with stress related behaviors. Under such physiological conditions enhanced excretion of adrenaline has been detected. Adrenaline is deaminated by MAO and methylamine is formed. This product can be further deaminated in the blood by SSAO and thus become harmful to the blood vessels (Yu and Zuo, submitted).

Chronic cytotoxic effect of SSAO-mediated deamination in rats

Although deamination of methylamine has been shown to be quite toxic towards the cultured endothelial cells (Yu and Zuo, 1993), whether or not methylamine indeed causes cytotoxicity in vivo is unclear. Recently, we have demonstrated that chronic administration of methylamine (4 mg/mL) via drinking water caused a significant increase of serum prorenin levels in rats three months after treatment (Yu and Zuo, 1996). High levels of plasma prorenin have been considered to be a marker of microvascular complications of diabetes (Halimi and Sealey, 1992; Frankenhaeuser, 1971; Luetscher et al., 1989). Kidneys are the main source of prorenin in the circulation. The significant increase of serum prorenin in the methylamine treated group may reflect microvascular damage related to its deamination, since methylamine is not so toxic unless in the presence of SSAO (Yu and Zuo, 1996).

Identification of the formaldehyde-protein adducts following administration of ^{14}C-methylamine

Following administration of ^{14}C-methylamine long lasting residual radioactive adducts are formed in the rat tissues (Yu and Zuo, 1996). Such reactions could be blocked by selective SSAO inhibitors. The protein adducts, such as from the kidney tissues, possess huge molecular weights (i.e. eluted in the void volume of Sephadex G-100 column), suggesting that these adducts are cross-linked protein complexes. When these labeled adducts in the kidney extracts were incubated with proteinase K, small proteolytic fragments of different molecular sizes were eluted in later fractions. The radioactive adducts are therefore proteins. It is also possible that formaldehyde in vivo may be converted to formate, which then enters the one-carbon metabolic poor, eventu-

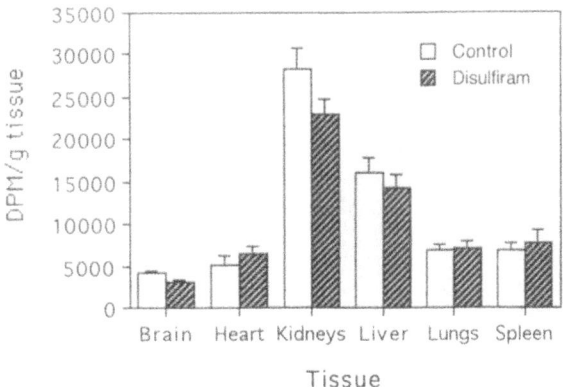

Fig. 1. Effect of alcohol dehydrogenase inhibitor on the radioactive labeled residual formation after intravenous administration of ^{14}C-methylamine. Disulfiram (200 mg/Kg) was orally administered daily for three days to the mice. ^{14}C-Methylamine (1 μCi/mouse) was injected intravenously one h after the last disulfiram treatment. The residual radioactivity in different tissues was analyzed 72 h after administration of ^{14}C-methylamine

ally metabolizing to amino acids (i.e. glycine), which are then incorporated into proteins. The radioactively labeled sites of these protein adducts have not yet been identified. Preliminary investigation showed that disulfiram, an alcohol dehydrogenase inhibitor, has very little effect on the residual formation after administration of ^{14}C-methylamine (Fig. 1).

In order to conclude whether or not SSAO-mediated production of formaldehyde is indeed involved in the adducts formation in vivo, it is necessary to detect N-hydroxymethyl-lysyl (derived from lysine-formaldehyde adduct) or thiazolidine-4-carboxylate (derived from cysteine-formaldehyde adduct) residues, etc. of the protein adducts.

Effect of formaldehyde, methylglyoxal and H_2O_2 on advanced glycation and AGEs formation in vitro

Nonenzymatic glycation involves the condensation of a sugar aldehyde or ketone with a free amino group by nucleophilic addition, resulting in a Schiff base, with subsequent rearrangement to the more stable Amadori products, which will be further degraded into a variety of highly reactive carbonyl compounds, such as 3-deoxyglucosone and the sugar fragmentation products to form a variety of intermediate and advanced glycation products (Brownlee, 1990), which can be determined by fluorescence. Formaldehyde causes protein crosslinking and H_2O_2 is involved in the AGE formation (Hunt et al., 1990). Both compounds are deaminated products of SSAO-mediated reactions. Recently, we found that both formaldehyde and methylglyoxal are unexpectedly capable of exacerbating glycation in vitro (see Fig. 2). Although it was reported that methylglyoxal by itself forms fluorescent adducts with albumin in vitro and this reaction can be blocked by aminoguanidine

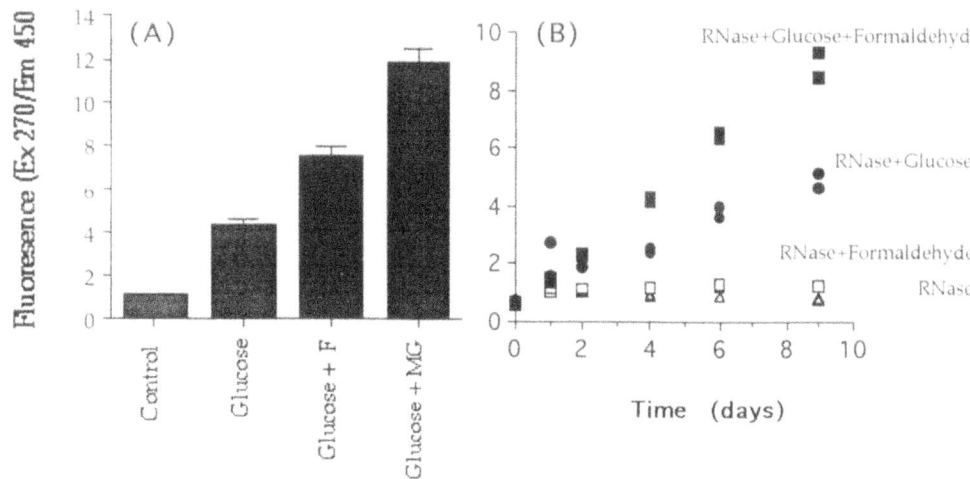

Fig. 2. Effect of formaldehyde and methylglyoxal on advanced glycation of RNase in vitro (Yu et al., unpublished). **A** RNase (20 mg/mL) was incubated with glucose (0.5 M) alone, and in the presence of glucose and 0.005% formaldehyde (F) and 0.005% methylglyoxal (MG) for 7 days; **B** time course of the effect of formaldehyde on RNase advanced glycation

(Selwood and Thornalley, 1993), formaldehyde does not form fluorescent adducts with RNase at all, yet it can substantially stimulate the advanced glycation of RNase.

This novel finding may be related to the advanced glycation theory for diabetic complications, since both methylglyoxal and formaldehyde seem to increase in diabetes. We also observed that the glycated RNase is more resistant towards proteolytic digestion by proteinase K in vitro. Interestingly, the formaldehyde-enhanced glycated RNase became even more resistant against the proteolytic hydrolysis under the same condition. This observation is consistent with the hypothesis that formaldehyde can further increase the degree of protein cross-linkage and unfolding, and thus make the protein molecules more rigid. The hardening of functional proteins could be related to complications associated with diabetes as well as other disorders.

Conclusion

This is a novel hypothesis with respect to clinical complications associated with diabetes. As can be seen in Scheme 1 deterioration of the vascular system in diabetes may be related to an excessive SSAO-mediated deamination of methylamine and/or aminoacetone. Uncontrolled production of toxic aldehydes, such as formaldehyde or methylglyoxal from methylamine or aminoacetone respectively, as well as hydrogen peroxide can either directly cause cytotoxicity (modification and crosslinking of functional or structural

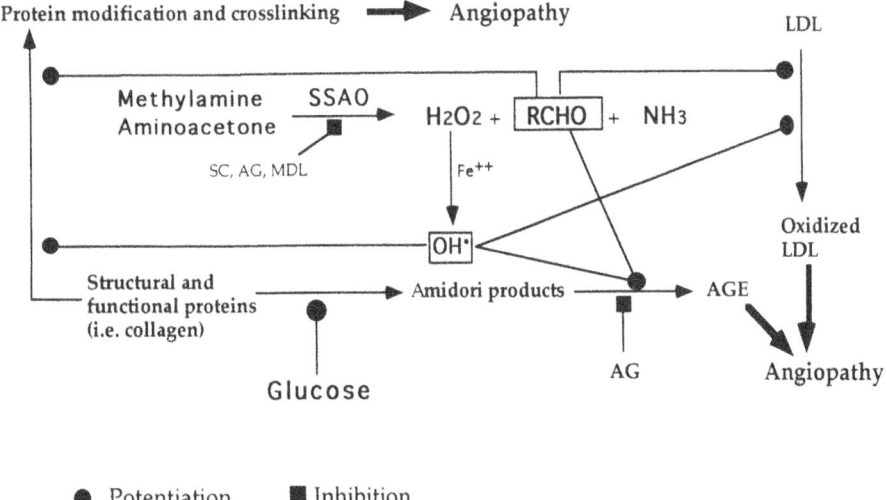

● Potentiation   ■ Inhibition

Scheme 1. Potential cytotoxic effects of SSAO-mediated deamination. *LDL* low density liproprotein, *AG* aminoguanidine, *SC* semicarbazide, *MDL* MDL-72974A

proteins) or indirectly enhance the oxidative glycation process, which is well known to be involved in the pathogenesis of complications associated with diabetes. Increased SSAO-mediated deamination may be a chronic risk factor of angiopathy, which is the leading cause of nephropathy, retinopathy and neuropathy in diabetes and possibly in other clinical disorders. Reduction of the SSAO-mediated deamination (i.e. inhibition of SSAO activity or reduction of available potentially harmful substrates) may be considered to be an alternative strategy to reduce the diabetic complications.

Acknowledgements

The author thanks Diabetes Association of Canada, Medical Research Council of Canada and Saskatchewan Health for their continuous financial support.

References

Asatoor AM, Kerr DNS (1961) Amines in blood and urine in relation to liver disease. Clin Chim Acta 6: 149–156

Auerbach O, Hammond EC, Garfinkel L (1965) Smoking in relation to atherosclerosis of the coronary arteries. N Engl J Med 273: 775–779

Baba S, Watanabe Y, Gejyo F, Arakwa M (1984) High-performance liquid chromatographic determination of serum aliphatic amines in chronic renal failure. Clin Chim Acta 136: 49–56

Baynes JW (1991) Role of oxidated stress in development of complications in diabetes. Diabetes 40: 405–412

Bierman EL (1992) Atherogenesis in diabetes. Arterioscler Thromb 12: 647–656

Blau K (1961) Chromatographic methods for the study of amines from biological material. Biochem J 80: 193–200

Bolt HM (1987) Experimental toxicology of formaldehyde. J Can Res Clin 13: 305–309

Boomsma FHM, van den Meiracker AH, Veld AJM, Schalekamp MADH (1995) Plasma semicarbazide-sensitive amine oxidase activity is elevated in diabetes mellitus and correlates with glycosylated haemoglobin. Clin Sci 88: 675–679

Boor PJ, Nelson TJ (1980) Allylamine cardiotoxicity. III. Protection by semicarbazide and in vivo derangements of monoamine oxidase. Toxicol 18: 87–102

Boor PJ, Hysmith RM (1987) Allylamine cardiovascular toxicity. Toxicol 44: 129–144

Boor PJ, Sanduja R, Nelson TJ, Ansari GAS (1987) In vivo metabolism of the cardiovascular toxin, allylamine. Biochem Pharmacol 36: 4347–4353

Boor PJ, Trent MB, Lyles GA, Tao M, Ansari GAS (1992) Methylamine metabolism to formaldehyde by vascular semicarbazide-sensitive amine oxidase. Toxicology 73: 251–258

Brett J, Ogawa S, Kirstein M, Radoff S, Vlassara H, Stern D (1990) Advanced glycosylation end products selectively attract monocytes to migrate across endothelial cell monolayers and induce activation and growth factor elaboration. Circulation 82 [Suppl 3]: 97

Brownlee M (1990) Advanced products of nonenzymatic glycoxylation and the pathogenesis of diabetic complications. In: Rifkin H, O'Porte D Jr (eds) Diabetes mellitus theory and practice. Elsevier, New York, pp 277–291

Brownlee M (1992) Non-enzymatic glycosylation of macromolecules: prospects of pharmacological modulation. Diabetes 41 [Suppl 2]: 57–60

Brownlee M, Vlassara H, Cerami A (1985) Nonenzymatic glycosylation products on collagen covalently trap low density lipoprotein. Diabetes 34: 938–941

Brownlee M, Cream A, Vlassara H (1988) Advanced glucosylation end products in tissue and the biochemical basis of diabetic complications. N Engl J Med 318: 1315–1321

Bucala R, Makita Z, Koschinsky T, Cerami A, Vlassara H (1993) Lipid advanced glycation: pathway for lipid oxidation in vivo. Proc Natl Acad Sci USA 90: 6434–6438

Callingham BA, Crosbie AE, Rous BA (1995) Some aspects of the pathophysiology of semicarbazide-sensitive amine oxidase enzymes. In: Current neurochemical and pharmacological aspects of biogenic amines: their function, oxidative deamination and inhibition. Prog Brain Res 106: 305–321

Cooper JR, Kini MM (1962) Editorial biochemical aspects of methanol poisoning. Biochem Pharmacol 11: 405–416

Courteix C, Eschalier A, Lavarenne J (1993) Streptozotocin-induced diabetic rats: behavioral evidence for a model of chronic pain. Pain 53: 81–88

Cryer PE, Haymond MW, Santiago JV, Shad SD (1976) Norepinephrine and epinephrine release and adrenergic mediation of smoking-associated haemodynamic and metabolic events. N Engl J Med 295: 573–577

Dar MS, Morselli PL, Bowman ER (1985) The enzymatic systems involved in the mammalian metabolism of methylamine. Gen Pharmacol 16: 557–560

D'Elia JA, Kaldany A, Miller DG, Abourizk NN, Weinrauch LA (1985) Diabetic nephropathy. In: Marble A, et al (eds) Joslin's diabetes mellitus. Lea and Febiger, Philadelphia, pp 635–663

Edelstein D, Brownlee M (1992) Mechanistic studies of advanced glycosylation end product inhibition by aminoguanidine. Diabetes 41: 26–29

Egyud LG, Szent-Gyorgyi A (1968) Cancer static action of methylglyoxal. Science 160: 1140

Elliott J, Callingham BA, Sharman DF (1989) The influence of amine metabolizing enzymes on the pharmacology of tyramine in the isolated perfused mesenteric arterial bed of rat. Br J Pharmacol 98: 515–522

Elliott J, Fowden AL, Callingham BA, Sharman DF, Silver M (1991) Physiological and pathological influences on sheep blood plasma amine oxidase: effect of pregnancy and experimental alloxan-induced diabetes mellitus. Res Vet Sci 50: 334–339

Epifano L, Di Vincenzo A, Fanelli C, Porcellati F, Perriello G, De Feo P, Motolese M, Brunetti P, Bolli CB (1992) Effect of cigarette smoking and of a transdermal nicotine delivery system on glucoregulation in type 2 diabetes mellitus. Eur J Clin Pharmacol 43: 257–263

Frankenhaeuser M (1971) Behavior and circulating catecholamines. Brain Res 31: 241–262

Fu MX, Knecht KJ, Thorpe SR, Baynes JW (1992) Role of oxygen in cross linking and chemical modification collagen by glucose. Diabetes 41 [Suppl 2]: 42–48

Gerich JE (1988) Role of insulin resistance in the pathogenesis of Type 2 (non-insulin-dependent) diabetes mellitus. In: Nattrass M, Halle PJ (eds) Bailliere's clinical endocrinology and metabolism 2: 307–326

Gibson JE (ed) (1983) Formaldehyde toxicity. Hemisphere Publ, Washington

Grafstrom RC, Curren RD, Yang LL, Harris CC (1985) Genotoxicity of formaldehyde in cultured human bronchial fibroblasts. Science 228: 89–91

Halimi JM, Sealey JE (1992) Prorenin in diabetes mellitus. Trends Endocrinol Metab 3: 270–275

Harker LA, Ross R, Slichter SJ, Scott CR (1976) Homocysteine induced arteriosclerosis. The role of endothelial injury and platelet response in its genesis. J Clin Invest 58: 731–741

Hayes BE, Clarke DE (1990) Semicarbazide-sensitive amine oxidase activity in streptozotocin diabetic rats. Res Comm Chem Pathol Pharmacol 69: 71–83

Helander A, Tottmar O (1987) Metabolism of biogenic aldehydes in isolated human blood cells, platelets and in plasma. Biochem Pharmacol 36: 1077–1082

Hunt JV, Smith CCT, Wolff SP (1990) Autooxidative glycosylation and possible involvement of peroxides and free radicals in LDL modification by glucose. Diabetes 39: 1420–1424

Hunt JV, Bottoms MA, Mitchinson MJ (1993) Oxidative alteration in the experimental glycation model of diabetes mellitus are due to protein-glucose adduct oxidation. Biochem J 291: 529–535

Jane SM, Mu D, Wemmer D, Smith JA, Kaur S, Maltby D, Burlingame AL, Klinman JP (1990) A new redox cofactor in eukaryotic enzymes: 6-hydroxydopa at the active site of bovine serum amine oxidase. Science 248: 981–987

Jennings PE, Jones AF, Florkouski CM, Lunic J, Barnett AH (1987) Increased diene conjugates in diabetic subjects with microangiopathy. Diabetic Med 4: 452–456

Kapeller-Adler R, Toda K (1932) Über das Vorkommen von Monomethylamine im Harn. Biochem Z 248: 403–425

Kalapos NP, Garzo T, Antoni F, Mardl S (1992) Accumulation of S-D-lactolylglutathione and transient decrease of glutathione level caused by methylglyoxal load in isolated hepatocytes. Biochim Biophys Acta 1135: 159–164

Lewinsohn R (1981) Amine oxidase in human blood vessels and non-vascular smooth muscle. J Pharm Pharmacol 33: 569–575

Lin SJ, Hong CY, Chang MS, Chiang BN, Chien S (1992) Long-term nicotine exposure increases aortic endothelial cell death and enhances transendothelial macromolecular transport in rats. Arterioscler Thromb 12: 1305–1312

Lo CS, Relf IRN, Myers KA, Wahlqvist ML (1986) Doppler ultrasound recognition of preclinical changes in arterial wall in diabetic subjects: compliance and pulse-wave damping. Diabetes Care 9: 27–31

Luetscher JA, Kraemer FB, Wilson DM (1989) Prorenin and vascular complications of diabetes. Am J Hypertens 2: 382–386

Lyles GA (1995) Substrate-specificity of mammalian tissue-bound semicarbazide-sensitive amine oxidase. Prog Brain Res 106: 293–303

Lyles GA, Singh I (1985) Vascular smooth muscle cells: a major source of the semicarbazide-sensitive amine oxidase of the rat aorta. J Pharm Pharmacol 37: 637–643

Lyles GA, Bertie KH (1987) Properties of a semicarbazide-sensitive amine oxidase in rat articular cartilage. Pharmacol Toxicol [Suppl] 1: 33

Lyles GA, McDougall SA (1989) The enhanced daily excretion of urinary methylamine in rats treated with semicarbazide or hydralazine may be related to the inhibition of semicarbazide-sensitive amine oxidase activities. J Pharm Pharmacol 41: 97–100

Lyles GA, Chalmers J (1992) The metabolism of aminoacetone to methylglyoxal by semicarbazide-sensitive amine oxidase in human umbilical artery. Biochem Pharmacol 31: 1417–1424

Malorny GN, Rietbrock N, Schneider M (1965) The oxidation of formaldehyde to formic acid in the blood. A contribution to the metabolism of formaldehyde. Schmiedebergs Arch Exp Path Pharmak 250: 419–436

McKennis HJr, Turnbull LB, Schwartz SL, Tamake E, Bowman ER (1962) Demethylation in the metabolism of (−)-nicotine. J Biol Chem 237: 541–546

McLellan AC, Phillips SA, Thornalley PJ (1992) The assay of methylglyoxal in biological systems by derivatization with diamno-4,5-dimethoxybenzene. Anal Biochem 206: 17–23

Moodonier VM, Kohn RR, Cerami A (1984) Accelerated age related browning of human collagen in diabetes mellitus. Proc Natl Acad Sci 81: 583–581

Morel DW, Hessler JR, Chisholm GM (1983) Low density lipoprotein cytotoxicity induced by free radical peroxidation of lipid. J Lipid Res 24: 1070–1076

More S (1981) Vascular injury and atherosclerosis. Marcel Dekker, New York

Mullarkey CJ, Edelstein D, Brownlee M (1990) Free radical generation by early glycation products: a mechanism for accelerated atherogenesis in diabetes. Biochem Biophys Res Comm 173: 932–939

Nelson TJ, Boor PJ (1982) Allylamine cardiotoxicity. IV. Metabolism to acrolein by cardiovascular tissues. Biochem Pharmacol 31: 509–514

Nilsson SE, Tryding N, Tufvesson G (1968) Serum monoamine oxidase in diabetes mellitus and some other internal diseases. Acta Med Scand 184: 105–108

Nixon R (1972) Volatile amines in mouse brain: a radioassay with picogram sensitivity. Anal Biochem 48: 460–470

O'Brien RE, Panangiotopoulos S, Cooper MS, Jerums G (1992) Anti-atherogenic effect of aminoguanidine, an inhibitor of advanced glycation. Diabetes 41 [Suppl 1]: 16A

Ohkuwa T, Sato Y, Naoi M (1995) Hydroxyl radical formation in diabetic rats induced by streptozotocin. Life Sci 56: 1789–1798

Picard S, Parathasarathy S, Fruebis J, Witztum JL (1992) Aminoguanidine inhibits oxidative modification of low density lipoprotein protein and the subsequent increase in uptake by macrophage scavenger receptors. Proc Natl Acad Sci USA 89: 6876–6880

Pirart J (1978) Diabetes mellitus and its degenerative complications: a prospective study of 4,400 patients observed between 1947 and 1973. Diabetes Care 1: 168–188

Precious E, Gunn CE, Lyles GA (1988) Deamination of methylamine by semicarbazide-sensitive amine oxidase in human umbilical artery and rat aorta. Biochem Pharmacol 37: 707–713

Rand LI (1991) Diabetic retinopathy: can we modify its course? Am J Med 90 [Suppl 2A]: 66–69

Reidy MA, Bowyer DE (1978) Distortion of endothelial repair. The effect of hypercholesterolaemia on regeneration of aortic endothelium following injury by endototoxin. Atherosclerosis 29: 459–466

Ross R (1986) The pathogenesis of atherosclerosis-an update. N Engl J Med 314: 488–500

Ross R, Harker L (1976) Hyperlipidemia and atherosclerosis. Chronic hyperlipidemia initiates and maintains lesion by endothelial cell desquamation and lipid accumulation. Science 193: 1094–1100

Regnstrom J, Nilsson J, Tornvall P, Landou C, Hamstein A (1987) Susceptibility of low density lipoprotein oxidation and coronary atherosclerosis in man. Lancet 339: 1183–1186

Ryder TA, Mackenzie ML, Pryse-Davies ML, Glover V, Lewinsohn R, Sandler M (1979) A coupled peroxidative oxidation technique for the histochemical localization of monoamine oxidase A and B and benzylamine oxidase. Histochem 62: 93–100

Sackett DL, Winkelstein W (1967) The relationship between cigarette usage and aortic atherosclerosis. Am J Epidemiol 86: 264–270

Sato Y, Hotta N, Sakamonto N, Matsuoka S, Ohishin N, Yagi IK (1979) Lipid peroxide levels in plasma of diabetic patients. Biochem Med 21: 104–107

Schayer RW, Smiley LR, Kaplan HE (1952) The metabolism of epinephrine containing isotopic carbon. J Biol Chem 198: 545–551

Selwood T, Thornalley PJ (1993) Binding of methylglyoxal to albumin and formation of fluorescent adducts. Inhibition by arginine, N_a-acetylarginine and aminoguanidine. Biochem Soc Trans 21: 170S

Smith AD, Jepson JB (1967) Chromatography of urinary and tissue amines and amino alcohols as 2,4-dinitrophenyl derivatives prepared with 2-nitrobenzene-sulfonic acid. Anal Biochem 18: 36–45

Sies H (1991) Oxidative stress; oxidants and antioxidants. Academic Press, London

Soulis-Liparota T, Cooper M, Papazoglou D, Clarke B, Jerums G (1991) Retardation by aminoguanidine of development of albuminuria, mesangial expansion and tissue fluorescence in streptozotocin induced diabetic rats. Diabetes 40: 1328–1335

Strong JP, Richards ML (1976) Cigarette smoking and atherosclerosis in autopsied men. Atherosclerosis 23: 451–476

Tesfamariam B, Cohen RA (1992) Free radicals mediate endothelial cell dysfunction caused by elevated glucose. Am J Physiol 263: H321–H326

Thomas PK (1991) Diabetic neuropathy: models, mechanisms and mayhem. Can J Neurol Sci 19: 1–7

Thornalley PJ (1993) The glyoxalase system in health and disease. Mol Asp Med 14: 287–371

Thornalley PJ (1994) Methylglyoxal, glyoxalases and the development of diabetic complications. Amino Acids 6: 15–23

Tilton RG, Chang K, Hasan KS, Smith SR, Petrash JM, Misko TP, Moore WM, Currie MG, Corbett JA, McDaniel ML, Williamson JR (1993) Prevention of diabetic dysfunction by guanidines. Inhibition of nitric oxide synthase versus advanced glycation end product formation. Diabetes 42: 221–232

Trézl L, Török G, Vasvári G, Pipek J (1992) Formation of burst chemiluminescence, excited aldehydes, and singlet oxygen in model reactions and from carcinogenic compound in rat liver S9 fractions. In: Tyihak E (ed) Role of formaldehyde in biological systems. Hungarian Biochem Soc, Sopron

Tryding N, Nilsson SE, Tufvesson G, Berg R, Carlstrom S, Elmfors B, Nilsson JE (1969) Physiological and pathological influences on serum monoamine oxidase level. Scan J Clin Lab Invest 23: 79–84

Tsuboi S, Kawase M, Takaka A, Hirmatus M, Wada Y, Kawakami Y, Ikeka M, Ohmori S (1992) Purification and characterization of formaldehyde dehydrogenase from rat liver cytosol. J Biochem 111: 465–471

US Dept Health and Human Services (1982) Constituents of tobacco smoke. USPHS Publication No 82-50179, pp 322

Uusitupa MIJ, Niskanen LK, Sittonen O, Voutilainen E, Pyorala K (1993) Ten-year cardiovascular mortality in relation to risk factors and abnormalities in lipoprotein composition in type 2 (non-insulin-dependent) diabetic and non-diabetic subjects. Diabetologia 36: 1175–1184

Vlassara H, Makita Z, Rayfield E, Freidman E, Cerami A, Morgelo S (1990) In vitro advanced glycation as a signal for monocyte migration in vessel wall: role in diabetes and aging. Circulation 82 [Suppl 3]: 92

Wibo M, Duong AT, Godfraind T (1980) Subcellular location of semicarbazide-sensitive amine oxidase in rat aorta. Eur J Biochem 112: 87–94

Wilens SL, Plair CM (1962) Cigarette smoking and atherosclerosis. Science 138: 975–977

World Health Organization (1989) Formaldehyde. Environmental Health Criteria 89, Geneva

Yu PH (1986) Monoamine oxidase. In: Boulton AA, Baker GB, Yu PH (eds) Neuromethods, vol V. Neurotransmitter enzymes. Humana Press, Clifton, New Jersey, pp 235–272

Yu PH (1990) Oxidative deamination of aliphatic amines by rat aorta semicarbazide-sensitive amine oxidase. J Pharm Pharmacol 42: 882–884

Yu PH, Zuo DM (1993) Methylamine, a potential endogenous toxin for vascular tissues: formation of formaldehyde via enzymatic deamination and the cytotoxic effects on endothelial cells. Diabetes 42: 594–603

Yu PH, Zuo DM (1995) Formaldehyde produced endogenously via deamination of methylamine; a potential risk factor for initiation of endothelial injury. Atherosclerosis 120: 189–197

Yu PH, Zuo DM (1997) Aminoguanidine inhibits semicarbazide-sensitive amine oxidase activity; implication for advanced glycation and angiopathy in diabetes. Diabetologia (in press)

Yu, PH, Lai CT, Zuo DM (1997) Evidence of formation of formaldehyde from adrenaline *in vivo*; a potential risk factor endothelial damage. Neurochem Res 22: 615–620

Yu PH, Zuo DM, Davis BA (1993) Human tissue and serum semicarbazide-sensitive amine oxidase: species heterogeneity. Biochem Pharmacol 47: 1055–1059

Yuen CT, Easton D, Misch KJ, Rhodes EL (1987) Increased activity of serum amine oxidase in granuloma annulare, necrobiosis lipoidica and diabetes. Br J Dermatol 116: 643–649

Zeisel SH, Dacosta KA (1986) Increase in human exposure to methylamine precursors of N-nitrosamines after eating fish. Cancer Res 46: 6136–6138

Zeisel SH, Wishnok JS, Blusztajn JK (1983) Formation of methylamines from ingested choline and lecithin. J Pharmacol Exp Ther 225: 320–324

Zuo DM, Yu PH (1994) Semicarbazide-sensitive amine oxidase and monoamine oxidase in rat brain microvessels, meninges, retina and eye sclera. Brain Res Bull 33: 307–311

Author's address: Dr. P. H. Yu, Neuropsychiatry Research Unit, Department of Psychiatry, University of Saskatchewan, Saskatoon, Saskatchewan, S7N 5E4 Canada

Monoamine oxidase activities in human cystic and colonic arteries — influence of age

I. V. Figueiredo[1], **A. Martinez Coscolla**[2], **M. D. Cotrim**[1], **M. M. Caramona**[1], and **B. A. Callingham**[2]

[1]Laboratório de Farmacologia, Faculdade de Farmácia, Universidade de Coimbra, Portugal
[2]Department of Pharmacology, Cambridge University, United Kingdom

Summary. The deamination of 5-hydroxytryptamine, phenylethylamine and benzylamine by monoamine oxidases (MAO-A and B) and semicarbazide sensitive amine oxidase (SSAO) respectively has been studied in homogenates of human cystic and colonic arteries by radiochemical assays. In cystic artery the deamination is mainly carried out by SSAO with a lower participation of MAO-B. The kinetic parameters were: to MAO-B the V_{max} = 15.11 ± 0.51 nmol/mg protein.h and the K_m = 78.51 ± 5.16 µM (±SE) and to SSAO the V_{max} = 211.70 ± 8.75 nmol/mg protein.h and the K_m = 211.51 ± 23.27 µM (±SE). We could not measure MAO-A activity in our experimental conditions and also the levels of catecholamines are very low and the histological studies show a poor innervation in these tissues. In colonic artery the kinetic parameters were: to MAO-B the V_{max} = 5.09 ± 0.31 nmol/mg protein.h and the K_m = 29.12 ± 4.55 µM (±SE) and to SSAO the V_{max} = 273.67 ± 8.35 nmol/mg protein.h and the K_m = 197.89 ± 21.81 µM (±SE). In this artery we could find MAO-A in five among the nine samples studied and the kinetic parameters were: the V_{max} = 14.48 ± 0.82 nmol/mg protein.h and the K_m = 136.40 ± 25.46 µM.

As we have performed the experiments with human vessels from donors with different age we could not find any relationship between the activity or affinity, in MAO-B and SSAO, with age. Nevertheless, the results show in cystic artery an increase in the affinity of MAO-B with age when we consider the female group which suggests a possible role of the hormonal condition in this behaviour.

Introduction

Catecholamines and serotonin play an important role as central and peripheral neurotransmitters which is specially significant in neuropsychiatry disorders. The oxidative deamination by monoamine oxidases (MAO) is one of the most important mechanisms of their inactivation.

A large amount of studies based on the selective inhibitors clorgyline and
(−) deprenyl has shown the existence of two types of enzymes which are
responsible for the activity of MAO in mammalian tissues (Johnston, 1968;
Fowler et al., 1978; Schurr, 1982). These two forms have a different behaviour
in relation with specific substrates. Thus, 5-hydroxytryptamine is only
metabolised by MAO-A Hall et al. (1969) while low concentrations of
phenylethylamine have been only metabolised by MAO-B Yang and
Neff (1973). Nevertheless, studies carried out on a wide range of tissues and
species (Callingham, 1983, 1986; Youdim et al., 1988) show results with
some apparent contradictions in substrate specificities and inhibitor sensitivities which show that this classification is too simple and any idea of ex-
trapolation from one species or tissue to another is not adequate for these
enzymes.

On the other hand, we have to consider a second group of amine oxidases,
SSAO (amine oxidase sensitive to semicarbazide) which has several differences from monoamine oxidases in relation with its biochemical identity
and distribution. Vascular wall and smooth muscle cells have been reported
as particularly rich in the enzyme (Lewinshon, 1984; Lyles and Fitzpatrick,
1985; Callingham and Barrand, 1987; Buffoni, 1993). Many papers show
an important role of this enzyme in mammalian tissues (Elliott et al., 1989,
1991; Banchelli et al., 1990) but its physiological importance still remains
unclear.

All these questions encourage us to characterise these three enzymes in
human tissues. The aim of the present work was to analyse the oxidative
deamination of different substrates by monoamine oxidases (MAO-A and
MAO-B) and semicarbazide sensitive amine oxidase (SSAO), trying to
stablish a relationship between these parameters and the catecholamine levels
and the density of innervation, in cystic and colonic arteries. As the donors
have different age we have related also the enzyme activities with age.

Materials and methods

Human arteries were provided by the University Hospital of Coimbra from patients
which had been submitted to surgery.

The human cystic and colonic arteries were isolated and transported to the laboratory
on ice in cold Krebs-Henseleit solution (Branco et al., 1992). In order to eliminate a
possible contamination from plasma amine oxidases activity, the arteries were stripped of
the surrounding tissues and washed to eliminate blood. Finally, they were stored at −80°C
in 0.01 M phosphate buffer solution (pH 7.4).

Prior to the enzymatic assay, the arteries were homogenated with 2 ml 0.01 M phosphate buffer pH 7.4 in a Duall homogeniser at 4°C.

MAO activity was determined using ^{3}H-5-hydroxytryptamine creatinine sulphate
(15.1 Ci/mmol) as a preferential substrate for MAO-A (in a concentration range
between 50 μM to 1,600 μM) and ^{14}C-phenylethylamine hydrochloride (56 mCi/mmol)
for MAO-B (between to 5–160 μM). The reaction mixture contained inhibitors in
aqueous solution (clorgyline 10^{-4}M as inhibitor of MAO-A and (−)-deprenyl 10^{-4}M for
MAO B) and protein homogenate (2 mg/ml). After the preincubation time (30 min at
37°C), the corresponding substrate were added and the solution was oxygenated. After

20 min of incubation at 37°C, the reaction was stopped by the addition of 10 μl HCl (3N). The deaminated product was extracted and measured by scintillation counting (Caramona, 1982).

Protein concentration of homogenates (mg/ml) were determinated by the method of Lowry et al. (1951) using bovine serum albumin as standard.

SSAO activity was determined as described for MAO (using both inhibitors) with times of preincubation and incubation of 30 and 5 min, respectively, using ^{14}C-benzylamine hydrochloride (59 mCi/mmol) as specific substrate (in a concentration range between 50 μM to 1,600 μM).

The levels of dopamine, norepinephrine and epinephrine in the arteries were measured by High Performance Liquid Chromatography (HPLC) with electrochemical detection (Warnhoff, 1984).

For the histological studies we used formalin-fixed, parafin-embedded blocks from the samples studied. Sections of the arteries were cut at 4 μm, dewaxed and rinsed in alcohol. A hematoxylin and eosin stain and the immunohystochemical study for neurone specific enolase (NSE) and S100 protein were performed for each section.

Endogenous peroxidase activity was blocked by 0.3% hydrogen peroxidase. After that, the tissue sections were incubated with the primary antibody, anti-NSE (DAKO-A598) or antibodies anti-S100 (DAKO-Z311) for 30 min at room temperature. The immunohystochemical method was an avidine-biotine complex immunoperoxidase method. The slices were developed with diaminobenzidine, rinsed with tap water and counterstained with hematoxylin, and mounted with entelan (Rode and Dhillon, 1984; Cabrita, 1986).

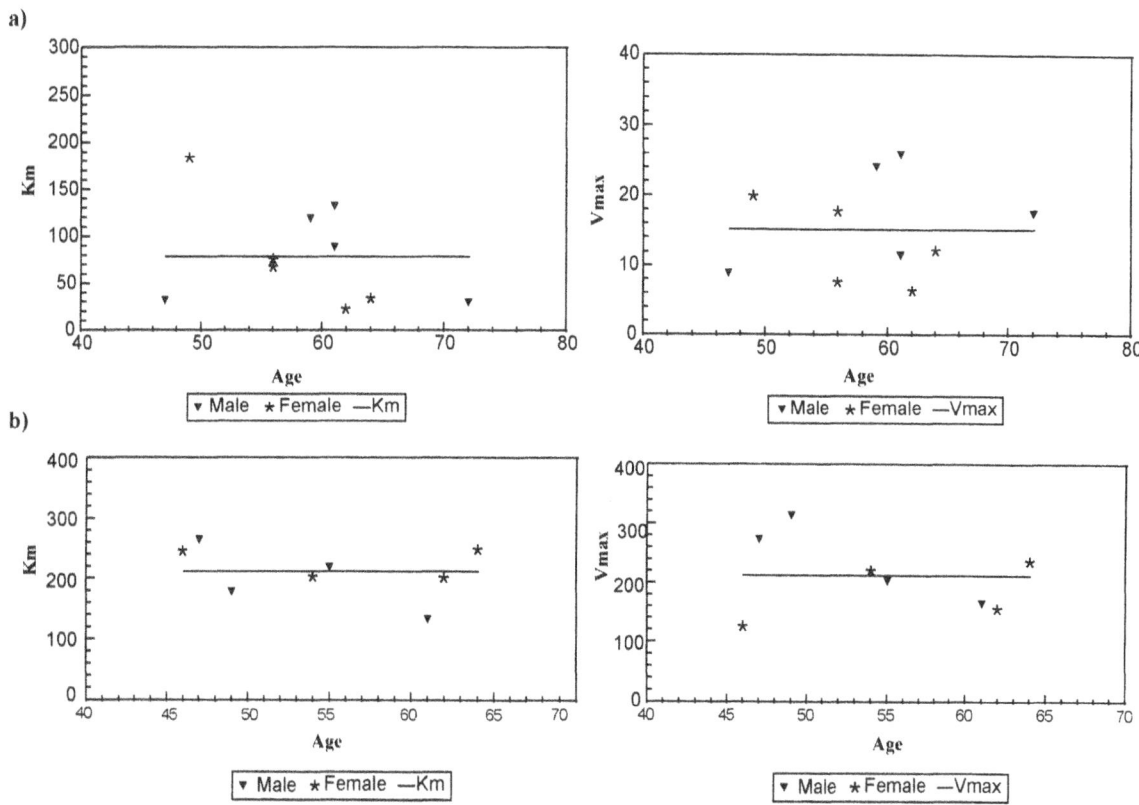

Fig. 1. Scatter diagrams for MAO B (**a**) and SSAO (**b**) to study the possibility of relationship between its Vmax or Km (y) and age (x) in cystic artery. The Vmax and Km lines (mean values of the population) are also drawn

Results

The results obtained for Vmax and Km in cystic arteries (n = 10) were the following: to MAO-B the Vmax = 15.11 ± 0.51 nmol/mg protein.h and the Km = 78.51 ± 5.15 μM (±SE); SSAO (n = 8) the Vmax = 211.70 ± 8.75 nmol/mg protein.h and the Km = 211.51 ± 23.27 μM (± SE). We could not measure any MAO-A activity in our experimental conditions. The values were according the Wilkinson Analysis (1961). Figures 1 and 2 show the scatter diagrams for MAO-B and SSAO in cystic and colonic arteries, respectively.

In colonic artery (n = 9), the results obtained were to MAO-B the Vmax = 5.09 ± 0.31 nmol/mg protein.h and the Km = 29.12 ± 4.55 μM (± SE) and to SSAO the Vmax = 273.67 ± 8.35 nmol/mg protein.h and the Km = 197.89 ± 21.81 μM (± SE). In this artery we could find MAO-A only in the five among the nine studied samples and the kinetic parameters were to the Vmax = 14.48 ± 0.82 nmol/mg protein.h and the Km = 136.40 ± 25.46 μM.

In both arteries the levels of catecholamines found were not statistically different, ranging from 200 to 300 pg/mg to norepinephrine, 7.5 to 10 pg/mg to epinephrine and 15 to 25 pg/mg to dopamine.

The histalogical and immunocytochemical studies showed a very low density of innervation that could be classified in very rare and rare neuronal

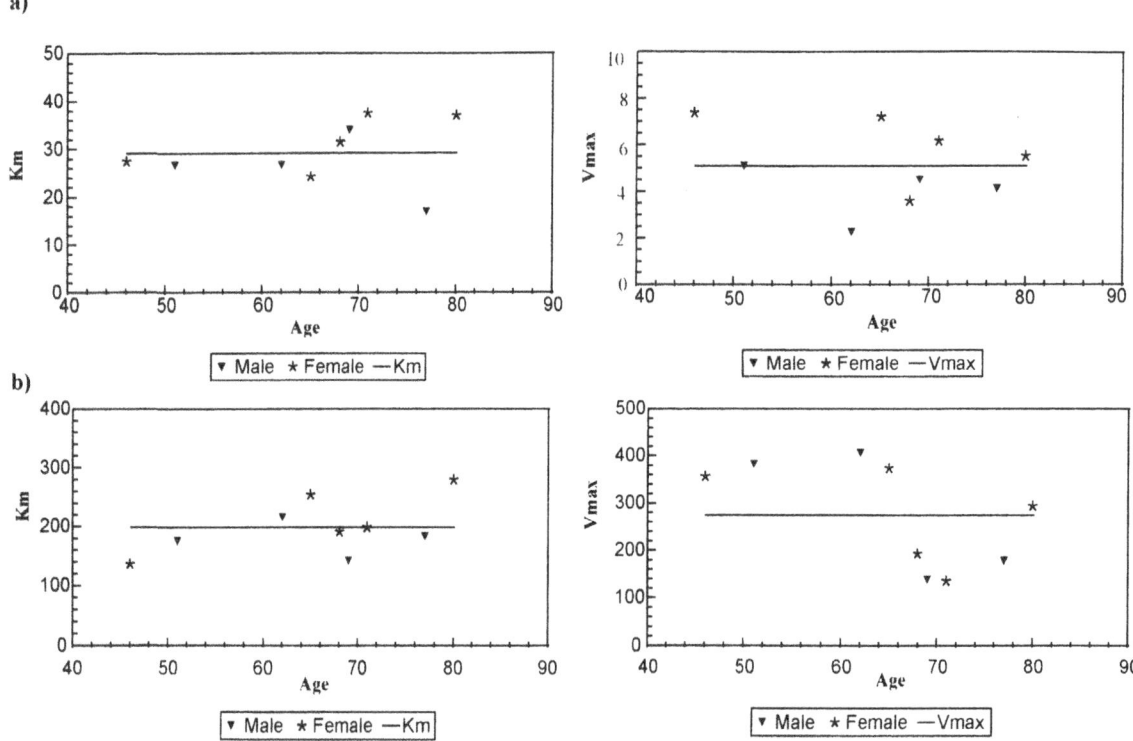

Fig. 2. Scatter diagrams for MAO B (**a**) and SSAO (**b**) to study the possibility of relationship between its Vmax or Km (y) and age (x) in colonic artery. The Vmax and Km lines (mean values of the population) are also drawn

structures in cystic and colonic arteries, respectively, compared to a positive control.

Discussion

The blood vessels are endowed with enzymatic and non-enzymatic mechanisms for the inactivation of catecholamines (Osswald and Guimarães, 1983; Kopin, 1985; Osswald, 1988). Although there is a wealth of information concerning the importance of the enzymes involved in the oxidative deamination of catecholamines in a number of isolated blood vessels of different animal species, little is known in the isolated human vascular tissue.

Therefore we feel that the results reported in this work allow us to compare the MAO (type A and B) and SSAO activity in two human blood vessels, the cystic and colonic arteries.

With the use of appropriate concentrations of selective substrates of both MAO-A and B and using the specific inhibitors clorgyline (MAO-A) and (−)-deprenyl (MAO-B) we study also the presence of a second group of amine oxidase, amine oxidase sensitive to semicarbazide (SSAO). In the cystic artery, the scatter diagrams (Fig. 1) show possible relationship between Vmax or Km and the age of the donors. The patterns made by the points plotted on the scatter diagrams suggest that there is no relation between activity (Vmax) or affinity (Km) and the age of patients for both enzymes (MAO-B and SSAO). Also, the points are scattered at random around the Vmax or Km line (mean values of the population tested) which suggests that all of them made a normal population. We could not measure MAO-A activity in our experimental conditions. This is in agreement with the low catecholamine levels and the hystological studies that show a poor innervation in these tissues, in contrast with other human vascular tissue, such as the uterine artery (Branco et al., 1992; Figueiredo et al., 1995). Nevertheless, if we pay attention to the relationship between Km and age for MAO B (Fig. 1a) when we consider the division of the patients into male or female the distribution of the points in the case of the female group does not follow an arbitrary way. In spite of a low number of values, their distribution shows an apparent increase in the affinity of the enzyme with age.

In the colonic artery, the diagrams in Fig. 2 do not show any change in the Km or the Vmax with age, either in MAO-B or in SSAO. In this artery we could find MAO-A only in the five among the nine studied samples and we can see its presence only in the old age.

According with our results the oxidative deamination is carried out by MAO-B in both arteries and this metabolism is also carried out by SSAO. MAO-A has a null (in cystic artery) or may be insignificant (colonic artery) participation in this enzymatic metabolism which is in relation with the low degree of innervation observed in histological or immunocytochemical assays.

Acknowledgements

We thank to the Surgery I Department from Coimbra University Hospital and also to Institute of Pathological Anatomy, Faculty of Medicine, University of Coimbra.

The authors are grateful to Prof. T. Macedo and Prof. F. Ribeiro from Institute of Pharmacology and Experimental Therapeutic, Faculty of Medicine, for their helpful collaboration.

References

Banchelli G, Buffoni F, Elliott J, Callingham BA (1990) A study of the biochemical pharmacology of 3,5-ethoxy-4-aminomethylpyiridine (B24), a novel amine oxidase inhibitor with selectivity for tissue bound semicarbazide-sensitive amine oxidase enzymes. Neurochem Int 17: 215–221

Branco D, Caramona MM, Martel F, Ferreira de Almeida JA, Osswald W (1992) Predominance of oxidative deamination in the metabolism of exogenous noradrenaline by the normal and chemically denervated human uterine artery. Naunyn Schmiedebergs Arch Pharmacol 346: 286–293

Buffoni F (1993) Properties, distribution and physiological role of semicarbazide-sensitive amine oxidase. Curr Topics Pharmacol 2: 33–50

Cabrita AMS (1986) A imunocitoquímica no estudo das lesões neoplásicas. Arq Patol Geral Anat Patol Univ Coimbra Vol XX (Nova série): 97–104

Callingham BA (1983) Comparative aspects of monoamine oxidase. Vet Res Commun 7: 325–330

Callingham BA (1986) Some aspects of monoamine oxidase pharmacology. Cell Biochem Funct 4: 99–108

Callingham BA, Barrand MA (1987) Some properties of semicarbazide-sensitive amine oxidases. J Neural Transm [Suppl] 23: 37–54

Caramona MM (1982) Monoamine oxidase of types A and B in the saphenous vein and mesenteric artery of the dog. Naunyn Schmiedebergs Arch Pharmacol 319: 121–124

Elliot J, Callingham BA, Sharman DF (1989) Semicarbazide-sensitive amine oxidase (SSAO) on rat aorta: interactions with some naturally occurring substrates and their structural analogues. Biochem Pharmacol 38: 1507–1515

Elliot J, Callingham BA, Sharman DF (1991) Amine oxidase enzymes of sheep blood vessels and blood plasma: a comparison of their properties. Comp Biochem Physiol 102: 83–89

Figueiredo IV, Caramona MM, Cotrim MD, Fontes Ribeiro CA, Macedo TRA, Pereira J, Silveira L, Falcão F (1995) The activity of semicarbazide sensitive amine oxidase in human uterine arteries. Pharmacol Res 31 [Suppl]: 240

Fowler CJ, Callingham BA (1978) Monoamine oxidase A and B: a useful concept? Biochem Pharmacol 27: 97–101

Hall DWR, Logan BW, Parsons GH (1969) Further studies on the inhibition of monoamine oxidase by M & B 9302 (clorgyline). I. Substrate specificity in various mammalian species. Biochem Pharmacol 18: 1447–1454

Johnston (1968) Some observations upon a new inhibitor of monoamine oxidase in brain tissue. Biochem Pharmacol 17: 1285–1297

Kopin IJ (1985) Catecholamines metabolism: basic aspects and clinical significance. Pharmacol Rev 37: 333–364

Lewinshon R (1984) Mammaliam monoamine-oxidising enzymes, with special references to benzylamine oxidase in human tissues. Braz J Med Biol Res 17: 223–256

Lowry OH, Rosebrough NJ, Farr AL, Randall RJ (1951) Protein measurement with the Folin phenol reagent. J Biol Chem 193: 265–275

Lyles GA, Fitzpatrick CMS (1985) An allylamine derivative (MDL 72145) with potent irreversible inhibitory actions on rat aorta semicarbazide-sensitive amine oxidase. J Pharm Pharmacol 37: 29–335

Osswald W (1988) Disposition of catecholamines blood vessel wall. In: Bevan JA, Majewski H, Maxwell RA, Story DF (eds) Vascular neuroeffector mechanisms. IRL Press, Oxford, pp 321–329

Osswald W, Guimaraes S (1983) Adrenergic mechanisms in blood vessels: morphological and pharmacological aspects. Rev Physiol Biochem Pharmacol 96: 53–122

Rode J, Dhillon AP (1984) Neuron specific enolase and S100 protein as possible prognostic indicators in melanoma. Histopath 8: 1041–1052

Schurr A (1982) Monoamine oxidase: to B or not to B? Life Sci 30: 1059–1063

Warnhoff M (1984) Simultaneous determination of Ne, dopamine, 5-HT and their main metabolits in rat brain using "HPLC-EC". J Chrom 307: 271–281

Wilkinson GN (1961) Statistical estimations in enzyme kinetic. Biocem J 80: 324–332

Yang HYT, Neff NH (1973) β-phenylethylamine, a specific substrate for type B monoamine oxidase of brain. J Pharmacol Exp Ther 187: 365–371

Youdim MBH, Finberg JPM, Tipton KF (1988) Monoamine oxidase. In: Trendelenburg U, Weiner N (eds) Catecholamines, vol I. Springer, Berlin Heidelberg New York Tokyo, pp 119–192

Authors' address: Dr. I. V. Figueiredo, Laboratório de Farmacologia, Faculdade de Farmácia da Universidade de Coimbra, Largo de D. Dinis, 3030 Coimbra, Portugal

Influence of maturation and ageing on the biotransformation of noradrenaline in the rat

J. T. Guimarães[2], M. Q. Paiva[1], D. Moura[1], and **S. Guimarães[1]**

[1] Institute of Pharmacology and Therapeutics, and [2] Department of Biochemistry,
Faculty of Medicine, Porto, Portugal

Summary. The present investigation was undertaken to study the influence of maturation and ageing on the disposition of noradrenaline by the aorta, heart (ventricle), liver and kidney of the rat. Slices of these tissues taken from rats aged less than 18h, 2.5–3 months or 18–24 months were incubated with $0.1\,\mu mol.l^{-1}$ ^{3}H-amine during 30min. At the end of this period, the accumulation of the intact amine in the tissue, as well as the ^{3}H-metabolites formed (3,4-dihydroxyphenylethylglycol, 3,4-dihydroxymandelic acid, normetanephrine and O-methylated deaminated metabolites) were determined by scintillation counting. The results obtained show that in the rat: 1) at any age, noradrenaline is preferentially deaminated; 2) while the capacity of the sympathetic nerve terminals in accumulating noradrenaline is rather well developed at birth, the metabolic system for its degradation is still immature; 3) aldehyde dehydrogenase activity or that of its co-factor (or both) of the heart is apparently missing at birth; 4) removal of noradrenaline by the liver and the kidney did not change with ageing, while that by the aorta decreased and that by the heart increased.

Introduction

The mechanisms for adrenergic function do not develop at the same time (Blatchford et al., 1976; Su et al., 1977). For example, the pathways for adrenergic transmitter inactivation, transmitter action on smooth muscle cells and neuronal transmitter release develop in that sequence in the fetal lamb (Su et al., 1977). On the other hand, ageing is associated with a variety of changes in cardiovascular function, as well as with an increased incidence of cardiovascular disorders (Docherty, 1990). The present investigation was undertaken to look for the influence of maturation and ageing on the role played by monoamine oxidase (MAO) and catechol O-methyl transferase (COMT) activities in the inactivation of noradrenaline in the rat.

Material and methods

Three groups of Wistar rats were used: one of animals less than 18 h old (body weight = 6.0 ± 0.3 g; n = 5), another of animals 2.5–3 months old (body weight = 351 ± 8 g; n = 4) and a third group of animals 18–24 months old (body weight = 841 ± 24 g; n = 6). The animals of either sex were anaesthetized with pentobarbital sodium (30 mg.kg^{-1}, i.p.) and the aorta, the heart, the liver and the kidney were dissected free and placed in aerated (95% O_2; 5% CO_2), modified Krebs-Henseleit solution (Guimarães et al., 1978) of the following composition (mmol.l^{-1}): NaCl, 118.6; KCl, 4.70; $CaCl_2$, 2.52; KH_2PO_4, 1.18; $MgSO_4$, 1.23; $NaHCO_3$, 25.0; glucose, 10.0. Then, slices of these organs were cut and placed for 30 min in small beakers containing 2.5 ml of modified Krebs-Henseleit aerated solution at 37°C. After this period of stabilization 0.1 μmol.l^{-1} ^{3}H-noradrenaline was added to the incubation medium for a further period of 30 min. To avoid autoxidation of the ^{3}H-amine, EDTA 0.027 and ascorbic acid 0.57 mmol.l^{-1} were added to the medium which was maintained at 37°C. At the end of this incubation, the tissues were immersed in 3 ml of 0.2 mol.l^{-1} perchloric acid and kept overnight for extraction of radioactivity.

Aliquots of the eluates and the tissue extracts were passed through alumina and Dowex 50WX4 columns, as described by Graefe et al. (1973). A second Dowex column was used to remove traces of ^{3}H-amine from the DOMA-containing alumina eluate (Trendelenburg et al., 1983). Five fractions were isolated: noradrenaline, 3,4-dihydroxyphenylglycol (DOPEG), 3,4-dihydroxymandelic acid (DOMA), normetanephrine (NMN) and O-methylated-deaminated metabolites (OMDA) (which represents 3-methoxy-4-hydroxyphenylglycol plus 3-methoxy-4-hydroxymandelic acid). The recovery of radioactivity in the chromatographic procedure (sum of radioactivity found in the five fractions/total radioactivity in the sample) was 0.92 ± 0.02 (n = 22). Results were not corrected for recovery.

Radioactivity was measured by liquid scintillation counting (liquid scintillation counter 1209 Rackbeta, LKB Wallac, Turku, Finland) in 2 ml aliquots of eluate (or 0.5 ml of tissue extract + 1.5 ml of Krebs solution) after addition of 8 ml of scintillation mixture (OptiPhase "HiSafe" 3, LKB, Loughborough, Leics, England).

Endogenous noradrenaline content of the tissue

The endogenous noradrenaline content of the tissues was measured by HPLC with electrochemical detection (Bioanalytical Systems, West Lafayette, Ind., USA). Immediately after removal of the tissues, a sample of each one was collected in perchloric acid (0.2 mol.l^{-1}) and kept overnight. Aliquots of the extract were directly injected in a HPLC system with a 5 μm ODS reverse-phase column. The mobile phase — which was a degassed solution of 50 mmol.l^{-1} KH_2PO_4, 1.7 mmol.l^{-1} sodium heptane sulphonate, 0.09 mmol.l^{-1} EDTA, 10% methanol v/v, pH 3.5 adjusted with perchloric acid — was pumped at a flow-rate of 1 ml.min^{-1}. Quantification was carried out with a carbon paste electrode at 0.75 V vs a Ag/AgCl reference electrode.

Determination of the extracellular space

Slices of heart (ventricle), liver and kidney or segments of aorta were incubated during 5 min in 3 ml of modified Krebs-Henseleit solution containing 0.2 μmol.l^{-} of ^{3}H-sorbitol and radioactivity was measured by liquid scintillation counting as described above.

Statistics

The results are presented as arithmetic means with standard errors. One-way analysis of variance was used to test differences between unpaired results. A probability level of 0.05 or less was considered statistically significant.

Drugs used

^{3}H-7-(−)-noradrenaline (10.4 Ci.mmol^{-1}, NEN Dupont, Dreieich, FRG); ^{3}H-sorbitol (12.9 Ci.mmol^{-1} NEN Dupont).

Results

1. Noradrenaline content of the different organs

The noradrenaline contents of the different organs at different ages are shown in Table 1. The liver, at any age, had the lowest noradrenaline content of all the organs studied. The noradrenaline contents of the aorta and liver in neonates were not different from those in adults. In contrast, there was a marked increase of noradrenaline contents in the heart and kidney during maturation. There was a marked reduction of the noradrenaline content of the aorta with ageing.

2. Metabolism of noradrenaline in the different organs

Aorta

As shown in Table 2, in the aorta of newborn rats the amount of noradrenaline accumulated was not significantly different from that which was metabolically degradated (182 vs. 234 pmol.g^{-1}; $P > 0.05$). Deamination was the mechanism largely predominant. There was practically no formation of O-methylated-deaminated metabolites. In adult rats, the total amount of metabolites formed from noradrenaline was more than twice that formed in the

Table 1. Noradrenaline content of some organs of the rat at different ages

	Aorta	Heart	Kidney	Liver
Newborns	1.16 ± 0.12	0.59 ± 0.06*	0.35 ± 0.08*	0.18 ± 0.05
Adults	1.67 ± 0.34	3.48 ± 0.06	1.88 ± 0.11	0.24 ± 0.05
Old rats	0.90 ± 0.07*	3.74 ± 0.18	2.15 ± 0.22	0.32 ± 0.01

Values are expressed as nmol.g^{-1} of tissue and represent mean ± s.e.m of 4–6 experiments. *Significantly different from the value obtained in adults in the same organ

Table 2. Metabolism of noradrenaline in the rat

	OMDA	NMN	DOPEG	DOMA	Noradrenaline	Total metabolites	Removal
Aorta							
Newborns	4 ± 2*	25 ± 9*	141 ± 35	64 ± 39	182 ± 21	234 ± 31*	456 ± 43*
Adults	256 ± 51	267 ± 93	124 ± 29	25 ± 20	198 ± 49	672 ± 169	870 ± 177
Old rats	138 ± 24	79 ± 3*	52 ± 16*	11 ± 8	83 ± 15*	280 ± 29*	363 ± 11*
Heart							
Newborns	3 ± 0*	15 ± 4*	41 ± 1*	6 ± 4*	87 ± 13*	65 ± 3*	152 ± 7*
Adults	62 ± 3	3 ± 1	361 ± 93	267 ± 115	260 ± 43	693 ± 52	953 ± 17
Old rats	105 ± 25	2 ± 0	375 ± 171	480 ± 120	203 ± 23	962 ± 143	1165 ± 142
Kidney							
Newborns	26 ± 2*	69 ± 4*	19 ± 5*	4 ± 4	54 ± 14	118 ± 7*	172 ± 18
Adults	88 ± 13	234 ± 26	38 ± 3	21 ± 9	99 ± 41	381 ± 46	480 ± 57
Old rats	102 ± 18	255 ± 23	39 ± 11	20 ± 9	148 ± 30	416 ± 45	564 ± 79
Liver							
Newborns	114 ± 7	37 ± 5*	35 ± 2	32 ± 13	1 ± 1	218 ± 14*	219 ± 12*
Adults	148 ± 32	4 ± 1	108 ± 42	62 ± 33	2 ± 1	322 ± 40	324 ± 43
Old rats	152 ± 23	5 ± 1	78 ± 35	128 ± 25	13 ± 3*	363 ± 33	376 ± 32

Shown are the metabolites formed during 30 min incubation with 0.1 $\mu mol.l^{-1}$ ^{3}H-noradrenaline in the four tissues indicated in the table. Values are expressed as $pmol.g^{-1}$ and represent mean ± s.e.m of 4–6 experiments. Removal represents accumulation plus total metabolites. * Significantly different from the value obtained in adults in the same organ

aorta of newborn animals, while the amine accumulated in newborn and adult rats was not different (182 vs. 198 $pmol.g^{-1}$; $P > 0.05$). Noradrenaline was both deaminated and O-methylated, the O-methylated metabolites predominating. While in newborns the ratio total metabolites/amine accumulated was 1.3, in adults it was 3.4.

In old rats, the amounts of the accumulated amine and of the total metabolites were only about 40% of those observed in adults. The amounts of O-methylated and of deaminated metabolites were not different. As in adults, O-methylation predominated over deamination (Table 2).

Heart

As shown in Table 2, the heart of newborn rats showed the lowest capacity to degradate noradrenaline among the organs studied. Noradrenaline was predominantly deaminated and the accumulation of intact amine was less than 50% that observed in aorta. In adults, deaminated metabolites (DOPEG + DOMA) represented more than 90% of the total metabolites formed. The total amount of metabolites formed was 11 times higher in adults than in newborns. In old rats, the cardiac capacity to degradate noradrenaline reached its maximum. The amount of metabolites formed was 15 times higher in old rats than in newborns. In old rats, the heart was the tissue possessing the highest capacity to degrade noradrenaline. Also in the heart of old rats, there was practically no O-methylation of noradrenaline. DOPEG + DOMA represented more than 90% of total metabolites formed and DOMA was the predominant metabolite, representing 50% of the total amount of metabolites formed.

Kidney

The kidney of newborn rats had a relatively low capacity to accumulate noradrenaline. The accumulation represented 31% of the removal. In contrast to what was observed in other tissues, the deamination of noradrenaline in the kidney was very low at any age, O-methylation predominating largely. These peculiar characteristics of the renal tissue were also observed in adult and old rats: a relative low accumulation of the intact amine and a large predominance of O-methylation over deamination. Both in adult and old rats, the total amount of metabolites formed was much higher than in newborns: 3.2 and 3.5 times higher in adult and old rats, respectively.

Liver

In newborn, adult and old rats the liver was the tissue with the lowest capacity to accumulate the intact amine (Table 2). As for the heart NMN practically did not appear in adult and old rats. On the contrary, similar amounts of NMN, DOPEG and DOMA were formed by the liver of newborns. As far as the metabolic degradation of noradrenaline is concerned, the liver was apparently the most mature tissue at birth. The ratio total metabolites formed in adults /total metabolites formed in newborns was: heart > kidney > aorta > liver (Table 2).

Discussion

The present results confirm that noradrenaline is preferentially inactivated by uptake into the sympathetic nerve terminals followed by accumulation or deamination. This has been already shown for adult animals in the canine saphenous vein (Guimarães, 1975; Guimarães and Paiva, 1977; Paiva and Guimarães, 1978) in the canine mesenteric artery (Garrett and Branco, 1977; Guimarães and Paiva, 1977) and in the rat heart (Fiebig and Trendelenburg, 1978). In vivo, it was also shown in adult rats, that noradrenaline is preferentially inactivated by neuronal pathways of metabolism (Eisenhofer, 1994; Eisenhofer and Finberg, 1994). The present results very clearly show that in the aorta (in neonates), in the heart (at any age), and in the liver (at any age), noradrenaline is preferentially deaminated at the concentration of $0.1\,\mu mol.l^{-1}$.

The amount of noradrenaline accumulated in the aorta, kidney and liver of newborns was not significantly different from that accumulated in the same organ of adult rats. This shows that at birth the sympathetic nerve endings are already developed and endowed with an active neuronal uptake. The presence of an active neuronal uptake at birth was previously reported for the carotid artery of the lamb (Su et al., 1977), for the rabbit aorta (Guimarães et al., 1991) and for the canine saphenous vein (Moura et al., 1993). However, the presence of neuronal uptake at birth does not necessarily means maturity of the sympathetic nerves since the development of the neuronal uptake precedes effective adrenergic transmission (Su et al., 1977). The transmembrane transport system is associated with the neuronal membrane of the new neurons, while the synthesis and storage capacities are linked to enzymes and vesicles which depend on the cell body (Dahlström and Häggendal, 1967; Laduron and Belpaire, 1968). According to the low noradrenaline content of the heart of newborn rats which was about 14% that of the heart of adult animals, the accumulation of noradrenaline was also lower in newborns than in adults, indicating that at birth the sympathetic nerves of the heart are either not yet fully present or not yet fully mature. Furthermore, in the heart, the enzymatic degradating capacity is even more underdeveloped at birth than the accumulating capacity, since the accumulation in the newborn represents about 35% that of the adult, while the total metabolites formation in the newborn reaches only about 10% of the value in the adult.

In sharp contrast to the capacity of the sympathetic nerve terminals in accumulating noradrenaline, the metabolic system for its degradation is still immature at birth. In all tissues, the amount of metabolites formed from both MAO and COMT activities is much smaller in newborns than in adults. The ratio metabolites formed/amine accumulated was about 2 times higher in adults than in newborns in the aorta, kidney and liver and about 3.5 times higher in adults than in newborns in the heart.

In adult and old rats, the predominant metabolites formed from noradrenaline in the heart were DOPEG and DOMA, while in newborns DOMA was practically not formed. The lack of DOMA formation at a moment at which DOPEG was the metabolite largely predominant favours the view that, at birth, there is a failure of the dehydrogenating process which can be due to either a failure of the aldehyde dehydrogenase or to a failure of its co-factor or to a failure of both the enzyme and the co-factor.

At any age, the accumulation of noradrenaline in the liver was very poor, as expected on the basis of the low noradrenaline content of this tissue. The low noradrenaline content and the very low amine accumulation in the liver indicate that the sympathetic innervation is very scarce and that deamination takes place predominantly at extraneuronal sites. In the isolated perfused rat liver, the extraneuronal uptake inhibitor corticosterone reduced the ^{3}H-metabolites without changing the accumulation of ^{3}H-noradrenaline (Steinberg et al., 1988) confirming that the metabolic degradation occurs mainly extraneuronally. While isolated hepatocytes possess a very high capacity to remove and metabolize both adrenaline and noradrenaline

(Martel et al., 1993), liver slices showed the lowest capacity to remove and degradate those amines among the four organs included in the present study.

At any age, the kidney showed an apparent lack of MAO activity, since noradrenaline was more O-methylated than deaminated and since more NMN than OMDA was formed. The ratio NMN/OMDA was about 3.

In adult and old rats, there was practically no NMN formation in the heart and in the liver. This unexpected finding might be ascribed to a high deaminating activity of the heart and the liver which so quickly transformed NMN into MOPEG or VMA that no NMN was accumulated. Alternatively one may admit that noradrenaline has a low affinity for COMT in these tissues.

With ageing (adult vs. old animals) there was a marked decline (by about 60%) in the capacity of the aorta to metabolize noradrenaline, while in the heart the capacity to degradate the same amine was markedly enhanced (a 2.5-fold increase). Most probably, the reduction in the metabolic capacity of the aorta is linked to degenerative changes occurred in the vessel wall dependent on age. Regarding the enhancement of this capacity in the heart it is tempting to speculate that this may be related to the high noradrenaline content of the hearts of old rats candidates to or already suffering from heart failure.

In summary, we conclude that in the rat: 1) at any age, noradrenaline is preferentially deaminated; 2) at birth, the capacity to metabolize noradrenaline is still underdeveloped, while that to accumulate it is already mature; 3) aldehyde dehydrogenase activity or that of its co-factor (or both) of the heart is apparently missing at birth; 4) removal of noradrenaline by the liver and the kidney did not change with ageing, while that by the aorta decreased and that by the heart increased.

Acknowledgements

This work was supported by JNICT (Junta Nacional de Investigação Cientifica), project 80/95, and by EC Biomed project EureCa.

References

Blatchford D, Holzbauer M, Grahame-Smith DG, Youdim MBH (1976) Ontogenesis of enzyme systems deaminating different monoamines. Br J Pharmacol 57: 279–293

Dahlström A, Häggendal J (1967) Studies on the transport and lifespan of amine storage granules in the adrenergic neuron system of the rabbit sciatic nerves. Acta Physiol Scand 69: 153–157

Docherty JR (1990) Cardiovascular responses in ageing: a review. Pharmacol Rev 42: 103–125

Eisenhofer G (1994) Plasma normetanephrine for examination of extraneuronal uptake and metabolism of noradrenaline in rats. Naunyn Schmiedebergs Arch Pharmacol 349: 259–269

Eisenhofer G, Finberg JPM (1994) Different metabolism of norepinephrine and epinephrine by catechol-O-methyltransferase and monoamine oxidase on rats. J Pharmacol Exp Ther 268: 1242–1251

Fiebig ER, Trendelenburg U (1978) The neuronal and extraneuronal uptake and metabolism of ^{3}H-(-)-noradrenaline in the perfused rat heart. Naunyn Schmiedebergs Arch Pharmacol 303: 21–35

Garrett J, Branco D (1977) Uptake and metabolism of noradrenaline by the mesenteric arteries of the dog. Blood Vessels 14: 43–54

Graefe K-H, Stefano F, Langer SZ (1973) Preferential metabolism of $(-)$-^{3}H-norepinephrine through deaminated glycol in the rat vas deferens. Biochem Pharmacol 22: 1147–1160

Guimarães S (1975) Further study of the adrenoceptors of the saphenous vein of the dog: influence of factors which interfere with the concentration of agonists at the receptor level. Eur J Pharmacol 34: 9–19

Guimarães S, Paiva MQ (1977) The role played by the extraneuronal system in the disposition of noradrenaline and adrenaline in vessels. Naunyn Schmiedebergs Arch Pharmacol 296: 279–287

Guimarães S, Brandão F, Paiva MQ (1978) A study of the adrenoceptor-mediated feedback mechanism by using adrenaline as a false transmitter. Naunyn Schmiedebergs Arch Pharmacol 305: 185–188

Guimarães S, Paiva MQ, Moura D, Vaz-da-Silva M (1991) Presynaptic α-autoadrenoceptors on peripheral noradrenergic neurones of newborn rabbits and dogs. In: Langer SZ, Galzin AM, Costentin J (eds) Presynaptic receptors and neuronal transporters. Pergamon Press, Oxford, pp 11–14 (Advances in the Biosciences 8)

Laduron P, Belpaire F (1968) Transport of noradrenaline and dopamine-β-hydroxilase in sympathetic nerves. Life Sci 7: 1–7

Martel F, Azevedo I, Osswald W (1993) Uptake and metabolism of ^{3}H-adrenaline and ^{3}H-noradrenaline by isolated hepatocytes and liver slices of the rat. Naunyn Schmiedebergs Arch Pharmacol 348: 450–457

Moura D, Vaz-da-Silva MJ, Azevedo I, Brandão F, Guimarães S (1993) Release and disposition of ^{3}H-noradrenaline in the saphenous vein of neonate and adult dogs. Naunyn Schmiedebergs Arch Pharmacol 347: 186–191

Paiva MQ, Guimarães S (1978) A comparative study of the uptake and metabolism of noradrenaline and adrenaline by the isolated saphenous vein of the dog. Naunyn Schmiedebergs Arch Pharmacol 303: 221–228

Steinberg P, Acevedo C, Masana MI, Rubio MC (1988) Uptake and metabolism of ^{3}H-($\pm$)-noradrenaline in the isolated perfused rat liver. Naunyn Schmiedebergs Arch Pharmacol 337: 392–396

Su C, Bevan JA, Assali NS, Brinkman CR (1977) Development of neuroeffector mechanisms in the carotid artery of the fetal lamb. Blood Vessels 14: 12–24

Trendelenburg U, Stefano E, Grohmann M (1983) The isotope effect of tritium in ^{3}H-noradrenaline. Naunyn Schmiedebergs Arch Pharmacol 323: 128–140

Authors' address: S. Guimarães, Institute of Pharmacology and Therapeutics, Faculty of Medicine, 4200 Porto, Portugal

The oxidation of dopamine and epinine by the two forms of monoamine oxidase from rat liver

M. Strolin Benedetti[1], **G. Sanson**[1], **L. Bona**[1], **M. Gallina**[1], **S. Persiani**[1], and **K. F. Tipton**[2]

[1] Department of Preclinical Development and Human Pharmacology, Zambon Group, Bresso, Italy
[2] Department of Biochemistry, Trinity College, Dublin, Ireland

Summary. Information on the "in vitro" oxidation of epinine by monoamine oxidase (MAO) compared to dopamine is very poor. The aim of this work was to study the oxidative deamination of epinine and dopamine by rat liver MAO-A and MAO-B.

The contributions of MAO-A and B to the metabolism of dopamine (55% and 45%, respectively) and epinine (70% and 30%, respectively) were similar. The results of this study show that epinine is a substrate for both forms of MAO in rat liver, although the contribution of MAO A to the deamination of this secondary amine appears to be slightly more important than that of MAO B.

Introduction

N-methyldopamine (epinine) is the active moiety of ibopamine, a do-paminergic agonist commonly used for the treatment of chronic heart failure. Indeed ibopamine is rapidly hydrolyzed to epinine by esterases both in laboratory animals and in humans (Pocchiari et al., 1986a,b; Lodola et al., 1986).

The comparison between the "in vitro" oxidation of adrenaline and noradrenaline by the two forms of monoamine oxidase (MAO) has been studied extensively (O'Carroll et al., 1986). In contrast, information on the "in vitro" oxidation of epinine by MAO compared to dopamine is very poor (Wiseman-Distler et al., 1965; Scriba and Borchardt, 1989) (Fig. 1). Only data comparing the behaviour of dopamine and epinine towards semicarbazide-sensitive amine oxidase in pig and human plasma have been published (Boomsma et al., 1993). The aim of this work was to study the oxidative deamination of epinine and dopamine by rat liver MAO-A and MAO-B.

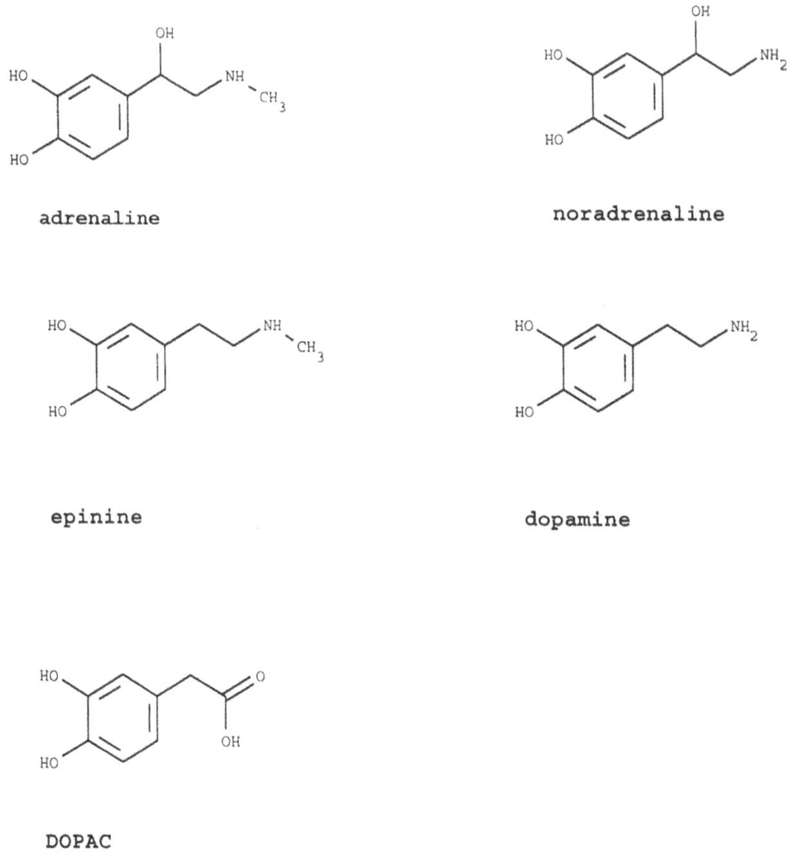

adrenaline

noradrenaline

epinine

dopamine

DOPAC

Fig. 1. Structures of biogenic amines and of their metabolite DOPAC

Materials and methods

Six Sprague Dawley male rats, 220–250 g body weight, were used. An homogenate of the six livers (5.33 g tissue/16 ml buffer) was prepared as previously described (Strolin Benedetti et al., 1983) and stored at $-196°C$ until use. The homogenate was further diluted 64 times. Aliquots of this liver homogenate (0.4 ml) were preincubated for 60 minutes at 37°C with concentrations of clorgyline in the range 3×10^{-10}–$3 \times 10^{-4} M$ (volume 0.45 ml), before being assayed for activity using either dopamine or epinine at the final concentration of 100 µM (final volume of incubation mixture 0.5 ml). Deamination of substrates was linear (Fig. 2a,b) with respect to both time (6 min) and enzyme concentration (1.54 mg fresh tissue/0.5 ml). After incubation with substrate, the 0.5 ml of incubation mixture was treated with 0.5 ml of 1N HCl and 100 µl of 2.2% sodium metabisulphite. Samples were centrifuged at 4,200 g for 15 minutes at +4°C and the supernatants were separated for the analysis of the end product 3,4-dihydroxyphenylacetic acid (DOPAC) by HPLC with electrochemical detection (Fig. 3), as reported by Mena et al. (1984), with some modifications. The chromatographic analysis was carried out on a Chrompack Nucleosil 100 C18 column (5 µm, 250 × 4.6 mm). The mobile phase was 0.02 M phosphate citrate buffer, pH 3.0/methanol (88:12, v:v) at a flow rate of 1 ml/min. The coulometric detector was an ESA 5100A model with an analytical cell model 5011 (oxidation potential +0.40 Volts). The injection volume was 25 µl and the retention time of DOPAC was about 15 minutes. Under these chromatographic conditions the limit of quantitation of the method was 0.1 ng/ml of supernatant, that is 2.5 pg of DOPAC injected.

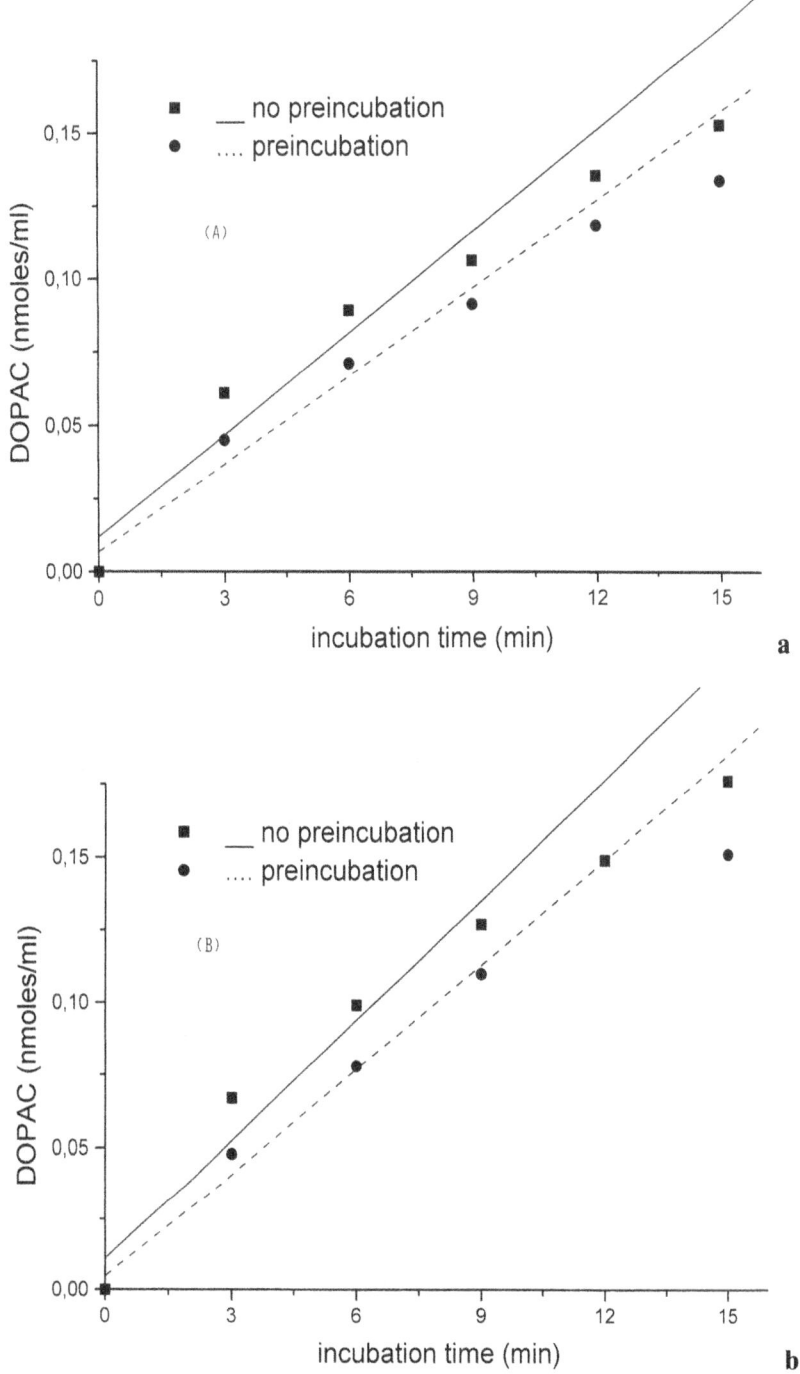

Fig. 2. Oxidative deamination of epinine (**A**) and dopamine (**B**) by MAOs of rat liver homogenates

Three separate experiments with the same homogenate were carried out with either epinire or dopamine in the presence of clorgyline. The enzyme activity was expressed as pmoles of DOPAC per mg of protein per minute, after having measured protein concentration in the initial liver homogenate by the method of Lowry et al. (1951).

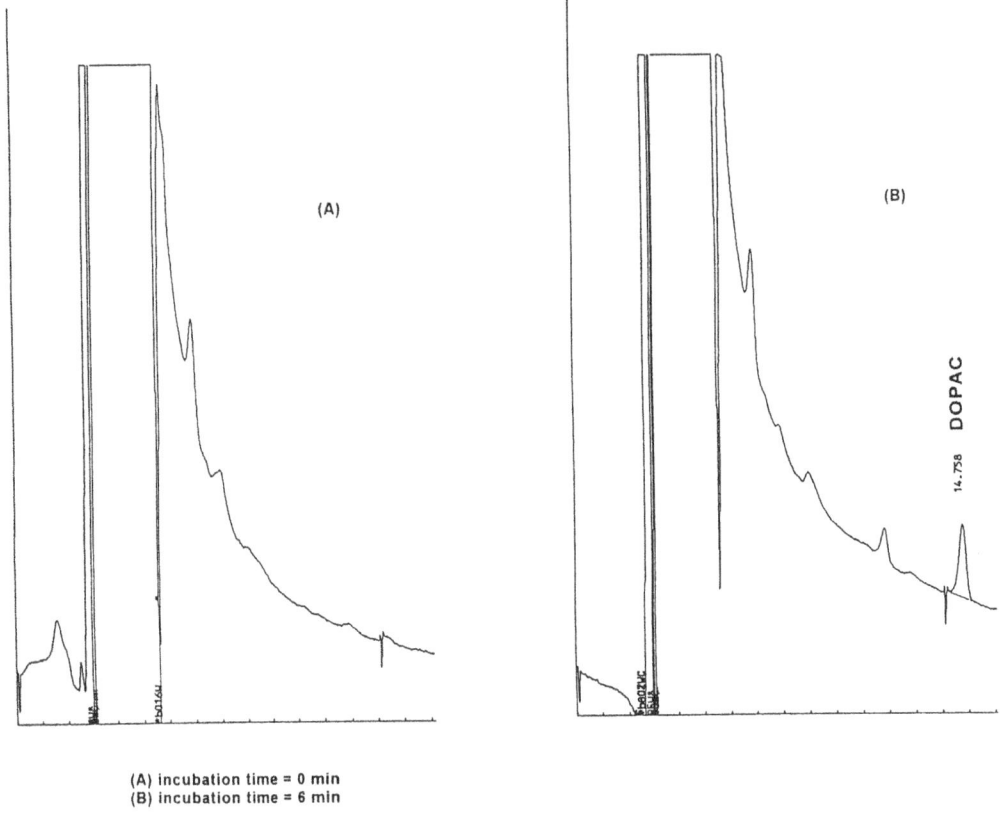

(A) incubation time = 0 min
(B) incubation time = 6 min

Fig. 3. Oxidation of epinine by rat liver monoamine oxidases

Endogenous level of DOPAC was approximately 0.005 nmoles/ml of supernatant (0.7 ng/ml). The DOPAC concentrations at each different incubation time reported in Fig. 2 are those obtained after subtraction of the endogenous DOPAC level.

Results and discussion

Recently, a similar methodology was used by Husseini et al. (1995) to measure the oxidative deamination of dopamine by human platelet MAO-B. The detection limit obtained by these authors was 8 pg of DOPAC injected compared to 2.5 pg in the present study. Husseini et al. used however an internal standard. The MAO activity of the homogenate (mean ± SD), after preincubation without clorgyline and 6 min incubation with either dopamine or epinine, was 45.1 ± 0.34 pmol·mg prot^{-1}·min^{-1} and 32.6 ± 7.46 pmol·mg prot^{-1}·min^{-1}, respectively. The loss of MAO activity due to homogenate preincubation was 20–22% with both substrates, in reasonable agreement with previous data on MAO from mouse brain and kynuramine as substrate (Hall et al., 1983). The contributions of MAO A and B to the metabolism of dopamine (55% and 45%, respectively) and epinine (70% and 30%, respectively) (Figs. 4 and 5) were similar. The data obtained for dopamine are in reasonable agreement

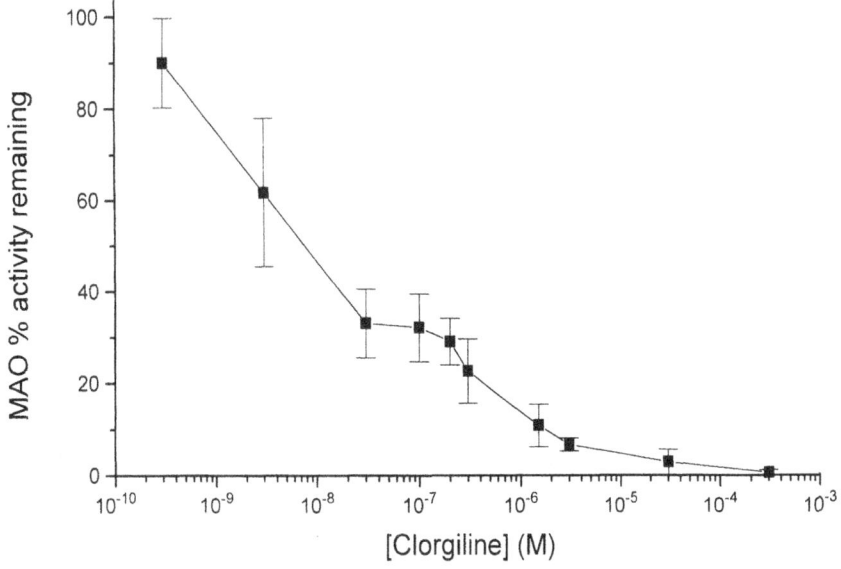

Fig. 4. Contribution of MAO-A and B to the metabolism of epinine

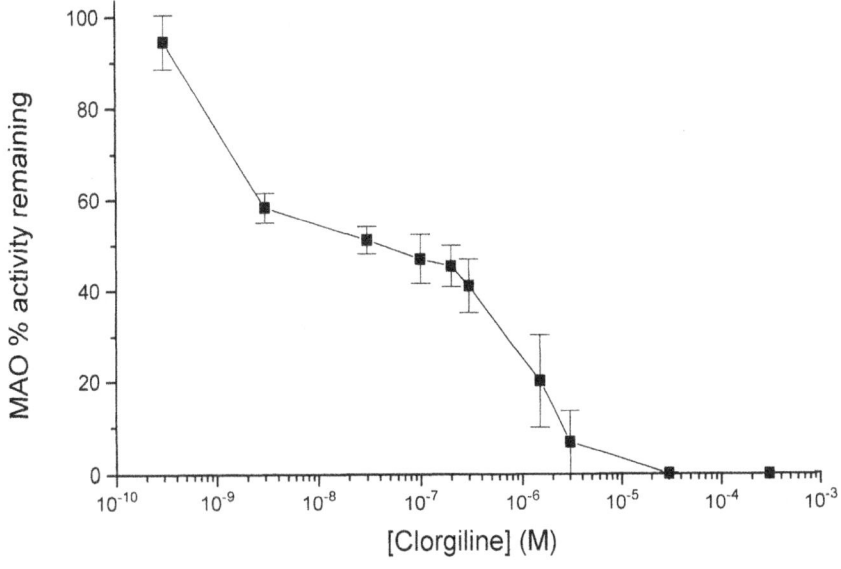

Fig. 5. Contribution of MAO-A and B to the metabolism of dopamine

with those of Fowler and Tipton (1984) who used however a radiochemical assay.

In conclusion, the results of this study show that epinine is a substrate for both forms of MAO in rat liver, although the contribution of MAO-A to the deamination of this secondary amine appears to be slightly more important than that of MAO-B. As regards the results obtained with epinine, it is interesting to recall that in cultured bovine brain microvessel endothelial cells, the MAO-mediated metabolism of epinine proved to be more sensitive to inhibition by clorgyline than by deprenyl (Scriba and Borchardt, 1989).

References

Boomsma F, van Woerkens LJ, Man in't Veld AJ, Verdouw PD, Schalekamp MADH (1993) High activity of semicarbazide-sensitive amine oxidase (SSAO): an important source of errors in the determination of the concentration of dopamine in pig plasma. J Cardiovasc Pharm 22: 198–202

Fowler CJ, Tipton KF (1984) On the substrate specificities of the two forms of monoamine oxidase. J Pharm Pharmacol 36: 111–115

Hall TR, Figueroa HR, Newton DK, Yurgens PB (1983) Thermal inactivation of mouse brain monoamine oxidase type A and type B. Gen Pharmacol 14: 681–684

Husseini H, Mitrovic V, Schlepper M (1995) Determination of monoamine oxidase B activity by high-performance liquid chromatography with electrochemical detection. J Chromatogr 672: 138–142

Lodola E, Borgia M, Longo A, Pocchiari F, Pataccini R, Sher D (1986) Ibopamine kinetics after a single oral dose in healthy volunteers. Arzneimittelforschung/Drug Res 36: 345–348

Lowry OH, Rosenbrough NJ, Farr AL, Randall RJ (1951) Protein measurement with the folin phenol reagent. J Biol Chem 193: 265–275

Mena MA, Aguado EG, de Yebenes JG (1984) Monoamine metabolites in human cerebrospinal fluid. HPLC/ED method. Acta Neurol Scand 69: 218–225

O'Carroll AM, Bardsley ME, Tipton KF (1986) The oxidation of adrenaline and noradrenaline by the two forms of monoamine oxidase from human and rat brain. Neurochem Int 8: 493–500

Pocchiari F, Pataccini R, Castelnovo P, Longo A, Casagrande C (1986a) Ibopamine, an orally active dopamine-like drug: metabolism and pharmacokinetics in rats. Arzneimittelforschung/Drug Res 36: 334–340

Pocchiari F, Pataccini R, Castelnovo P, Longo A, Paro M, Casagrande C (1986b) Ibopamine, an orally active dopamine-like drug: metabolism and pharmacokinetics in dogs. Arzneimittelforschung/Drug Res 36: 341–344

Scriba GKE, Borchardt RT (1989) Metabolism of catecholamine esters by cultured bovine brain microvessel endothelial cells. J Neurochem 53: 610–615

Strolin Benedetti M, Dostert P, Guffroy C, Tipton KF (1983) Partial or total protection from long-acting monoamine oxidase inhibitors (MAOIs) by new short-acting MAOIs of type A MD780515 and type B MD780236. Mod Probl Pharmacopsychiat 19: 82–104

Wiseman-Distler MH, Sourkes TL, Carabin S (1965) Precursors of 3,4-dihydroxyphenylacetic acid and 4-hydroxy-3-methoxyphenylacetic acid in the rat. Clin Chim Acta 12: 335–339

Authors' address: M. Strolin Benedetti, Preclinical Development and Human Pharmacology, Zambon Group S.p.A., Via Lillo del Duca, 10, I-20091 Bresso (MI), Italy

Properties and functions of tissue-bound semicarbazide-sensitive amine oxidases in isolated cell preparations and cell cultures

G. A. Lyles and **R. Pino***

Department of Pharmacology and Clinical Pharmacology, University of Dundee,
Ninewells Hospital and Medical School, Dundee, Scotland, United Kingdom

Summary. The demonstration of semicarbazide-sensitive amine oxidase
(SSAO) activity in some freshly-dispersed cell preparations and in particular
types of cells grown in culture, provides increasing opportunities for investigating the importance of SSAO in various aspects of cellular function. Assays
of benzylamine and methylamine metabolism in homogenates of cultured
cells have established clearly that SSAO is expressed in rat and pig vascular
(aortic) smooth muscle cells, as well as in rat non-vascular (anococcygeus,
trachea) smooth muscle, brown and white adipocytes. However, to date little
or no SSAO activity has been detected in cultures of human vascular smooth
muscle cells grown from blood vessels (e.g. umbilical artery) known to contain
the enzyme, and the reason for this is not yet apparent. However, those cell
cultures expressing SSAO are offering useful experimental models for studying biochemical and toxicological consequences upon cellular function which
may result from the metabolism of various aromatic and aliphatic amines
suggested to be possible physiological and xenobiotic substrates of the
enzyme.

Introduction

Although progress continues to be made, a clearer understanding of the
role of the tissue-bound semicarbazide-sensitive amine oxidase (SSAO) enzymes in cellular amine metabolism, and the physiological and/or toxicological effects of this metabolism upon cell and organ function, remains a primary
objective in studies upon this group of enzymes (see Lyles, 1996 for review).
While a central feature characterizing these enzymes is a relatively high
deaminating activity towards the synthetic amine benzylamine (BZ), there is
currently no evidence that BZ is likely to exist as an endogenously-occurring
physiological substrate for SSAO. Consequently, some support for the
idea that these enzymes may regulate the physiological and pharmacological

---

*Current address: Department of Pharmacology, University of Florence, Florence, Italy

actions of certain endogenous aromatic amines (e.g. tyramine, tryptamine, β-phenylethylamine, dopamine, histamine) has come from the finding that in various tissue homogenates, the metabolism of these amines is in part sensitive to inhibition by semicarbazide and other SSAO inhibitors (see Lyles, 1995). However, one major complication associated with the proposal that these aromatic amines are physiological substrates for SSAO enzymes is that their ability to be metabolised is very species-dependent, such that in vitro they are very poor substrates, if at all, for the human enzyme.

A more promising avenue has opened up with an increasing number of studies suggesting that SSAO may be involved in the metabolism of various endogenous aliphatic amines, with particular attention having been focussed upon methylamine and aminoacetone, which are good SSAO substrates in a variety of tissue homogenates from several species including man (Lyles, 1995; Callingham et al., 1995). Methylamine is derived from adrenaline, sarcosine and creatinine metabolic pathways, whereas aminoacetone is a product of glycine and threonine metabolism (Precious et al., 1988; Lyles and Chalmers, 1992). In addition, the fact that these amines are metabolized to potentially toxic aldehydes (formaldehyde and methylglyoxal) has suggested a possible toxicological aspect to aliphatic amine metabolism by SSAO, which was first highlighted by work from Boor's group showing that pathological lesions resembling atherosclerotic disease in cardiovascular tissues can be produced by experimental administration of the xenobiotic aliphatic amine allylamine to laboratory animals. These actions are generally attributed to the metabolism of allylamine to the cytotoxic aldehyde acrolein, by the action of SSAO, especially in vascular smooth muscle cells (Boor et al., 1990).

Over the last few years, there has been an increasing use of isolated cells, either immediately after dissociation from parent tissues, or after proliferation in cell cultures, in order to investigate the cellular distribution, as well as the properties and possible functions of SSAO. Here, we review some of the results which have emerged from these studies, as a step to assessing if the properties of SSAO in these cells are representative of those in the intact tissues from which they are derived.

Localization of SSAO in various cell types

Vascular smooth muscle and endothelial cells

It has been evident for many years, based upon studies of amine metabolism by blood vessel homogenates from a variety of animal species, that the vasculature contains particularly high activity of SSAO. The application of light microscopy-based histochemical methods for detecting hydrogen peroxide formation resulting from benzylamine (BZ) metabolism in human and rat vascular sections indicated that this SSAO was associated predominantly with smooth muscle cell-containing layers of the tunica media (Lewinsohn, 1984;

Lyles and Singh, 1985; Precious and Lyles, 1988). In a general review of SSAO, Lewinsohn (1984) reported that enzyme activity (towards BZ) was present in cultures of rat aortic smooth muscle cells, although experimental details of the methods used were rather sketchy. Lyles and Singh (1985) similarly found SSAO activity associated with freshly-dispersed smooth muscle cells obtained after collagenase treatment of the medial layers of the rat aorta.

More detailed studies on the activity of SSAO in cultured vascular smooth muscle cells have come from assays (using BZ) upon homogenates of pig aortic cells (Hysmith and Boor, 1987). At confluence, these cells expressed a similar level of SSAO activity in primary culture as in subcultures up to the sixth passage. However, the specific enzyme activity then gradually increased (approximately doubling) between the 6th and 15th passage, and remained stable thereafter (up to 21 passages). Interestingly, this increase in cell-associated SSAO activity mirrored the appearance of an ability of the cells to secrete a soluble form of SSAO into their culture medium. This soluble SSAO, subsequently purified from the conditioned medium, could be immunoprecipitated by an antiserum raised in rabbits against the purified cell-associated SSAO (Hysmith and Boor, 1988a). In these studies both purified forms of SSAO had a subunit molecular weight of 130 kDa determined by sodium dodecyl sulphate — polyacrylamide gel electrophoresis. Furthermore, the fact that the antiserum also recognized a soluble SSAO prepared from pig plasma raised the possibility that vascular smooth muscle cells may be one source of the plasma form of the enzyme. In relation to this, other studies showing that a cDNA isolated from a bovine liver cDNA library encodes the complete primary sequence of bovine plasma SSAO, has led to the suggestion that the liver may also participate in synthesis and secretion of the plasma enzyme (Mu et al., 1994).

Table 1. Kinetic constants for amine metabolism in homogenates of rat aortic cultured smooth muscle cells

Substrate	Culture	K_m (μM)	V_{max} (nmol/h/mg protein)
Benzylamine	Primary	8.9 ± 1.3	4.0 ± 2.3
Benzylamine	First passage	6.7 ± 1.3	2.6 ± 0.7
Methylamine	Primary	254 ± 59	3.6 ± 1.2
Methylamine	First-passage	347 ± 31	1.6 ± 0.2

Cells were homogenized in 1 mM potassium phosphate buffer and the metabolism of benzylamine and methylamine was measured by different radiochemical assay methods previously described (Precious et al., 1988; Lyles et al., 1990). Kinetic constants were determined in homogenates preincubated for 20 min at 37°C with 100 μM clorgyline (to inhibit any MAO activity), and are given as mean ± s.e. from 3–5 cultures. Benzylamine data from Blicharski and Lyles (1990), methylamine data previously unpublished by same authors

In our laboratory, we have determined SSAO activity towards both BZ and methylamine in homogenates of primary and first-passage cultures of rat aortic smooth muscle cells. Kinetic constants shown in Table 1 indicated that the K_m values for BZ (around 6–9 μM) and for methylamine (250–350 μM) were fairly close to values previously obtained with these substrates with SSAO in rat aortic homogenates (see Lyles, 1996). Thus by these particular criteria, the SSAO activity expressed in the cultures seems to have similar, if not identical properties, to the activity present in the original tissue.

Notwithstanding the success to date in detecting and studying SSAO activity in pig and rat aortic cultured cells, there appears at present to be no convincing demonstration of easily measurable SSAO activity expressed in human cultured vascular smooth muscle cells. In this respect, we have grown primary cultures of smooth muscle cells (verified by anti-[smooth muscle-actin] immunofluorescence) from explant tissue of human umbilical artery, a blood vessel which we have previously used extensively to characterize the properties of human SSAO (Precious et al., 1988; Precious and Lyles, 1988; Lyles and Chalmers, 1992). These cultures, grown in explant-seeded six-well tissue culture trays, proliferated very slowly by outgrowth from the explants, and required from 4–6 weeks to reach confluence. After combining and homogenizing cells scraped from several wells, we found little if any deaminating activity towards 1 mM [^{14}C] BZ. In a few cultures allowed to grow into dense overlapping layers of cells, there sometimes appeared to be a very small amount of BZ metabolism which was sensitive to inhibition by semicarbazide, although this activity was only marginally detectable above the assay blank values. Interestingly, there was no difficulty in measuring SSAO activity in homogenates made from the tissue explants from which the cultured cells had migrated and proliferated. It is not clear why SSAO activity appears to be either absent or poorly expressed by these cells. However, a similar inability to detect SSAO activity in smooth muscle cells grown from several adult human blood vessels has been found by other workers (Boor, personal communication). Perhaps the phenotypic and functional properties of the cells when grown in culture differ in various respects from those of the smooth muscle cells organized within the extracellular matrix of the parent tissue. Further investigation of such possibilities may provide important clues about the role of SSAO in human vascular smooth muscle.

The presence of SSAO in vascular tissue homogenates from various species raises the question as to whether the enzyme is also present in endothelial cells, in order to metabolize amines at the interface between the circulation and the vascular wall. Hysmith and Boor (1987) found no detectable enzyme activity in pig aortic endothelial cells cultured up to 21 passages. In contrast, Ramos et al. (1988) proposed that rat aortic endothelial cells may contain SSAO on the basis of the finding that cell suspensions could metabolize allylamine to acrolein, albeit with very low activity compared with smooth muscle cells. However, the lack of suitable inhibitor studies here leaves this interpretation as to the nature of the enzyme involved rather unclear.

Non-vascular smooth muscle cells

As well as showing that SSAO is associated with vascular smooth muscle, Lewinsohn (1981) also proposed that SSAO is found in non-vascular smooth muscle, since high enzyme activity was present in homogenates of smooth muscle-rich tissues such as the human uterus, ureter and vas deferens. Support for this also came from the demonstration that SSAO is present in homogenates of the rat anococcygeus, another tissue predominantly containing smooth muscle (Barrand et al., 1981).

In order to investigate this further, we have recently developed methods to grow cultures of smooth muscle cells dissociated with collagenase from the rat anococcygeus and trachea (in the latter case by using the strip of smooth muscle-rich tissue found along the length of the trachea at the regions of incomplete closure of the cartilage rings). Cells were identified as smooth muscle again by anti-[smooth muscle actin] immunofluorescence. Metabolism of BZ ($10\mu M$) determined in homogenates of first-passage cells was predominantly sensitive to inhibition by $100\mu M$ semicarbazide presumably corresponding to SSAO activity, although a small component inhibited by $100\mu M$ pargyline suggested that some monoamine oxidase (MAO) activity was also present, especially in the anococcygeus smooth muscle cells (Table 2). K_m values for BZ metabolism (Table 3), determined with relatively low substrate concentrations in these cells, are very similar to those values

Table 2. Inhibition of $10\mu M$ benzylamine (BZ) metabolism in homogenates of rat tracheal and anococcygeal cultured smooth muscle cells

Inhibitor	BZ metabolism (% of control)	
	Trachea	Anococcygeus
Semicarbazide ($100\mu M$)	18 ± 2	24 ± 6
Pargyline ($100\mu M$)	89 ± 4	76 ± 9

Smooth muscle cells were dissociated from tissues by collagenase treatment ($2\,mg/ml$) in Medium 199, and grown in six-well culture dishes containing Medium 199, foetal calf serum (10% v/v), antibiotics (penicillin/streptomycin) and amphotericin. At confluence, they were passaged by trypsinization, and further grown to confluence before collection by scraping from dishes and use in assays. Homogenates were prepared in $10\,mM$ potassium phosphate buffer pH 7.8 and preincubated with inhibitors for $20\,min$ at $37°C$ before determination of [^{14}C] BZ metabolism as previously described (Precious et al., 1988). Control samples were preincubated without inhibitors. Results are mean $\pm$ s.e. of 4 (trachea) or 3 (anococcygeus) cultures, each from different rats

Table 3. Kinetic constants for benzylamine (BZ) metabolism in homogenates of rat tracheal and anococcygeal cultured smooth muscle cells

Tissue	K_m (μM)	V_{max} (nmol/h/mg protein)
Trachea (7)	5.6 ± 0.4	2.1 ± 0.4
Anococcygeus (4)	7.9 ± 4.0	2.5 ± 0.8

Homogenates were prepared as in Table 2. [^{14}C] BZ metabolism was determined at 2, 5, 10 and 20 μM. Numbers of cultures (from different rats) are given in parentheses. Values for kinetic constants are mean ± s.e.

usually found for BZ metabolism by SSAO in rat tissue homogenates (see Lyles, 1996). Consequently, it is clear that cultured non-vascular smooth muscle cells, from the rat at least, can express SSAO activity and this should provide further opportunities for investigating the role of SSAO in smooth muscle function.

White and brown adipocytes

The other major cell type in which SSAO activity has been studied fairly extensively is the adipocyte. SSAO activity was first described in homogenates of rat brown adipose tissue by Barrand and Callingham (1982), and histochemical studies with the electron microscope subsequently localized the enzyme to the plasma membrane of intact brown adipocytes (Barrand et al., 1984). Freshly isolated white adipocytes from the rat were subsequently shown to contain a membrane-bound SSAO activity which metabolized not only BZ, but also β-phenylethylamine and histamine (Raimondi et al., 1991).

Raimondi et al. (1990) also showed that preadipocytes from both white and brown adipose tissue can be isolated and maintained in culture, and that these cells transform over several days to a committed adipocyte phenotype if grown in a lipogenic culture medium containing isobutylmethylxanthine, dexamethasone and insulin. SSAO activity (towards BZ) which was barely detectable in the preadipocytes increased markedly during cell transformation and the accumulation of lipid, suggesting that there may be some link between SSAO and these processes. These authors have also indicated that this transformation can be delayed by the SSAO inhibitors semicarbazide and B24 (Raimondi et al., 1993). In addition to metabolizing BZ, it has been shown that white and brown mature adipocytes freshly dissociated from intact adipose tissue, as well as white and brown adipocytes obtained by the transformation process in culture, are capable of metabolizing methylamine (Conforti et al., 1993) with a K_m (around 250–300 μM), similar to values for methylamine metabolism found in other rat tissues (Lyles, 1996). An immunofluorescence-based demonstration of SSAO on pig white adipocytes has also been reported, with the use of a rabbit antiserum prepared against pig plasma SSAO which recognized the adipocyte-associated enzyme (Raimondi et al., 1992).

Other cell types

SSAO activity has been localized to fibroblasts in guinea pig skin by immunochemical methods (Buffoni et al., 1993) and in chondrocytes in sections of rat articular cartilage by peroxidase-based histochemical methods (Lyles and Bertie, 1987). However, whether or not these cell types also express SSAO if grown under culture conditions remains to be established.

Amine metabolism by SSAO in cultured cells: physiological or toxicological consequences?

Subcellular fractionation studies carried out with both pig aortic cultured smooth muscle cells and rat brown adipocytes freshly dissociated from adipose tissue, have led to the conclusion that a major part of the SSAO activity in these cells is associated with the plasma membrane (Barrand et al., 1984; Hysmith and Boor, 1988b). This is consistent with other studies based on subcellular fractionation of whole tissues (e.g. rat aorta — Wibo et al., 1980). Furthermore, it has been suggested that SSAO activity may face outwards from the plasmalemma, such that it could metabolize extracellular amines and regulate their functional effects upon cells (Barrand et al., 1984; Holt and Callingham, 1994). Some experimental support for this hypothesis is available with freshly dissociated white adipocytes. Raimondi et al. (1993) demonstrated that histamine can stimulate lipolysis in these cells principally by activating histamine-H_2 receptors. Histamine is metabolized by SSAO in homogenates of these cells, and thus it is of interest that the SSAO inhibitors semicarbazide and B24 were capable of potentiating the lipolytic effects of histamine.

By far the major use of cultured cells to investigate functional aspects of amine metabolism by SSAO has been in studies of possible toxicological effects which may arise from the metabolism of certain aliphatic amines, principally focussing to date upon allylamine and methylamine. The administration of the xenobiotic and industrially-used compound allylamine to various laboratory species produces necrotic and fibrotic lesions of myocardial and vascular tissue, in the latter case often accompanied by smooth muscle cell proliferation, which under some circumstances resembles similar pathological changes seen in atherosclerotic disease (Boor and Hysmith, 1987). Allylamine is also cytotoxic towards cultured smooth muscle cells from rat and pig aorta, causing inhibition of cell growth and also inducing cell lysis. It has been proposed that this is as a consequence of the bioconversion of allylamine to the toxic aldehyde acrolein, by the action of SSAO associated with the cells, since these cytotoxic effects (and the pathological changes in vivo) could be prevented by SSAO inhibitors (Hysmith and Boor, 1988b; Ramos et al., 1988). Indeed, the direct detection of acrolein formation by cultured vascular smooth muscle cells has been described (Ramos et al., 1988; Boor et al., 1990) Amine metabolism by SSAO is also accompanied by hydrogen peroxide formation, another potential cytotoxic agent, and the possibility that this contributes to the effects of allylamine was suggested by studies showing that the toxicity of the latter upon cultured cells was partially prevented by addition of catalase (Ramos et al., 1988).

Cellular detoxification of acrolein involves its conjugation with reduced glutathione (GSH), and this can lead to the depletion of cellular GSH, as has been observed in allylamine-treated smooth muscle cells (Ramos and Thurlow, 1993). GSH-dependent peroxidase activity is also involved in hydrogen peroxide metabolism (Shan et al., 1990), and thus cellular GSH levels probably play a crucial role in determining the extent of the cytotoxic insult produced by allylamine (and acrolein). In support of this, we have previously shown that allylamine-induced cell lethality is potentiated in rat aortic smooth muscle cells pretreated with buthionine sulphoximine, an inhibitor of GSH synthesis (Blicharski and Lyles, 1991).

In addition to cytotoxic effects upon smooth muscle cells, allylamine has also been shown to have similar actions usually at fairly high concentrations on cultured pig and rat endothelial cells (Hysmith and Boor, 1986; Ramos et al., 1988), as well as on myocytes and fibroblasts from neonatal rat heart (Toraason et al., 1989). Although these effects also could be blocked by SSAO inhibitors, it has not been established in these cases to what extent this might represent a capacity of the cells themselves to metabolize allylamine weakly, or alternatively be due to the metabolic activity of a soluble SSAO activity in the bovine serum component of the medium in which cells are usually cultured. We recently showed that cytotoxic actions of allylamine ($50\,\mu M$) upon cultures of human umbilical vein endothelial cells were greatly increased by adding samples of human umbilical artery homogenates to the cells, and these effects could be prevented by SSAO inhibitors, thus demonstrating that SSAO in the human blood vessel is also capable of metabolizing allylamine (Pino and Lyles, submitted).

A similar approach to the latter was originally used by Yu and Zuo (1993) who showed that methylamine is also cytotoxic to cultured endothelial cells from calf pulmonary artery and human umbilical vein, particularly in the presence of SSAO activity provided by adding samples of human umbilical artery homogenate or human plasma to the culture dishes. These cytotoxic effects were substantially inhibited by the SSAO inhibitors semicarbazide and MDL 72974A. Formaldehyde and hydrogen peroxide, two of the metabolic products of methylamine deamination by SSAO were also shown to be directly cytotoxic in this study.

To our knowledge, there have not been any investigations to date of the possible toxicological consequences of aminoacetone metabolism (to methylglyoxal) by SSAO in cultured vascular cells or tissues. Such studies would be of interest in relation to increasing evidence that aminoacetone is a very good substrate for SSAO in a number of species (see Lyles, 1996). However, we have found recently that formaldehyde (200μM) and methylglyoxal (500μM), which by themselves have relatively little effects upon the viability of human umbilical vein cultured endothelial cells, have their cytotoxicity greatly enhanced, leading to death of almost all cells, after depletion of GSH levels in the cells with buthionine sulphoximine (Pino and Lyles, submitted).

Many of the studies described above have utilised assays based upon cell death as the end point to assess the cytotoxic effects of aliphatic amines and their SSAO-derived aldehyde metabolites. Although such studies have often used relatively high concentrations of these agents, exceeding those normally likely to occur endogenously, it is probable that more subtle actions to produce important alterations in normal cellular functions, but without necessarily killing cells, may occur at lower physiological and/or pathological concentrations of these compounds and further investigation of these possibilities would appear to be worthwhile.

Conclusions

Studies with homogeneous populations of cultured cells offer a variety of opportunities for investigating factors which may regulate the expression of SSAO enzymes, their biochemical properties and the effects upon cellular function resulting from the metabolism of various amines by these enzymes. Of course, if cultured cells are to provide suitable models for SSAO activities found in vivo, the properties of the enzymes expressed in both cases should ideally be identical. Where attempts have been made to confirm this, for example, in studies of subcellular localization and substrate specificity (including determinations of K_m values towards substrates), there appears to be a good correspondence between SSAO properties expressed in cultured cells, compared with those of the enzyme in the parent tissue from which the cells were derived. The techniques of molecular biology, as they develop in this area, should also be capable of probing and comparing the molecular properties of enzyme structure in these cases.

It is envisaged that culture systems may also be useful in examining the role that specific growth factors or hormones have upon SSAO expression, in a similar manner to the intriguing experiments demonstrating an increase in SSAO activity occurring at the same time as lipogenesis in rat transformed preadipocytes (Raimondi et al., 1990). Studies following the time course of SSAO recovery in cultures treated with irreversible SSAO inhibitors should allow information about the processes of enzyme turnover (synthesis and degradation rates) to be investigated. In this respect, Hysmith and Boor (1988b) have demonstrated that SSAO activity in pig aortic smooth muscle cells returned completely to control levels, 6h after complete inhibition with 100μM semicarbazide, whereas recovery appeared to be rather slower, with approximately 50% recovery occurring 8h after inhibition with 100μM phenelzine.

Since the substrate specificity of SSAO varies quite markedly between species (See Lyles, 1995), studies on tissues from laboratory animals may have limited predictive value in defining the role of SSAO in amine metabolism in man. Although SSAO activity has been found, using conventional assay methods, in homogenates prepared from a wide range of human tissues, to date we are unaware of any human cell line(s) expressing significant amounts of easily detectable enzyme activity, and this includes cultured smooth muscle cells grown from the human umbilical artery, a blood vessel known to possess SSAO in vivo. The discovery of an appropriate cell line, or suitable culture conditions, which permit a more ready demonstration of SSAO expression in human cells would be of considerable importance for promoting the investigation of possible physiological and toxicological influences of SSAO upon cellular function in man.

Acknowledgements

Recent work in this laboratory has been supported by a European Union Research Contract (CHRX-CT93-0256) under the Human Capital and Mobility Programme.

References

Barrand MA, Callingham BA, Lyles GA (1981) Monoamine deamination in the anococcygeus muscle of the rat. Br J Pharmacol 74: 198P

Barrand MA, Callingham BA (1982) Monoamine oxidase activities in brown adipose tissue of the rat: some properties and subcellular distribution. Biochem Pharmacol 31: 2177–2184

Barrand MA, Fox SA, Callingham BA (1984) Amine oxidase activities in brown adipose tissue of the rat: identification of semicarbazide-sensitive (clorgyline-resistant) activity at the fat cell membrane. J Pharm Pharmacol 38: 288–293

Blicharski JRD, Lyles GA (1990) Semicarbazide-sensitive amine oxidase activity in rat aortic cultured smooth muscle cells. J Neural Transm [Suppl] 32: 337–339

Blicharski JRD, Lyles GA (1991) D,L-buthionine sulphoximine, a glutathione depleting agent, potentiates allylamine-induced cytotoxicity in rat aortic smooth muscle cultures. Br J Pharmacol 102: 184P

Boor PJ, Hysmith RM (1987) Allylamine cardiovascular toxicity. Toxicology 44: 129–145

Boor PJ, Hysmith RM, Sanduja R (1990) A role for a new vascular enzyme in the metabolism of xenobiotic amines. Circ Res 66: 249–252

Buffoni F, Banchelli G, Cambi S, Ignesti G, Pirisino R, Raimondi L, Vannelli G (1993) Skin would healing: some biochemical parameters in guinea pig. J Pharm Pharmacol 45: 784–790

Callingham BA, Crosbie AE, Rous BA (1995) Some aspects of the pathophysiology of semicarbazide-sensitive amine oxidase enzymes. In: Yu PM, Tipton KF, Boulton AA (eds) Current neurochemical and pharmacological aspects of biogenic amines. Elsevier, Amsterdam, pp 305–321 (Prog Brain Res 106)

Conforti L, Raimondi L, Lyles GA (1993) Metabolism of methylamine by semicarbazide-sensitive amine oxidase in white and brown adipose tissue of the rat. Biochem Pharmacol 46:603–607

Holt A, Callingham BA (1994) Location of the active site of rat vascular semicarbazide-sensitive amine oxidase. J Neural Transm [Suppl] 41: 433–437

Hysmith RM, Boor PJ (1986) Comparative toxicity of the cardiovascular toxin allylamine to porcine aortic smooth muscle and endothelial cells. Toxicology 38: 141–150

Hysmith RM, Boor PJ (1987) In vitro expression of benzylamine oxidase activity in cultured porcine smooth muscle cells. J Cardiovasc Pharmacol 9: 668–674

Hysmith RM, Boor PJ (1988a) Purification of benzylamine oxidase from cultured porcine aortic smooth muscle cells. Biochem Cell Biol 66: 821–829

Hysmith RM, Boor PJ (1988b) Role of benzylamine oxidase in the cytotoxicity of allylamine toward aortic smooth muscle cells. Toxicology 51: 133–145

Lewinsohn R (1981) Amine oxidase in human blood vessels and non-vascular smooth muscle. J Pharm Pharmacol 33: 569–575

Lewinsohn R (1984) Mammalian monoamine-oxidizing enzymes, with special reference to benzylamine oxidase in human tissues. Braz J Med Biol Res 17: 223–256

Lyles GA (1995) Substrate-specificity of mammalian tissue-bound semicarbazide-sensitive amine oxidase. In: Yu PM, Tipton KF, Boulton AA (eds) Current neurochemical and pharmacological aspects of biogenic amines. Elsevier, Amsterdam, pp 293–303 (Prog Brain Res 106)

Lyles GA (1996) Mammalian plasma and tissue-bound semicarbazide-sensitive amine oxidases: biochemical, pharmacological and toxicological aspects. Int J Biochem Cell Biol 28: 259–274

Lyles GA, Singh I (1985) Vascular smooth muscle cells: a major source of the semicarbazide-sensitive amine oxidase of the rat aorta. J Pharm Pharmacol 37: 637–643

Lyles GA, Bertie KH (1987) Properties of a semicarbazide-sensitive amine oxidase in rat articular cartilage. Pharmacol Toxicol [Suppl 1] 60: 33

Lyles GA, Chalmers J (1992) Aminoacetone metabolism by semicarbazide-sensitive amine oxidase in human umbilical artery. Biochem Pharmacol 43:1409–1414

Lyles GA, Holt A, Marshall CMS (1990) Further studies on the metabolism of methylamine by semicarbazide-sensitive amine oxidase activities in human plasma, umbilical artery and rat aorta. J Pharm Pharmacol 42: 332–338

Mu D, Medzihradsky KF, Adams GW, Mayer P, Hines WM, Burlingame AL, Smith AJ, Cai D, Klinman JP (1994) Primary structures for a mammalian and serum copper amine oxidase. J Biol Chem 269: 9926–9932

Precious E, Lyles GA (1988) Properties of a semicarbazide-sensitive amine oxidase in human umbilical artery. J Pharm Pharmacol 40: 627–633

Precious E, Gunn CE, Lyles GA (1988) Deamination of methylamine by semicarbazide-sensitive amine oxidase in human umbilical artery and rat aorta. Biochem Pharmacol 37: 707–713

Raimondi L, Pirisino R, Banchelli G, Ignesti G, Conforti L, Buffoni F (1990) Cultured preadipocytes produce a semicarbazide-sensitive amine oxidase (SSAO) activity. J Neural Transm [Suppl] 32: 331–336

Raimondi L, Pirisino R, Ignesti G, Capecchi S, Banchelli G, Buffoni F (1991) Semicarbazide-sensitive amine oxidase activity (SSAO) of rat epididymal white adipose tissue. Biochem Pharmacol 41: 467–470

Raimondi L, Pirisino R, Banchelli G, Ignesti G, Conforti L, Romanelli E, Buffoni F (1992) Further studies on semicarbazide-sensitive amine oxidase activities (SSAO) of white adipose tissue. Comp Biochem Physiol 102B: 953–960

Raimondi L, Conforti L, Banchelli G, Ignesti G, Pirisino R, Buffoni F (1993) Histamine lipolytic activity and semicarbazide-sensitive amine oxidase (SSAO) of rat white adipose tissue (WAT). Biochem Pharmacol 46: 1369–1376

Ramos KS, Thurlow CH (1993) Comparative cytotoxic responses of cultured avian and rodent aortic smooth muscle cells. J Toxicol Environ Health 40: 61–76

Ramos K, Grossman SL, Cox LR (1988) Allylamine-induced vascular toxicity in vitro: prevention by semicarbazide-sensitive amine oxidase inhibitors. Toxicol Appl Pharmacol 95: 61–71

Shan X, Aw TY, Jones DP (1990) Glutathione-dependent protection against oxidative injury. Pharmacol Ther 47: 61–71

Toraason M, Luken ME, Breitenstein M, Krueger JA, Biagini RE (1989) Comparative toxicity of allylamine and acrolein in cultured myocytes and fibroblasts from neonatal rat heart. Toxicology 56: 107–117

Wibo M, Duong AT, Godfraind T (1980) Subcellular location of semicarbazide-sensitive amine oxidase in rat aorta. Eur J Biochem 112: 87–94

Yu PH, Zuo D-M (1993) Oxidative deamination of methylamine by semicarbazide-sensitive amine oxidase leads to cytotoxic damage in endothelial cells. Diabetes 42: 594–603

Authors' address: Dr. G. A. Lyles, Department of Pharmacology and Clinical Pharmacology, University of Dundee, Ninewells Hospital and Medical School, Dundee, DD1 9SY, United Kingdom

Studies on the time-dependent activation of microsomal semicarbazide-sensitive amine oxidase

J. M. Lizcano[1], **A. Fernández de Arriba**[1], **K. F. Tipton**[2], and **M. Unzeta**[1]

[1]Department of Bioquimica i Biologia Molecular, Universitat Autonoma de Barcelona, Barcelona, Bellaterra, Spain
[2]Biochemistry Department, Trinity College, Dublin, Ireland

Summary. The semicarbazide-sensitive amine oxidase (SSAO) from bovine lung microsomes was activated in a temperature- and time-dependent process. This behaviour was observed when the enzyme was preincubated at 25°C, 37°C and 50°C but not at 4°C. This activation was only observed when benzylamine was used as substrate but not when methylamine, histamine or 2-phenylethylamine were used. The activation was independent of pH, ionic strength and the nature of the buffer used. At 37°C the specific activity had risen to a value that was about 7 times higher than that of the starting material after 120 min. This process affected only the maximum velocity of the reaction with the K_m value remaining essentially unchanged. Treatment of SSAO with phospholipases and detergents did not affect this behaviour. Incubation of the enzyme with serine proteases, metal chelating agents, reducing agents or protease inhibitors, had no effect on the activation. The fact that both forms of the enzyme (activated and non-activated), showed the same Mr values on gel filtration chromatography excluded the possibility of an enzyme aggregation and/or degradation being involved in this process.

Introduction

Physiologically-active endogenous and xenobiotic amines can be metabolized by a semicarbazide-sensitive amine oxidase (SSAO) activity (EC 1.4.3.6), which is distinct from the more widely-studied monoamine oxidase (EC 1.4.3.4. MAO).

SSAO is present in many different cell types, with activity being particularly high in vascular smooth muscle tissue. It is characterized by its high affinity towards benzylamine, although, like MAO, it can also catalyze the oxidative deamination of dopamine, tyramine and 2-phenylethylamine.

The tissue enzyme has been shown to be membrane-bound and Barrand et al. (1982) demonstrated a plasma membrane localization in adipocytes. This suggests that a physiological role of the enzyme could be to scavenge circulating toxic amines, without the necessity for a reuptake process. Furthermore the generation of H_2O_2 by SSAO in the membrane could act as a second messenger which might play an important physiological role in the control of the cellular activity (see Schreck et al., 1991; Meyer et al., 1993; Feng et al., 1995). Nevertheless its true physiological role remains to be fully elucidated.

In our studies on purification of SSAO from microsomes of bovine lung (Lizcano et al., manuscript in preparation), we have observed an anomalous activation behaviour towards benzylamine as substrate. The aim of the present work was to study this phenomena from a kinetic point of view.

Materials and methods

Bovine lung microsomes were prepared essentially as described previously (Lizcano et al., 1994). Microsomes were solubilized by mixing with an equal volume of Triton X-100, sodium cholate or β-octylglucoside at different concentrations, and shaking at 4°C for 15 min folowed by centrifugation for 1 hour at 105,000 g. The supernatant containing the solubilized enzyme was stored at −20°C. SSAO activity was determined radiochemically using ^{14}C-benzylamine, ^{14}C-2-phenylethylamine (Fowler et al., 1981) or ^{14}C-methylamine (Precious et al., 1988) as substrates. In the case of benzylamine, SSAO activity was also determined spectrophotometrically by continously monitoring the formation of benzaldehyde at 250 nm (Tabor et al., 1954). For determining histamine (20 mM) oxidation the fluorimetric method of Crosbie et al. (1994), was used. Protein concentration was determined by the method of Hartree et al. (1972).

Mixed substrate experiments were carried out with benzylamine and 2-phenylethylamine at their K_m concentration (40 and 312 μM, respectively). They were assayed separately and as a mixture.

Activation experiments were done by preincubating the enzyme in 50 mM potassium phosphate buffer, pH 7.2, for different times before starting the reaction with substrate.

Partially pure SSAO (after hydroxyapatite, DEAE-Sephacel, and Lens-culinaris agarose chromatography columns) was chromatographed through FPLC-Superdex 200 HR (exclusion limit 30–600 kDa). The enzyme was eluted with 150 mM NaCl, 20 mM potassium phosphate buffer, pH 7.2, containing 0.1% Triton X-100, at a flow rate of 0.3 ml/min and 0.5 ml fractions were collected (Lizcano et al., manuscript in preparation). The enzyme was purified 46 fold in a yield of 13%, by this procedure to yield a preparation with a specific activity of 46 nmol·min^{-1}·mg protein^{-1}.

Results

Incubation of the microsomal fraction from bovine lung in 50 mM potassium phosphate buffer, pH 7.2, at 37°C induced a time-dependent activation of SSAO when benzylamine was used as substrate. This behaviour was not observed when 2-phenylethylamine, methylamine or histamine were used (Fig. 1a).

This phenomena was also observed when the enzyme was incubated at 25 and 50, but not at 4°C (Fig. 1b). However once the activation has been started by incubation for a period at 37°C it continued to progress when the incubation temperature was reduced at 4°C (Fig. 2).

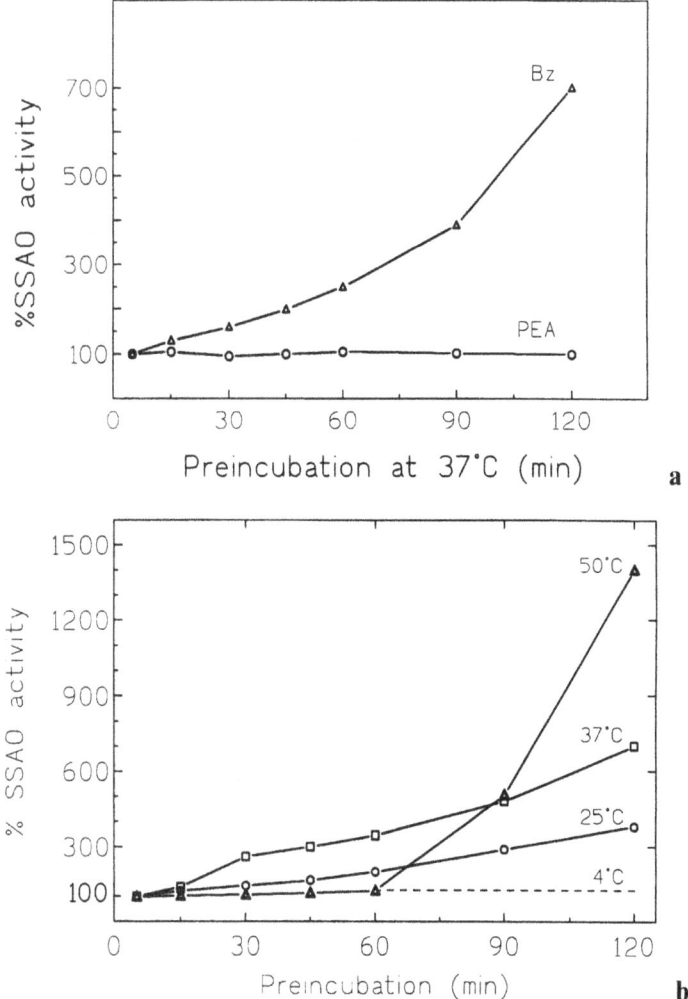

Fig. 1. The effects of preincubation on SSAO activity. **a** The enzyme was preincubated with buffer at 37°C for different times before the reaction was started by adding the substrate, benzylamine (40 μM) (Bz), methylamine (100 μM), histamine (2 mM), or 2-phenylethylamine (PEA 312 μM). The behaviour with the last three substrates was not significantly different from one another and for clarity only the data obtained for PEA are shown. **b** The time-courses of SSAO activation were determined during preincubation at 25, 37, and 50°C with benzylamine as substrate. Each point is the mean of triplicate determinations. Ranges of individual values were less than 5% in all cases. Errors bars are omitted for clarity

The activation was due solely to an increase in V_{max} value, with the K_m value not being significantly affected (Fig. 3).

This activation phenomena was independent of the pH in the range (6 to 10), ionic strength (50 mM–400 mM potassium phosphate buffer, pH 7.2) and the nature of the buffer utilized (Tris-HCl, Hepes, or phosphate; data not shown).

The elution profile on gel filtration through Superdex 200HR was not significantly affected by the enzyme activation (Fig. 4). This rules out the

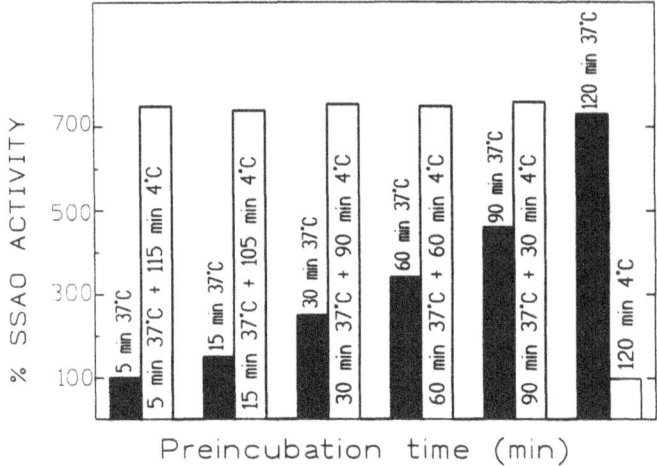

Fig. 2. The effects of altering the incubation temperature on the activation of SSAO. The enzyme was incubated at 37°C for the times indicated after which the mixture was cooled to 4°C and the incubation was continued for the stated time before the activity was determined with benzylamine as substrate. Each value is the mean of triplicate determinations. Ranges of individual values were less than 5% in all cases. Errors bars are omitted for clarity

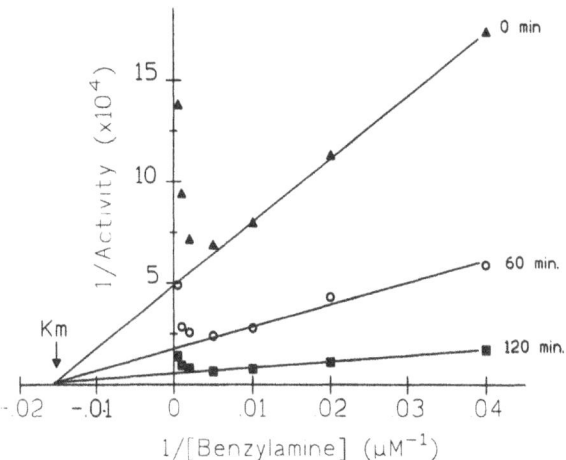

Fig. 3. Double-reciprocal plots of the initial velocities of benzylamine (0.025–2 mM) oxidation by SSAO after different preincubation times at 37°C. Each point is the mean of triplicate determination. Ranges of individual values were less than 5% in all cases. Errors bars are omitted for clarity

possibility of an aggregation process or extensive degradation of the enzyme being involved.

The absence of changes on the SDS-PAGE pattern, of the partially purified enzyme, after the activation and the fact that different protease inhibitors (1 mM PMSF, 5 mM EDTA, 5 mM EGTA, 0.1 μM pepstatin, and 1 mM N-ethylmaleimide) plus a mixture of each of these did not affect this process (data not shown) is also consistent with the conclusion that it is not a result of

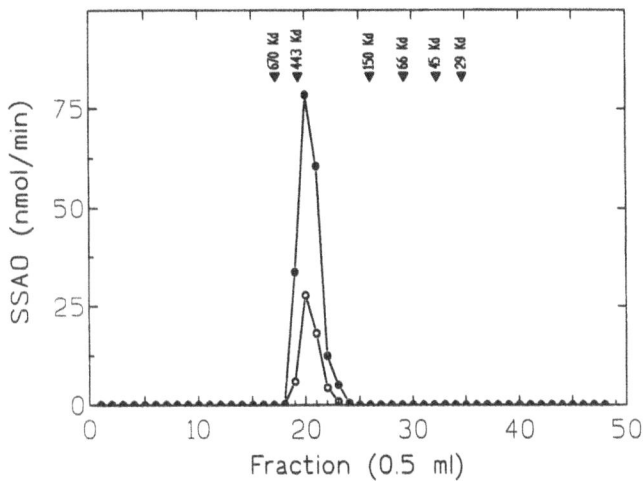

Fig. 4. Gel filtration on Superdex 200 HR (FPLC) of partially purified SSAO before (○) and after (●) preincubation at 37°C for 120 min. Activity was determined with benzylamine as the substrate

a proteolytic cleavage. Furthermore inclusion of either trypsin or chymotrypsin at a ratio of 1:5 (protease:protein) neither affected the enzyme activity not the activation profile over a 2 h incubation period.

To examine the possibility that the activation might be due to a specific interaction with a membrane component, the microsomal SSAO was treated with different detergents (0.6% Triton X-100, 0.75% Na-cholate, or 0.75% β-octyl-glycoside) or with phospholipases (3 units C, 0.5 units C-phosphoinolsitol specific, 30 μg A_2 or 90 units D). In all cases the activation was similar to that observed in the untreated control.

When SSAO was assayed at K_m concentrations of benzylamine and 2-phenylethylamine (40 μM and 312 μM respectively), the respective activities obtained were 1,130 ± 32 and 1,410 ± 35 dpm·min^{-1}. Determination of the activity towards a mixture of these substrates, each at the same concentration, gave an initial velocity of 1,680 ± 90 dpm·min^{-1}. Thus the activity with the mixture was 66.6% of the sum of activities with the two substrates assayed separately, a value not significantly different from the 66.7% that would be expected for two substrates competing for the same active site. Thus these results are not consistent with the existence of two enzymes, one specific for the benzylamine that is susceptible to activation and another, with a wide substrate specificity, that is unaffected.

Mu et al. (1992), have reported that the topa-quinone cofactor of some copper containing amine oxidases is generated by oxidation of a tyrosine residue present in the polypeptide chain. Preincubation of the SSAO either in an atmosphere of N_2 or O_2, or in presence of tyrosinase, did not significantly change the activation pattern with respect to the control. Thus it is unlikely that the formation of this cofactor plays a role in the activation process.

Discussion

A time-dependent activation phenomena was observed when microsomal SSAO from bovine lung was preincubated at temperatures above 25°C and the activity was measured towards benzylamine as substrate. Such behaviour was not observed using the physiological substrates 2-phenylethylamine, histamine and methylamine. This activation which was also observed for preparations of the enzyme from human lung (data not shown) does not appear to involve aggregation or substantial degradation of the enzyme.

Although the fact that this phenomenon has only been observed with the non-physiological substrate benzylamine might suggest a limited metabolic significance, further work will be necessary to investigate whether it does occur with any naturally occurring amines.

The results from the mixed substrate experiments are consistent with a single enzyme active-site being involved in the oxidative deamination of benzylamine and 2-phenylethylamine. This observation might therefore imply that different reaction pathways are followed for the oxidation of these two substrates and that the activation involves a differential effect on them. Further work is in progress to investigate this possibility.

Acknowledgements

This work was partially supported by the European Community by a grant from the Human Capital and Mobility Programme, contract n°CHRX-CT93-0256.

References

Barrand MA, Callingham BA (1982) New amine oxidases activities in brown adipose tissue of the rat: some properties and subcellular distribution. Biochem Pharmacol 31: 2177–2184

Crosbie A, Callingham B (1994) Semicarbazide sensitive amine oxidase in sheep plasma: interaction with some substrates and inhibitors. J Neural Transm [Suppl] 41: 427–432

Feng L, Xia Y, Garcia GE, Hwang D, Wilson CB (1995) Involvement of reactive oxygen intermediates in cyclooxygenase-2 expression induced by interleukin-1, tumor necrosis factor-alpha, and lipopolysaccharide. J Clin Invest 95: 1669–1675

Fowler CJ, Tipton KF (1981) Concentration dependence of the oxidation of tyramine by the two forms of rat liver mitochondrial monoamine oxidase. Biochem Pharmacol 30: 3329–3332

Hartree EP (1972) Determination of protein: a modification of the Lowry method that gives a linear photometric response. Anal Biochem 48: 422–427

Lizcano JM, Fernández de Arriba A, Lyles GA, Unzeta M (1994) Several aspects of the amine oxidation by semicarbazide-sensitive amine oxidase (SSAO) from in bovine lung. J Neural Transm [Suppl] 41: 415–420

Meyer M, Schreck R, Baeuerle PA (1993) H₂O₂ and antioxidants have opposite effects on activation of NF-kappa B and AP-1 in intact cells: AP-1 as secondary antioxidant-responsive factor. EMBO J 12: 2005–2015

Mu D, Janes SM, Smith AJ, Brown DE, Dooley DM, Klinman JP (1992) Tyrosine codon corresponds to topa quinone at the active site of copper amine oxidase. J Biol Chem 267: 7979–7982

Precious E, Gunn CE, Lyles GA (1988) Deamination of methylamine by semicarbazide-sensitive amine oxidase in human umbilical artery and rat aorta. Biochem Pharmacol 37: 707–713

Schreck R, Rieber P, Baeuerle PA (1991) Reactive oxygen intermediates as apparently widely used messengers in the activation of the NF-kappa B transcription factor and HIV-1. EMBO J 10: 2247–2258

Tabor CW, Tabor H, Rosenthal SM (1954) Purification of amine oxidase from beef plasma. J Biol Chem 208: 644–661

Authors' address: Dr. M. Unzeta, Departament de Bioquímica i de Biologia Molecular, Facultad de Medicina, E-08193 Bellaterra (Barcelona), Spain

Studies on the behaviour of semicarbazide-sensitive amine oxidase in Sprague-Dawley rats treated with the monoamine oxidase inhibitor tranylcypromine

D. H. Fitzgerald[1], **K. F. Tipton**[1], and **G. A. Lyles**[2]

[1]Department of Biochemistry, Trinity College, Dublin, Ireland
[2]Department of Pharmacology and Clinical Pharmacology, Ninewells Hospital and Medical School, University of Dundee, Dundee, Scotland, United Kingdom

Summary. The possibility that increased levels of the activity of the semicarbazide-sensitive amine oxidase (SSAO) might, to some extent, compensate for the loss of monoamine oxidase (MAO) activity in the atypical form of Norrie Disease, was examined using the rat as a model. Long-term treatment with the MAO inhibitor tranylcypromine (1 mg/kg/day) resulted in sustained inhibition of MAO-A and MAO-B activities in liver and brain. After one week, the SSAO activity in heart had increased by 79% above the control levels. This increase was maintained for 3 weeks. Since such alterations might result from enzyme induction, the turnover of the enzyme was studied in cultured cells from rat aortic smooth muscle. The time-course of recovery of enzyme activity following irreversible inhibition by MDL 72145 corresponded to a half-life of approximately 6 days for this process.

Introduction

Despite the deletion of both MAO-A and MAO-B activities in individuals suffering from the atypical form of Norrie Disease (Warburg, 1996; Norrie, 1927), their ability to metabolise dopamine and 5-HT appears to be impaired to a much lower extent than that towards 2-phenylethylamine, tyramine or noradrenaline (Murphy et al., 1990). This raises the possibility that another enzyme may metabolise these amines in a MAO-deficient state.

The semicarbazide-sensitive amine oxidases (EC 1.4.3.6; SSAO) are a group of non-FAD-dependant, copper-containing, enzymes which may be encoded by a gene, distant from the genes for MAO-A and MAO-B. The SSAO from human plasma has been reported to deaminate dopamine (McEwen, 1972; Murphy et al., 1976). It has also been reported (Zis et al., 1980) that the level of human plasma SSAO increases in response to the infusion of dopamine. It is therefore possible that SSAO may be activated in MAO-deficient individuals, thus permitting semi-normal function of this neurotransmitter.

The present studies, in which the inhibitor tranylcypromine was used to cause substantial long-term inhibition of MAO in the rat, were aimed at investigating whether changes of SSAO levels could represent an adaptive response to the high levels of neurotransmitter amines, such as occurs in Norrie disease.

A rat aortic smooth muscle cell-culture system, which expresses detectable quantities of SSAO, was also used to estimate the rate of turnover of SSAO since this would be a major factor in any induction of the enzyme in response to metabolite disturbances.

Materials and methods

Tranylcypromine, dissolved in 0.9% NaCl solution, was administered to 15 female Sprague-Dawley rats by intraperitoneal injection at a dose of 1 mg/kg/day (determined from preliminary experiments to give 70–95% inhibition of both MAO A and B). At the same time, 15 control rats were injected with 0.9% NaCl. At various times after injection (30 min, 7, 14, 21 and 28 days), 4 ml samples of blood was drawn into heparinised tubes by cardiac puncture under halothane anaesthesia from 3 control and 3 experimental animals. They were then sacrificed by cervical dislocation. SSAO and MAO activities were measured at 37°C in tissue homogenates and in the blood plasma samples, using a modification of the radiochemical assay described by Tipton (1985) with $10 \mu M$ [^{14}C]-benzylamine (specific radioactivity, 12.09 Ci/mol) as substrate. Samples were preincubated for 30 min at 37°C in the presence or absence of 1 mM clorgyline to allow the activity due to SSAO and the total (MAO plus SSAO) activity, respectively, to be determined. For each time point, assays were performed on control and experimental samples in parallel. A reaction time of 30 min was used. Control experiments indicated initial-rate conditions to hold for this reaction time.

The activities of MAO-A and -B in liver and brain homogenates were measured throughout the study to ensure these enzymes remained inhibited to the required extent. MAO-A activity towards $100 \mu M$ [^{14}C]-5-hydroxytryptamine (specific radioactivity, 4.5 Ci/mol) and MAO-B activity towards $20 \mu M$ [^{14}C]-2-phenylethylamine (specific radioactivity, 2.39 Ci/mol) were determined as described by Tipton (1985).

Rat aortic smooth muscle cells were isolated from adult male rats using a method modified from that of Campbell and Campbell (1993). Confluent cells from 3rd and 4th passage were allowed to grow in dense layers in 6-well plates to ensure adequate material for the detection of SSAO activity. For turnover experiments, experimental wells were exposed for 2 h, to $10 \mu M$ MDL 72145 which had been dissolved in the cell medium, after which time, all wells were washed with 0.9% saline and fresh medium was applied. Control wells were given cell medium without inhibitor. After 2 h, cells from 2 control and 2 experimental wells were harvested, homogenised in 1.5 ml 1 mM potassium phosphate buffer, pH 7.8, and assayed for SSAO activity towards $5 \mu M$ [^{14}C]-benzylamine (specific radioactivity 10 Ci/mol) using the radiochemical method described by Lyles and Callingham (1982). A 30 min preincubation with $100 \mu M$ clorgyline was used to inhibit any MAO present. On day 3, 6, 9 and 11, control wells and experimental wells were assayed for benzylamine deamination to monitor the return of SSAO activity to control levels.

Results

The level of SSAO activity in tranylcypromine-treated rat heart homogenates was increased relative to controls after 1, 2 and 3 weeks but appeared to fall

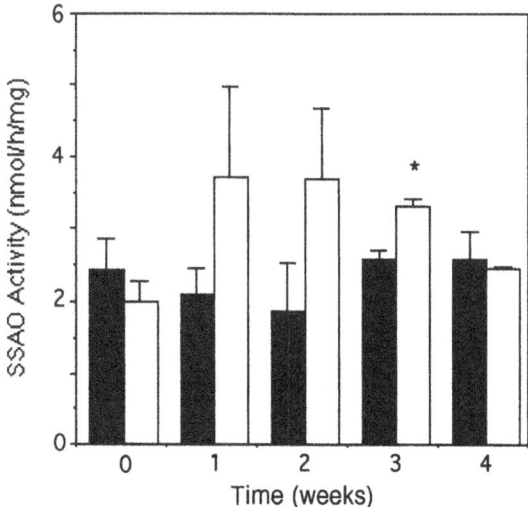

Fig. 1. Changes in SSAO activity in heart homogenates from female Sprague-Dawley rats treated with tranylcypromine. Homogenates from control (■) and tranylcypromine-treated animals (□) were assayed simultaneously. Values are the means ± s.e.m. from three animals; *p < 0.0085 by Student's t-test

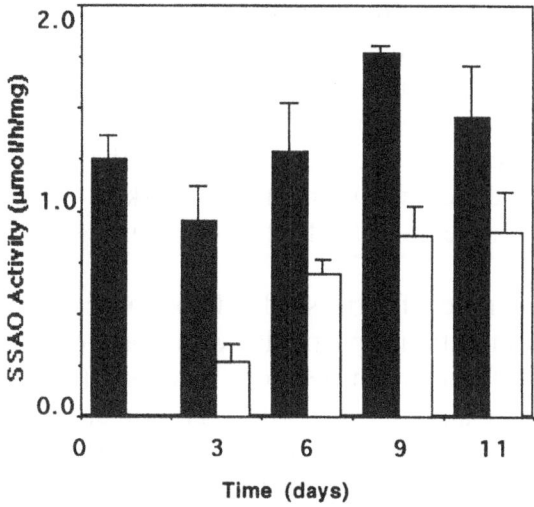

Fig. 2. The return of SSAO activity in rat aortic smooth muscle cell homogenates following exposure to 10 µM MDL72145. Values shown are the means of triplicate determinations ± s.e.m. Homogenates were prepared from cells of 2 cell-culture wells. Control wells (■) and wells exposed to MDL72145 for 2 h (□) were assayed simultaneously

back to control levels after 4 weeks (Fig. 1). The enzyme activity towards 10 µM benzylamine that remained in rat heart following preincubation with 1 mM clorgyline, was sensitive to inhibition by 1 mM semicarbazide, confirming it to represent SSAO.

The levels of SSAO in rat plasma were found to be present only in trace amounts (pmol range) and were too small to permit reliable activity determinations for the comparison purposes, required in this study.

The MAO-A activity determined towards 5-HT and MAO-B activity determined towards 2-phenylethylamine, in the rat liver and brain homogenates remained inhibited to an extent of 70–95%, in the tranylcypromine-treated groups, for the duration of these experiments.

Following its complete inhibition with 10 µM MDL72145, the return of SSAO activity, in homogenates prepared from the rat aortic smooth muscle cell cultures, was followed over 11 days. At 6 days, SSAO activity had returned to about 50% of the control levels (Fig. 2).

Discussion

In the rat heart, SSAO levels increase by 79% above the controls within 7 days of continual MAO inhibition. The level of SSAO remained elevated, compared to controls, for up to three weeks of tranylcypromine treatment after which it declined to control levels. At three weeks the increase in SSAO activity was clearly significant. It is not yet clear whether this observed increase results from an induction of enzyme synthesis, an intrinsic change in the enzyme itself or from changes in the levels of an endogenous effector. In the last of these cases, the dilution of the samples in the preparation and the assay of the homogenates might, however, have been expected to result in the dissociation of any effector that bound reversibly to the enzyme. The specific activity of the SSAO in brain homogenate and plasma was found to be very low. In the former case it, presumably, represented that associated with vascular tissue. Such low activities precluded studies of the possible effects of long-term tranylcypromine treatment. In further work, it might be of interest to investigate changes of the SSAO activities in brain regions and particularly in the microvessels.

The use of the rat model for the study of possible changes in human Norrie disease has some limitations, particularly in view of the fact that SSAO has considerably different affinities for its substrates in rat compared to human tissues (see Clarke et al., 1982; Hayes et al., 1983; Suzuki and Matsumoto, 1984). However, the ability of SSAO from rat deferens to metabolise dopamine has been demonstrated (Lizcano et al., 1991). Although dopamine may be metabolised by SSAO in Norrie disease little kinetic information is available on the metabolism of this amine in human tissue. Clearly, a full kinetic profile of the interaction of human SSAO with dopamine will be necessary in any further studies of this nature and work on this aspect is currently in progress.

The turnover experiments, indicated that the half-life of SSAO was approximately 6 days in the rat aortic smooth muscle cell culture. This suggests that, if SSAO is induced in this system in response to exogenous dopamine, it would take several days for the induction to become appreciable. An accurate assessment of the turnover of the human enzyme would be important since Zis et al. (1980) has reported rapid changes in the activity of human plasma SSAO in response to the infusion of dopamine. Furthermore, a detailed correlation between the time-course of SSAO activity increase in the rat heart following MAO inhibition and the turnover rate of the enzyme in that tissue would be important in understanding the mechanisms involved.

SSAO activity has been reported to alter in a number of conditions; including human development (Lewinsohn et al., 1980), diabetes mellitus, pregnancy in sheep (Elliott et al., 1991) cancer, and also in burn victims (Lewinsohn, 1977, 1984). These have been interpreted as indicating some role for the enzyme in cell development and repair. An understanding of the factors involved in the regulation of enzyme activity in the tissues would clearly be of considerable importance in connection with these phenomena.

Acknowledgements

We are grateful to the EC HCM Programme (Contract No ERBCHRXCT 93 0256), the Health Research Board of Ireland and the Trinity Trust Research Fund for support.

References

Campbell JH, Campbell GR (1993) Culture techniques and their applications to studies of vascular smooth muscle. Clin Sci 85: 501–513

Clarke DE, Lyles GA, Callingham BA (1982) A comparison of cardiac and vascular clorgyline resistant amine oxidase and monoamine oxidase. Biochem Pharmacol 31: 27–35

Elliott J, Fowden AL, Callingham BA, Sharman DF, Silver M (1991) Physiological and pathological influences on sheep plasma amine oxidase: effects of pregnancy and experimental allotoxan-induced diabetes mellitus. Res Vet Sci 50: 334–339

Hayes BE, Ostrow PT, Clarke DE (1983) Benzylamine oxidase in normal and atherosclerotic aortae. Exp Mol Pathol 38: 243–254

Lizcano JM, Balsa D, Tipton KF, Unzeta M (1991) The oxidation of dopamine by the semicarbazide-sensitive amine oxidase (SSAO) from rat vas deferens. Biochem Pharmacol 41: 1107–1110

Lewinsohn R (1977) Human serum amine oxidase, enzyme activity in severely burnt patients and in patients with cancer. Clin Chim Acta 81: 247–256

Lewinsohn R (1984) Mammalian monoamine oxidising enzymes, with special reference to benzylamine oxidase in human tissues. Brazil J Med Biol Res 17: 223–256

Lewinsohn R, Glover V, Sandler M (1980) Development of benzylamine oxidase and monoamine oxidase A and B in man. Biochem Pharmacol 29: 1221–1230

Lyles GA, Callingham BA (1982) In vitro and in vivo inhibition by benserazide of clorgyline-resistant amine oxidases in rat cardiovascular tissues. Biochem Pharmacol 31: 1417–1424

McEwen CM Jr (1972) The soluble monoamine oxidase of human plasma and sera. Adv Biochem Psychopharmacol 5: 51–165

Murphy DL, Wright C, Buchsbaum M, Costa J, Nichols A, Wyatt R (1976) Platelet and plasma amine activities in 680 normals; sex and age differences and stability over time. Biochem Med 16: 254–265

Murphy DL, Sims KB, Karoum F, de la Chapelle A, Norio R, Sankila E-M, Breakfield XO (1990) Marked amine and metabolite changes in Norrie disease patients with an X-chromosomal deletion affecting monoamine oxidase. J Neurochem 54: 242–247

Norrie G (1926) Causes of blindness in children. Acta Opthalmol (Copenhagen) 5: 357–386

Suzuki O, Matsumoto T (1984) Some properties of benzylamine oxidase in human aorta. Biogenic Amines 1: 249–257

Tipton KF (1985) Determination of monoamine oxidase. Meth Find Exp Clin Pharmacol 7: 361–367

Warburg M (1966) Norries disease: a congenital, progressive, occuloacoustic, cerebral degeneration. Acta Opthalmol (Copenh) [Suppl] 89: 1–47

Zis AP, Gold PW, Paul SM, Goodwin FK, Murphy DL (1980) Elevation of human plasma and platelet amine oxidase activity in response to intravenous dopamine. Life Sci 28: 371–376

Authors' address: D. H. Fitzgerald, Department of Biochemistry, Trinity College, Dublin 2, Ireland

Semicarbazide-sensitive amine oxidase in pig heart

T. G. Dowling*, S. Cambi, and **F. Buffoni**

Department of Pharmacology, University of Florence, Italy

Summary. A semicarbazide-sensitive amine oxidase (SSAO) (E.C.1.4.3.6) has been purified from pig heart. Western blot analysis showed that the enzyme reacts with a polyclonal antibody raised against homogeneous crystalline pig plasma benzylamine oxidase (BAO).

A subunit molecular mass of 97 KDa obtained by SDS electrophoresis is identical to the plasma enzyme. The purification procedure consisted of sequential DEAE cellulose, octyl-Sepharose, Con A-Sepharose and hydroxyapatite columns. Two peaks of activity were obtained on octyl-Sepharose which were found to be kinetically and immunologically indistinguishable. The specific activity of the purified enzyme was 0.045 μmol/min/mg of protein at 37°C and the Km for benzylamine was estimated to be 63 μM. The enzyme was inhibited by carbonyl reagents such as semicarbazide but was insensitive to the effect of pargyline.

Introduction

The presence of a SSAO activity with high affinity for benzylamine in the heart has been described by Callingham and Fowler (1977) in developing chick heart, by Lyles and Callingham (1979) and by Clarke et al. (1982) in rat heart, by Buffoni et al. (1989) in rabbit heart and by Puliti and Buffoni (1996) in guinea pig heart. SSAO's with high affinity for benzylamine have been described in blood vessels of many different species (reviewed by Buffoni, 1995) and are expressed by vascular smooth muscle cells in culture (Lyles and Singh, 1985; Hysmith and Boor, 1988) and by adipocytes (Raimondi et al., 1991).

Few attempts at the purification of the tissue-bound SSAO with high affinity for benzylamine have been carried out. Mehrabian and Nalbandyan (1983) have isolated a benzylamine oxidase from bovine microvessels which apparently contains 1 g-atom of copper per 60 kDa subunit. Tipnis et al. (1992) have isolated an enzyme from porcine aorta which has many properties in common with the benzylamine oxidase characterised from cultured porcine aortic smooth muscle cells (Hysmith and Boor, 1988). Guinea pig

---

* Research fellow of Europe (grant CHRXCT93-0256). Present address: Department of Biochemistry, Trinity College, Dublin, Ireland

skin SSAO has been isolated by Buffoni et al. (1994) and both bovine and human lung SSAO have been recently studied by Lizcano et al. (1990). The lung enzyme seems to differ quite considerably from pig benzylamine oxidase.

Pig plasma BAO has been subject of previous studies from this laboratory. Pig heart has some SSAO enzymic activity that we considered useful to purify in order to compare the properties of tissue and plasma enzymes.

Materials and methods

Materials

[^{14}C]-Benzylamine was obtained from Amersham Int., Buckinghamshire, U.K.. Microgranular DEAE-cellulose (DE 32) was obtained from Whatman, Maidstone, Kent, U.K., Con A Sepharose and octyl-Sepharose was obtained from Pharmacia Biotech, Cologno Monzese, Italy. The protease inhibitors: phenylmethylsulphonyl-fluoride (PMSF), l-tosyl-l-phenyl-alanylchlormethane (TPCK) and N-tosyl-l-lysylchlormethan hydrochloride (TLCK) were obtained from Merck Laborchimica, Milan, Italy. Pargyline hydrochloride was obtained from Sigma Chemical Co., St. Louis, Mo., USA and catalase was obtained from Boehringer, Mannheim, Germany. Biotinylated alkaline phosphatase was supplied by Bio-Rad Laboratories, Milan, Italy. All other reagents were of analytical grade. Hydroxyapatite was prepared in the laboratory according to Tiselius et al. (1956).

General methods

A polyclonal antibody to homogenous crystalline pig plasma BAO was produced in rabbit as previously described (Buffoni et al., 1977). SDS electrophoresis in 7.5% resolving gels was performed as described by Laemmli (1970). Protein were transferred to nitrocellulose by the method of Towbin et al. (1979). Visualisation of immunoblotted proteins was performed using a modification the streptavidin/biotinylated alkaline phosphatase method of Coughtrie et al. (1988). This procedure allowed 50 ng of pure BAO to be visualised. Protein was estimated using the method of Lowry et al. (1951) with bovine serum albumin as the standard, or by measurement of the absorbance either at 280 nm or at 215–225 nm.

Enzyme assay

Due to the low activity of BAO in pig heart the purification was monitored radiochemically using 333 µM [^{14}C]-benzylamine according to Buffoni et al. (1989). Purified BAO was assayed spectrophotometrically at 37°C using a modification of the method of Tabor et al. (1954), and spectrofluorimetrically using the method described by Matsumoto et al. (1982).

Enzyme purification

All operations were carried out at 4°C.

Pig hearts, obtained not more than 10 min after slaughter, were immediately transported to the laboratory where they were kept at −20°C until used. The thawed material was cleaned of excess fat and blood vessels and minced in a commercial mincer. The resulting material (2 kg) was homogenised 1:3 (w:v) using a Waring blender (3x 15s bursts) in Na_2HPO_4-KH_2PO_4 (10 mM), EDTA (2 mM), Sucrose (0.25 M), pH 7.0 containing the following protease inhibitors: PMSF (200 mg/6L), TLCK (100 mg/6L), TPCK (150 mg/6L).

The resulting homogenate was spun at 4,500 g for 1 h at 4°C. Ammonium sulphate was added to give a final degree of saturation of 30% (first purification procedure), 35% (second purification procedure). The pellet obtained by centrifugation was discarded and additional ammonium sulphate was added to give a final degree of saturation of 70% (first purification procedure), 55% (second purification procedure). The pellet obtained by centrifugation was fully dialysed against 10 mM phosphate buffer pH 7.0 and applied to a DEAE-cellulose column equilibrated in the above buffer. The enzyme was eluted by increasing the ionic strength of the buffer. Activity was precipitated with ammonium, sulphate (70%) and applied to an octyl-Sepharose column equilibrated in 10 mM phosphate, 2 mM EDTA, pH 7.0 containing 1.2 M ammonium sulphate. The enzyme was eluted with a negative ammonium sulphate gradient (1.2–0 M). Two peaks of activity were found. The active fractions from the major peak were pooled, precipitated with ammonium sulphate and dialysed against 50 mM phosphate buffer pH 7.0 containing 1 M NaCl and applied to a Con A- Sepharose column equilibrated with the same buffer. The enzyme was eluted with buffer containing increasing concentrations of alpha-methyl-D-glucoside. The pooled active fractions were again precipitated with ammonium sulphate, dialysed against 6 mM phosphate buffer pH 6.8 and applied to a hydroxyapatite column equilibrated with the same buffer. The enzyme was eluted by increasing the buffer concentration to 60 mM. The pooled active fractions were concentrated by ammonium sulphate precipitation. The precipitate was dialysed against 50 mM phosphate buffer and kept frozen at −20°C until use.

In the second purification procedure octyl-Sepharose was omitted.

Results

SDS electrophoresis under reducing conditions showed that the isolated enzyme had a subunit relative molecular mass of 97 kDa, in exactly the same position as that of the pig plasma enzyme. It also cross reacted in western blots with a polyclonal antibody raised against pig plasma BAO.

Direct fluorimetric measurement of H_2O_2 formation demonstrated that histamine was also a substrate for pig heart BAO and, at a concentration of 1 mM, was oxidized at a rate that was 25% the rate of oxidation of benzylamine. In the radiochemical assay, 1 mM histamine reduced the rate of oxidation of benzylamine (0.6 mM) by 21% as would be expected if both substrates compete for the same active site.

The purified enzyme was totally inhibited by 0.01 mM 3,5-diethoxy-4-amino-methylpyridine (B24) which is a specific inhibitor of SSAO enzymes with high affinity for benzylamine, and by 0.01 mM semicarbazide. Alpha-aminoguanidine (0.01 mM) inhibited the activity by 29% under similar conditions.

The enzyme was also totally inhibited by 0.01 mM cuprizone which is both a cupric copper chelating agent and a carbonyl reagent, but was not inhibited by 0.1 mM o-phenantroline, 0.1 mM neocuproine, both cuprous copper chelating agents. The enzyme was not inhibited by 0.01 mM pargyline.

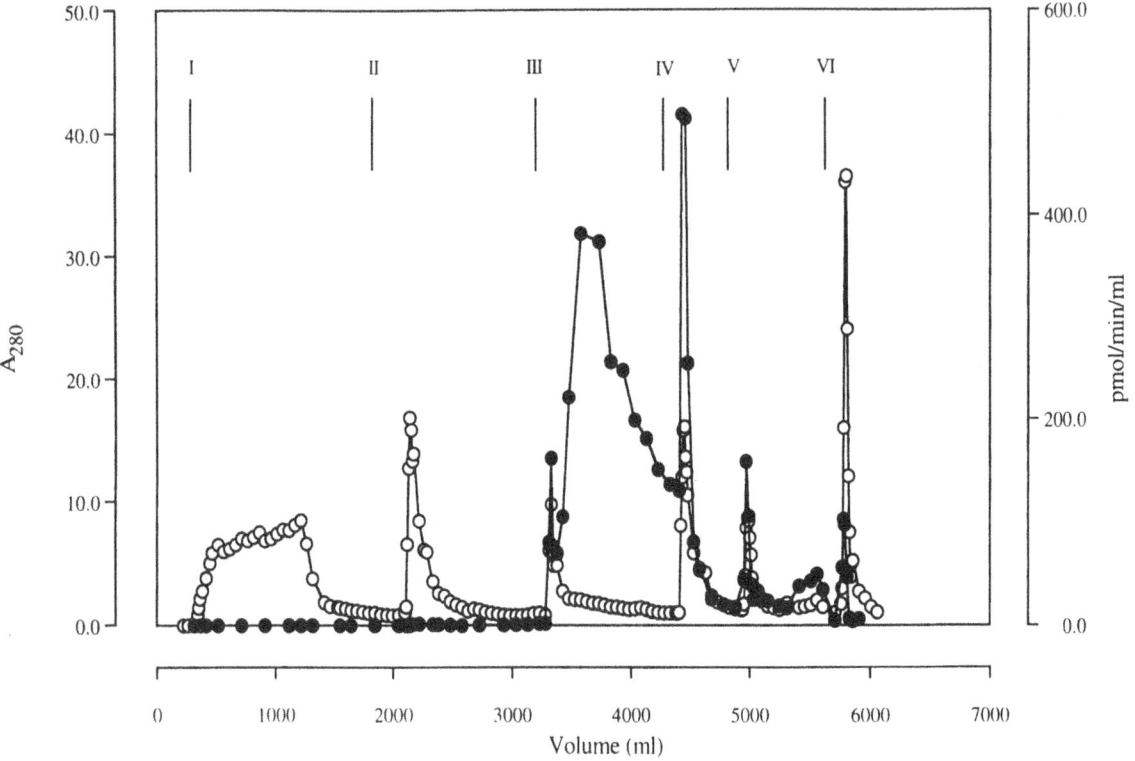

Fig. 1. Elution profile of pig heart BAO on DEAE-Cellulose (–•–pmol/min/ml; –○–
A280). At points I, II, II, IV, V and VI the column was washed successively with 10, 30,
50, 75, 100 and 200 mM phosphate, pH 7.0

The second purification procedure gave an enzyme preparation with a
higher specific activity. Non-linear regression of initial velocities obtained
radiochemically and spectrophotometrically gave similar values (63 ± 4 μM)
for the Michaelis constant. A Vmax of 45 ± 1 nmol/min/mg was obtained.
Enzfitter (Leatherbarrow, 1987) was used for non-linear regression analysis.

Discussion

The preliminary experiments reported here demonstrate that a soluble amine
oxidase is present in pig heart which is inhibited by the specific SSAO inhibitor B24, by semicarbazide and by alpha-aminoguanidine. Like the plasma
enzyme it has histaminase activity and on SDS electrophoresis appears identical to pig plasma BAO with a relative molecular mass of 97 kDa. It cross-
reacts with the polyclonal antibody raised against pure pig plasma BAO. The
Km of the heart enzyme (63 μM) is similar to that of pig plasma BAO, but the
Vmax (0.045 μmol/min/mg) is lower than that found by Buffoni and Blashko
(1964) with homogenous crystalline pig plasma benzylamine oxidase. Partially
purified guinea pig skin BAO also has a higher Vmax (0.075 μmol/min/mg;
Buffoni et al., 1994). SDS electrophoresis revealed that the purified pig heart
enzyme still contained some contaminants, and this may partially account for

the lower specific activity. The enzyme may also be less stable during purification than the pig plasma enzyme studied by Buffoni and Blaschko (1964).

Acknowledgements

This paper was supported by a grant from European Community (Human Capital and Mobility CHRX-CT93-0256), by the Italian MURST and by BIOMED I programme (contract No. BMH1-CT94-1402).

References

Buffoni F (1995) Semicarbazide-sensitive amine oxidases: some biochemical properties and general considerations. In: Yu PM, Tipton KF, Boulton AA (eds) Current neurochemical and pharmacological aspects of biogenic amines. Their function, oxidative deamination and inhibition. Elsevier, Amsterdam, pp 323–331 (Prog Brain Res 106)

Buffoni F, Blaschko H (1964) Benzylamine oxidase and histaminase: purification and crystallization of an enzyme from pig plasma. Proc Roy Soc B 161: 153–167

Buffoni F, Della Corte L, Hope DB (1977) Immunofluorescence histochemistry of porcine tissue using antibodies to pig plasma amine oxidase. Proc Roy Soc B 195: 417–423

Buffoni F, Banchelli G, Bertocci B, Raimondi L (1989) Effect of pyridoxamine on semicarbazide-sensitive amine oxidase activity of rabbit lung and heart. J Pharm Pharmacol 41: 469–473

Buffoni F, Cambi S, Banchelli G, Ignesti G, Pirisino R, Raimondi L (1994) Semicarbazide-sensitive amine oxidase activity of guinea pig dorsal skin. J Neural Transm [Suppl] 41: 421–426

Callingham BA, Fowler CJ (1977) Monoamine oxidation in tissues of the developing chick. Br J Pharmacol 60: 306P

Clarke DE, Lyles GA, Callingham BA (1982) A comparison of cardiac clorgyline resistant amine oxidase and monoamine oxdidase. Biochem Pharmacol 31: 27–35

Coughtrie MWH, Burchell B, Leakey JEA, Hume R (1988) The inadequacy of perinatal glucuronidation immunoblot analysis of the developmental expression of individual UDP-glucuronyltransferase isoenzymes in rat and human liver microsomes. Mol Pharmacol 34: 729–735

Hysmith RM, Boor PJ (1988) Purification of benzylamine oxidase from cultured porcine aortic smooth muscle cells. Biochem Cell Biol 66: 821–829

Laemmli UK (1970) Cleavage of structural proteins during the assembly of the head of bacteriophage T4. Nature 227: 680–685

Leatherbarrow RJ (1987) Enzfitter. A non-linear regression data analysis program for IBM PC. Elsevier Biosoft

Lizcano JM, Balsa D, Tipton KF, Unzeta M (1990) Amine oxidase activities in bovine lung. J Neural Transm [Suppl] 32: 341–344

Lowry OH, Rosebrough NJ, Farr AL, Randall RJ (1951) Protein measurement with the Folin phenol reagent. J Biol Chem 193: 265–275

Lyles GA, Callingham BA (1979) Selective influences of age and thyroid hormones on type A monoamine oxidase of the rat heart. J Pharm Pharmacol 31: 755–760

Lyles GA, Singh I (1985) Vascular smooth muscle cells: a major source of the semicarbazide-sensitive amine oxidase of the rat aorta. J Pharm Pharmacol 37: 637–643

Matsumoto T, Furuta T, Nimura Y (1982) Increased sensitivity of the fluorometric method of Snyder and Hendley for oxidase assay. Biochem Pharmacol 31: 2307–2309

Mehrabian ZB, Nalbandyan RM (1983) Benzylamine oxidase from brain microvessels. FEBS Lett 164: 89–92

Puliti M, Cambi S, Buffoni F (1996) Evidence of semicarbazide-sensitive amine oxidase activity in guinea pig tissues. Comp Biochem Physiol 115B: 159–165

Raimondi L, Pirisino R, Ignesti G, Capecchi S, Banchelli G, Buffoni F (1991) Semicarbazide-sensitive amine oxidase activity (SSAO) of rat epididymal white adipose tissue. Biochem Pharmacol 41: 467–470

Tabor H, Tabor CW, Rosenthal SM (1954) Purification of amine oxidase from beef plasma. J Biol Chem 208: 645–661

Tipnis UR, Tao M, Boor PJ (1992) Purification and characterization of semicarbazide-sensitive amine oxidase from porcine aorta. Cell Mol Biol 38: 575–584

Tiselius A, Hjerten S, Levin O (1956) Protein chromatography on calcium phosphate columns. Arch Biochem Biophys 65: 132–155

Towbin H, Staehelin T, Gordon J (1979) Electrophoretic transfer of proteins from polyacrylamide gels to nitrocellulose sheets: procedures and some applications. Proc Natl Acad Sci USA 76: 4350–4354

Authors' address: Dr. T. G. Dowling, Department of Biochemistry, Trinity College, Dublin 2, Ireland

Chronic TVP-1012 (rasagiline) dose — activity response of monoamine oxidases A and B in the brain of the common marmoset

M. E. Götz[1], W. Breithaupt[1], J. Sautter[2], A. Kupsch[2,3], J. Schwarz[3], W. H. Oertel[3], M. B. H. Youdim[4], P. Riederer[1], and M. Gerlach[1,5]

[1] Division of Clinical Neurochemistry, Department of Psychiatry, Julius-Maximilians-University, Würzburg, [2] Institute of Physiology, Ludwig-Maximilians-University, Munich, and [3] Department of Neurology / Hospital Grosshadern, Ludwig-Maximilians-University, Munich, Federal Republic of Germany
[4] Department of Pharmacology, Rappaport Familiy Research Institute, US National Parkinson Foundation, Faculty of Medicine, Technion, Haifa, Israel
[5] Division of Clinical Neurochemistry, University St. Joseph's Hospital for Neurology, Ruhr-University, Bochum, Federal Republic of Germany

Summary. The stereospecific form of the known acetylenic mechanism-based MAO-inhibitor AGN1135 (Rasagiline, TVP-1012) is devoid of sympathomimetic amphetamine-like properties.

To evaluate the efficiency and selectivity of subcutaneous injections of TVP-1012 (dose range from 0.01 up to 10 mg/kg for 7 days) the activities of monoamine oxidases A and B (MAO-A,-B) were determined in different brain regions of the common marmoset.

At a dose of 0.1 mg/kg TVP-1012, almost 80% of MAO-B activity is inhibited in all brain regions investigated (prefrontal and occipital cortex, cerebellum, caudate nucleus, putamen, nucleus accumbens). In contrast, MAO-A is not inhibited in putamen and nucleus accumbens. However, by increasing the TVP-1012 dose to 0.5 mg/kg, MAO-A is inhibited to a significant extent as well, concomitant to total inhibition of MAO-B.

The results obtained indicate that TVP-1012 irreversibly inhibits both types of MAO in the common marmoset with selectivity for MAO-B at doses less than 0.5 mg/kg. TVP-1012 could be useful in studies requiring selective MAO-B inhibition without concomitant sympathomimetic amphetamine-like effects and could thus be of therapeutic interest for Parkinson's disease and retarded depression.

Introduction

Like selegiline, rasagiline (TVP-1012) belongs to the acetylenic monoamine oxidase inhibitors (Fig. 1) and produces a selective and irreversible inhibition of monoamine oxidase type B in vitro similar to that produced by selegiline

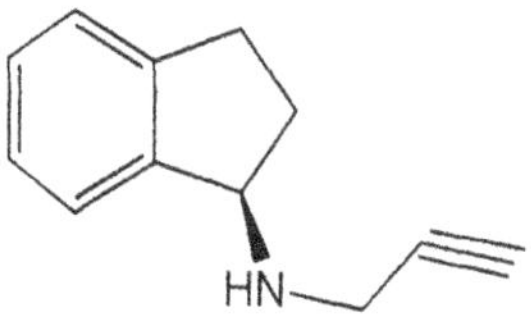

Fig. 1. Structure of rasagiline

(Knoll et al., 1978). In addition, they have similar tyramine antagonistic properties in rat vas deferens (Finberg et al., 1981). Selegiline is metabolized to methamphetamine and amphetamine in man. Therefore, it has been suggested that an intrinsic amphetamine-like effect contributes to the mechanisms responsible for selegiline's clinical effects in Parkinson's disease (e.g. see Gerlach et al., 1992). However, rasagiline which was developed from AGN 1135 (Kalir et al., 1981) cannot possess such an endogenous amphetamine-like effect (Fig. 1) and therefore may be a good candidate for clinical use in the treatment of Parkinson's disease (Youdim and Finberg, 1987; Finberg et al., 1991a,b). To determine the dose of rasagiline which selectively inhibits the type B form of brain monoamine oxidase for use in such clinical trials, we have mapped the regional activities of both forms of the enzyme towards their specific substrates serotonin and phenylethylamine in the brain of the common marmoset (Callithrix jacchus) ex vivo following the subchronic administration of various doses of rasagiline. We have used common marmosets, a non-human primate species, because differences between rodents and primates in the substrate specifities were reported. For example, in rat brain homogenates dopamine appears mainly to be deaminated by the type A form of the enzyme, whereas the type B form predominates in the human brain. Furthermore, the half-lives of the enzymes from the rat brain have been reported to be relatively short, whereas that for the type B enzyme in primate brain has been reported to be about 30 days (Maitre et al., 1976; Arnett et al., 1987).

Materials and methods

Animal experiments

All experiments on common marmosets have been approved by the Government of Upper Bavaria/Institutional Animal Care and Use Committee. Marmosets of either sex were housed under identical conditions and treated according the following scheme:

Experiment I: Four marmosets (300 g–450 g body weight, average age of 6 years) received daily subcutaneous injections with 0.01, 0.1, 0.5 or 1.0 mg/kg rasagiline (dissolved in 0.9% sodium chloride in a volume of 1 ml/100 g body weight) for seven days (n = 1 per group). Monkeys were sacrificed by decapitation under deep anaesthesia with Saffan^R (4 ml/kg body weight) 24 h following the last drug injection.

Experiment II: Two groups of marmosets (200 g–400 g body weight, average age of 3 years) received daily subcutaneous injections with 10 mg/kg rasagiline (dissolved in 0.9% sodium chloride in a volume of 1 ml/100 g body weight) or with 0.9% sodium chloride alone (1 ml/100 g) for seven days. Monkeys were sacrificed according to experiment I but 7 days after the end of treatment.

The brains were quickly removed from the skull, rinsed in ice cold saline and dissected on an ice cold plastic dish. Prefrontal cortex, nucleus accumbens, putamen, caudate nucleus, occipital cortex and cerebellum were separated according to the stereotaxic atlas of Stephan et al. (1980). Brain samples were immediately frozen and stored in liquid nitrogen pending biochemical analysis for MAO-A and -B activities.

Biochemical analysis

Prior to the assay the brain tissue was disrupted by 30 strokes at 4°C in a 1 ml Dounce homogenizer. The amount of protein was assayed according to the method of Bradford (1976) with bovine serum albumin as a standard (Bio-Rad, Munich, F.R.G.) MAO activity of different primate brain areas was assayed according to a modification of the method described by Wurtman and Axelrod (1963) and evaluated according to Tipton and Youdim (1976). Later reevaluations of this radiochemical discontinuous assay were considered (Tipton and Youdim, 1983; Yu, 1986; Tipton and Singer, 1993; Krueger and Singer, 1993).

An aliquot of brain homogenate (300 μg of protein) was preincubated in 200 μl 0.1 M phosphate buffer, pH 7.4 for 7 min at 37°C by shaking in a 6 ml borosilicate open glass tube.

In order to measure MAO-A activity the reaction was started by the addition of 50 μl 0.1 M phosphate buffer PH 7.4 containing [2-^{14}C]-5-hydroxytryptamine creatinine sulfate (specific activity 2.04 GBq/mmol, 1 mM, Amersham-International, England) and incubation continued by shaking for 30 min at 37°C. Following the addition of 250 μl citric acid (2 M) and 5 ml of an extraction mixture (toluene/ethylacetate 1/1 v/v containing 0.6% 2,5-diphenyloxazole, Merck, Darmstadt, F.R.G.) the reaction mixture was shaken vigorously for 30 min and centrifuged at 1,000 g for 5 min. The hydrophilic phase was frozen out at −80°C. The lipophilic upper phase was counted for ^{14}C-activity by liquid scintillation counting (β-counter type LS 5000 TD, Beckman Instruments, Munich, F.R.G.). For 5-HT metabolites recoveries of labeled products in human cortex amounted to 33% of total radioactivity used.

For the measurement of MAO-B activity 25 μl 0.1 M phosphate buffer pH 7.4 containing [1-^{14}C-ethyl] phenylethylamine hydrochloride (100 μM; specific activity 1.16 GBq/mmol, New England Nuclear, Boston, U.S.A.) was added to 225 μl of brain homogenate (300 μg protein) in 0.1 M phosphate buffer (pH 7.4). After 10 min of incubation (continuously shaken) 50 μl of 1 M HCl was added. The acidified solution was then extracted with 2 ml of ethyl acetate by vigorous shaking for 10 min. The two phases were separated by centrifugation at 1,000 g for 5 min and 1 ml of the organic phase containing the deaminated metabolites were combined with 4 ml biosolve cocktail (Roth, Karlsruhe F.R.G.) and counted in a liquid scintillation counter (Beckmann Instruments, dt.). Recovery of labeled products amounted to 98% of total radioactivity used.

All assays were performed in triplicate. Values are expressed as means ± SEM. Values were corrected for blank activity (radioactivity extracted from acid pre-deactivated reaction mixtures) and counting was quench corrected. MAO-activity is given as nmol of products formed per mg protein per min. 5-HT has been used at the final concentration of 200 μM (actual activity 7.2 kBq/assay) whereas PEA has been used at the final concentration of 10 μM with an activity of 4.2 kBq/assay.

Choosing these conditions MAO-A or -B activities were linear with respect to time with 300 μg marmoset brain protein, up to 45 min or 20 min, respectively.

Table 1. MAO-A/B activities and activity ratios in different brain regions of the common marmoset (nmol products formed per mg protein per min)

Brain region	MAO-A	MAO-B	MAO-A/B ratios
prefrontal cortex	0.71 ± 0.10	0.28 ± 0.01	2.54
occipital cortex	0.64 ± 0.03	0.11 ± 0.01	5.82
cerebellum	0.79 ± 0.07	0.20 ± 0.01	3.95
caudate nucleus	1.83 ± 0.02	0.53 ± 0.02	3.21
putamen	0.91 ± 0.01	0.28 ± 0.01	3.25
nucleus accumbens	1.18 ± 0.08	0.57 ± 0.01	2.07

Listing of activities and activity ratios of means of MAO-A and -B activities measured in triplicate ex vivo in different brain regions of four common marmosets (controls) in the presence of 200 µM 5-HT (incubation for 30 min at 37°C) or 10 µM PEA (incubation for 10 min at 37°C), respectively, 7 days after the last administration of placebo

Table 2. Percentages of inhibition of MAO activity following cumulative doses of 0.07 up to 7 mg of the irreversible MAO-inhibitor rasagiline in brain of the common marmoset

MAO-A inhibition

[rasagiline] mg/kg s.c. daily	0.01	0.1	0.5	1.0
prefrontal cortex	3 ± 0.1	7 ± 3.4	40 ± 0.5	54 ± 2.6
occipital cortex	22 ± 1.2	47 ± 3.5	56 ± 5.4	67 ± 4.9
cerebellum	57 ± 4.6	57 ± 2.3	74 ± 2.4	80 ± 1.9
caudate nucleus	47 ± 0.2	61 ± 3.2	71 ± 1.1	77 ± 0.8
putamen	—	—	16 ± 1.3	31 ± 0.5
nucleus accumbens	—	—	45 ± 4.4	53 ± 1.6

MAO-B inhibition

[rasagiline] mg/kg s.c. daily	0.01	0.1	0.5	1.0
prefrontal cortex	—	76 ± 1.9	95 ± 6.8	96 ± 11.2
occipital cortex	47 ± 9.9	88 ± 1.9	97 ± 14.5	99 ± 4.9
cerebellum	28 ± 0.9	84 ± 2.3	96 ± 4.2	97 ± 6.3
caudate nucleus	18 ± 0.5	78 ± 3.5	96 ± 0.5	96 ± 4.9
putamen	—	66 ± 0.8	93 ± 1.8	93 ± 6.3
nucleus accumbens	11 ± 2.4	72 ± 1.5	95 ± 1.4	96 ± 2.5

Listing of mean values of % inhibition of MAO-A and -B activities ± SEM measured in triplicate ex vivo in different brain regions of the common marmoset in the presence of 200 µM 5-HT (incubation for 30 min at 37°C) or 10 µM PEA (incubation for 7 min at 37°C), respectively, 24 h after the last administration of rasagiline. Marmosets received doses mentioned. — indicates no inhibition versus controls

We checked our assay for recovery of 5-HT and PEA metabolites including extraction efficiency and quench correction in liquid scintillation counting. PEA metabolites could be almost quantitatively recovered (98%). In contrast, 5-HT recovery is only 33%. High amounts of oxidation products formed are retained within the aqueous phase

(Huether et al., 1990). We could not increase the yield of 5-HT metabolites by additional enzyme or by increasing time of incubation longer than 90 min above 33% of total substrate added. Thus, this correction factor has been included in activity calculations for MAO-A.

Results

The interregional distribution pattern of MAO-A and -B are similar in the examined brain regions. However, the activities toward the substrates 5-HT and PEA do not show a 1:1 relationship (Table 1). The highest specific activities were found in the nucleus accumbens and the caudate nucleus.

Following rasagiline administration brain MAO-A and -B activities are dose dependently inhibited. Table 2 demonstrates that the selectivity for inhibition of MAO-B depends on the brain region investigated and the dose of rasagiline. MAO-B inhibition is already complete with s.c. 0.5 mg/kg for 7 days rasagiline. Even with a dose of rasagiline of 0.1 mg/kg for 7 days marmoset brain MAO-B was inhibited to a mean of 80%, while MAO-A was only moderately or not inhibited.

In the second experiment 4 marmosets received on 7 consecutive days 10 mg/kg rasagiline. With this drug regimen almost complete inhibition of MAO-A and MAO-B (>95% in all cases) was found in all examined brain regions ex vivo (data not shown).

Discussion

MAO-A and B are flavoproteins localized in the outer mitochondrial membrane in neuronal, glial and endothelial cells in brain catalyzing the oxidative deamination of monoaminergic neurotransmitters and xenobiotic amines (Youdim et al., 1988). The existence of two subtypes A and B was established pharmacologically by the use of selective irreversible inhibitors (selegiline for MAO-B and clorgyline for MAO-A) and has been confirmed immuno-histochemically using monoclonal antibodies specific to either forms (Denney et al., 1982; Konradi et al., 1988, 1989). The expression of the MAO subtypes is species and tissue specific (Stenström et al., 1987; Shih et al., 1990; Saura et al., 1992; Zhu et al., 1992) and does not reflect the distribution of the presumed natural substrates (Westlund et al., 1988).

This is the first comprehensive study describing the distribution of MAO activities toward 5-HT and PEA and MAOA/B activity ratios in several brain regions of the common marmoset. In our investigation the activities of MAO-A in cortical areas are somewhat higher compared to that previously reported in the cerebral cortex of the common marmoset (Ueki et al., 1989).

The highest specific activity of MAO-A and -B in the human brain was found in the caudate nucleus, the putamen and the globus pallidus (Riederer and Youdim, 1986). The lowest activities were measured in cortical areas. This agrees with data presented here for the common marmoset.

The brain regions affected first by rasagiline (at the daily dose of 0.01 mg/kg) are cortical areas, the cerebellum and the caudate nucleus. Whether there is a selective uptake mechanism for rasagiline into these brain regions has not yet been investigated.

Daily doses of 0.1 mg/kg rasagiline s.c. inhibit 80% of MAO-B whilst MAO-A is only moderately inhibited. At this dose selectivity of rasagiline seems to be optimal. Exceeding the dose of 0.5 mg/kg/day s.c. rasagiline becomes increasingly unselective. At the dose of 10 mg/kg/day MAO-A and MAO-B are completely inhibited by rasagiline in all brain regions investigated (data not presented).

Even 7 days after cessation of administration of subchronic doses (10 mg/kg) rasagiline, nearly complete inhibition of both enzyme types were found in all examined brain regions. This result confirms the reported long half-life of MAO. In comparison, selegiline is a relatively selective inhibitor of MAO-B following a daily s.c. dose of 0.05–0.25 mg/kg for 3 weeks to rodents (Ekstedt et al., 1978; Knoll et al., 1978).

In summary, a selective inhibition of marmoset brain MAO-B activity was found following daily s.c. doses of 0.1 mg/kg rasagiline for seven days. Thus, rasagiline could be useful in studies requiring selective MAO-B inhibition without concomitant sympathomimetic amphetamine-like effects and could be of therapeutic interest in Parkinson's disease since it has neuroprotective activities (Yoles and Schwartz, 1995; Finberg et al., 1996).

Acknowledgements

This work was supported by grants from the Bundesministerium für Forschung und Technologie (BMFT) grant numbers 01KL 9001; 01KL9013; 01KL9405; the TEVA Pharmaceutical Company Israel; Golding Parkinson's disease research fund and BIOMED I, European Community Brussels and National Parkinson Foundation, USA.

References

Arnett CD, Fowler JS, MacGregor RR, Schlyer DJ, Wolf AP, Langström B, Halldin C (1987) Turnover of brain monoamine oxidase measured in vivo by positron emission tomography using L-[^{11}C]deprenyl. J Neurochem 49: 522–527

Bradford MM (1976) A rapid and sensitive method for the quantitation of microgram quantities of protein utilizing the principle of protein-dye binding. Anal Biochem 72: 248–254

Denney RM, Patel NT, Fritz RR, Abell CW (1982) A monoclonal antibody elicited to human platelet monoamine oxidase. Isolation and specificity for human monoamine oxidase B but not A. Mol Pharmacol 22: 500–508

Ekstedt B, Magyar K Knoll J (1978) Does the B form selective monoamine oxidase inhibitor deprenyl lose selectivity by long term treatment? Biochem Pharmacol 28: 919–923

Finberg JP, Tenne M, Youdim MBH (1981) Tyramine antagonistic properties of AGN 1135, an irreversible inhibitor of monoamine oxidase type B. Br J Pharmacol 73: 65–74

Finberg JPM, Youdim MBH, Levy R, Sterling S, Eliash S (1991a) Pharmacology of TVP 1012, a selective inhibitor of MAO type B lacking sympathomimetic action. In: Proceedings of the Tenth International Symposium on Parkinson's Disease. University Press, Tokyo, p 108

Finberg JPM, Youdim MBH, Levy R, Sterling S, Eliash S (1991b) TVP 1012, a potent and selective irreversible inhibitor of MAO type B. In: Proceedings of the Tenth International Symposium on Parkinson's Disease. University Press, Tokyo, p 134

Finberg JPM, Lamensdorf I, Commissiong JW, Youdim MBH (1996) Pharmacology and neuroprotective properties of rasagiline. J Neural Transm [Suppl 48]: 95–103

Gerlach M, Riederer P, Youdim MBH (1992) The molecular pharmacology of L-deprenyl. Eur J Pharmacol Mol Pharmacol Sect 226: 97–108

Huether G, Reimer A, Schmidt F, Schuff-Werner P, Brudny MM (1990) Oxidation of the indole nucleus of 5-hydroxytryptamine and formation of dimers in the presence of peroxidase and H_2O_2. J Neural Transm [Suppl] 32: 249–257

Kalir A, Sabbagh A, Youdim MBH (1981) Selective acetylenic "suicide" and reversible inhibitors of monoamine oxidase types A and B. Br J Pharmacol 73: 55–64

Knoll J, Ecsery Z, Magyar K, Satory E (1978) Novel (−)deprenyl-derived selective inhibitors of B-type monoamine oxidase. The relation of structure to their action. Biochem Pharmacol 27: 1739–1747

Konradi C, Svoma E, Jellinger K, Riederer P, Denney R, Thibault J (1988) Topographic immunocytochemical mapping of monoamine oxidase-A, monoamine oxidase-B and tyrosine hydroxylase in human post mortem brain stem. Neuroscience 26: 791–802

Konradi C, Kornhuber J, Frölich L, Fritze J, Heinsen H, Beckmann H, Schulz E, Riederer P (1989) Demonstration of monamine oxidase-A and -B in the human brainstem by a histochemical technique. Neuroscience 33: 383–400

Krueger MJ, Singer TP (1993) An examination of the reliability of the radiochemical assay for monoamine oxidases A and B. Anal Biochem 214: 116–123

Maitre L, Delini-Stula A, Waldmeier PC (1976) Relations between the degree of monoamine oxidase inhibition and some psychopharmacological responses to monoamine oxidase inhibitors in rats. In: Wolstenholme GEW, Knight J (eds) Monoamine oxidase and its inhibition. Elsevier, Amsterdam, pp 247–267 (Ciba Fdn Symp)

Riederer P, Youdim MBH (1986) Monoamine oxidase activity and monoamine metabolism in brains of parkinsonian patients treated with L-deprenyl. J Neurochem 46: 1359–1365

Saura J, Kettler R, Da Prada M, Richards JG (1992) Quantitative enzyme radioautography with ^{3}H-Ro 41-1049 and ^{3}H-Ro 19-6327 in vitro: localization and abundance of MAO-A and MAO-B in rat CNS, peripheral organs and human brain. J Neurosci 12: 1977–1999

Shih JC, Grimsby J, Chen K (1990) The expression of human MAO-A and B genes. J Neural Transm [Suppl] 32: 41–47

Stenström A, Hardy J, Oreland L (1987) Intra- and extra-DA-synaptosomal localization of monoamine oxidase in striatal homogenates from four species. Biochem Pharmacol 18: 2931–2935

Stephan H, Baron G, Schwerdtfeger WK (1980) The brain of the common marmoset (Callithrix jacchus)-a stereotaxic atlas. Springer, Berlin Heidelberg New York

Tipton KF, Youdim MBH (1976) Assay of monoamine oxidase. In: Wolstenholme GEW, Knight J (eds) Monoamine oxidase and its inhibition. Excerpta Medica, North-Holland, pp 393–403

Tipton KF, Youdim MBH (1983) The assay of monoamine oxidase. In: Parvez S, Nagatsu T, Nagatsu I, Parvez H (eds) Methods in biogenic amine research. Elsevier, Amsterdam, pp 441–465

Tipton KF, Singer TP (1993) The biochemical assay for monoamine oxidase activity. Problems and pitfalls. Biochem Pharmacol 46: 1311–1316

Ueki A, Willoughby J, Glover V, Sandler M (1989) Distribution of phenolsulfo-transferase and monoamine oxidase in the common marmoset. Biochem Pharmacol 38: 2383–2385

Westlund KN, Denney RM, Rose RM, Abell CW (1988) Localization of distinct monoamine oxidase A and monoamine oxidase B cell populations in human brainstem. Neurosci 25: 439–456

Wurtman RJ, Axelrod J (1963) A sensitive and specific assay for the estimation of monoamine oxidase. Biochem Pharmacol 12: 1439–1441

Youdim MBH, Finberg JPM (1987) MAO type B inhibitors as adjunct to L-DOPA therapy. Adv Neurol 45: 127–136

Youdim MBH, Finberg JPM, Tipton KF (1988) Monoamine oxidase. In: Trendelenburg U, Weiner N (eds) Handbook of experimental pharmacology, vol 90/I. Springer, Berlin Heidelberg New York Tokyo, pp 119–192

Yoles E, Schwartz M (1995) N-Propargyl-1 (R)-Aminoindan (TVP-1012), a putative neuroprotective agent, enhances in vitro neuronal survival after glutamate toxicity. Soc Neurosci Abstr 230.18: 562

Yu PH (1986) Monoamine oxidase. In: Boulton AA, Baker GB, Yu PH (eds) Neuromethods, vol 5. Humana Press, Clifton, pp 235–272

Zhu Q-S, Grimsby J, Chen K, Shih JC (1992) Promoter organization and activity of human monoamine oxidase (MAO) A and B genes. J Neurosci 12: 4437–4446

Authors' address: Dr. M. E. Götz, Department of Psychiatry, Division of Clinical Neurochemistry, University of Würzburg, Füchsleinstrasse 15, D-97080 Würzburg, Federal Republic of Germany

Increased striatal dopamine production from L-DOPA following selective inhibition of monoamine oxidase B by R(+)-N-propargyl-1-aminoindan (rasagiline) in the monkey

J. P. M. Finberg, J. Wang, K. Bankiewicz, J. Harvey-White, I. J. Kopin, and **D. S. Goldstein**

Clinical Neurosciences Branch and Surgical Neurology, NINDS, NIH, Bethesda, MD, USA

Summary. Striatal extracellular fluid concentrations of dopamine and metabolites in response to direct striatal administration of two L-DOPA boluses administered sequentially were determined in three rhesus monkeys during halothane anesthesia. Whereas in an initial microdialysis run, generation of dopamine was less following the second L-DOPA bolus than the first, in a subsequent run, in which the selective MAO-B inhibitor R(+)-N-propargyl-1-aminoindan (rasagiline) was administered systemically (0.2 mg/kg s.c.) between the two L-DOPA boluses, generation of dopamine was greater following the second bolus.

Introduction

Dopamine (DA) is a substrate for both types of monoamine oxidase (MAO) A and B, but its in vivo metabolism by MAO depends on the subtype of the enzyme present within the neuron and at the site of uptake following neuronal release. Following synthesis of DA from exogenous L-DOPA in the brain, metabolism will depend on the type of enzyme present at sites of decarboxylation of the L-DOPA, and at the site of uptake of the DA formed.

Selective inhibition of MAO type B by (−)-deprenyl is effective in potentiating the anti-Parkinsonian efficacy of L-DOPA in human patients who are experiencing loss of response to L-DOPA (Birkmayer et al., 1975). Brain tissue from such deprenyl-treated patients shows increased levels of dopamine (DA) and reduced levels of deaminated DA metabolites (Riederer and Youdim, 1986), indicating that in the human, DA formed from exogenous L-DOPA is metabolised by MAO-B. Used in monotherapy in parkinsonian subjects, (−)-deprenyl possesses a symptomatic effect (Parkinson's Study Group, 1993), indicating that endogenously released DA is also metabolised by MAO-B, although deprenyl may also have DA releasing properties which could be responsible for this effect (Knoll, 1978).

In human brain, about 75% of the MAO activity towards a joint substrate such as tyramine is type B, but the dopaminergic cells of the nigro-striatal pathway contain only MAO-A, as shown by immunohistochemical and ligand binding techniques (Westlund et al., 1985; Richards et al., this volume). In the rat, in which brain MAO activity is about 55% MAO-A (Tipton et al., 1976), striatal extracellular fluid DA levels are increased by inhibition of MAO-A but not MAO-B (Colzi et al., 1990). No comparable data for the effect of selective MAO inhibition on striatal DA levels exist in primate species. We therefore carried out an in vivo microdialysis study in the rhesus monkey to study the effect of selective inhibition of MAO-B on DA produced from L-DOPA. The L-DOPA was administered directly into the striatum, in order to study striatal generation of DA without modification of its release by activation of somatodendritic receptors, as may follow systemic L-DOPA (Finberg et al., 1995). We used the selective MAO-B inhibitor R(+)-N-propargyl-1-aminoindan (rasagiline; Finberg et al., 1995), which is not metabolised to amphetamine, and is therefore a better tool than deprenyl for study of MAO-B inhibition uncomplicated by effects of dopamine-releasing metabolites. This drug is currently in phase 3 clinical trial for treatment of Parkinson's disease. The racemic form of this compound, known as AGN-1135, was described by Kalir et al. (1981), and Finberg et al. (1981).

Methods

Subjects were three adult rhesus monkeys (Macaca mulatta). The animals were housed with a 12h light-dark cycle, and fed Purina monkey chow twice daily with free access to water. The experimental protocol was approved by the NINDS Animal Care and Use Committee, and the animals were treated in accordance with NIH guidelines.

Each animal underwent 2 microdialysis runs, separated by a 2 week recovery period. In each run, the animal was anesthetised with halothane in room air, intubated, and microdialysis probes were inserted bilaterally into the caudate nucleus and putamen, using co-ordinates from the superior sagittal sinus, and entry angles previously determined under NMR (Wang et al., 1990).

The probes were manufactured as descibed by Wang et al. (1990), from fused silica and stainless steel tubing. The dialysis membrane was 4mm of AN 69 polyacetonitrile membrane with 40,000MW cut-off (Hospal Medical Co), and was changed for each experiment.

In vitro recovery for DA, HVA[1] and DOPAC is between 25–30%. In each animal 4 probes with similar in vitro recovery rates were used.

The probes were perfused with Ringer's solution (composition: Na^+ 147, Ca^{++} 2.2, K^+ 4, Cl^- 155mmol/L, pH 6.0) at 2μl/min. Baseline collections of microdialysate (each 20min) were commenced 60min after probe insertion. After 4–5 baseline collections, an L-DOPA bolus was administered via the microdialysis probe, by perfusion with Ringer's solution containing L-DOPA (0.5mM) for 15min. The collection immediately after the bolus was rejected, and then post-DOPA collections were made for 4h. A second L-DOPA bolus was then administered in an identical manner to the first, and a further 7 collections of dialysate were made.

In the first run, which was carried out to establish the characteristics of dopamine metabolism in response to the double L-DOPA bolus, saline (0.2ml/kg s.c.) was ad-

---

[1] *HVA* homovanillic acid, *DOPAC* dihydroxyphenylacetic acid

ministered to the animal in collection 11. In the second run, rasagiline (0.2 mg/kg s.c.) was administered instead of saline. Rasagiline was used as hydrochloride salt (TVP-101).

Microdialysate was collected into vials containing 1 μl of 1 M perchloric acid, and deep frozen until analysis. For HPLC determination of DA and metabolites, the volume of each collection was measured, and an appropriate aliquot injected directly into the solvent stream of a Waters HPLC apparatus. Amines and metabolites were determined as described by Finberg et al. (1995).

Results

Resting levels of DA and metabolites in caudate nucleus microdialysate are shown in Table 1. Basal DA levels were significantly higher in the second run. The local infusion of L-DOPA caused a rapid generation of DA and DOPAC in the striatum, with no increase in HVA levels. Although an equal amount of drug was contained in the infusion solutions, the amount of L-DOPA recovered from the striatum was not exactly equivalent in the two bolus infusions. When the total efflux of L-DOPA was calculated, the cumulative efflux in the first bolus was 987 ± 118, and in the second 808 ± 115 pmol ($P < 0.05$) yielding an average in vivo recovery of 6%. This slight difference in L-DOPA levels

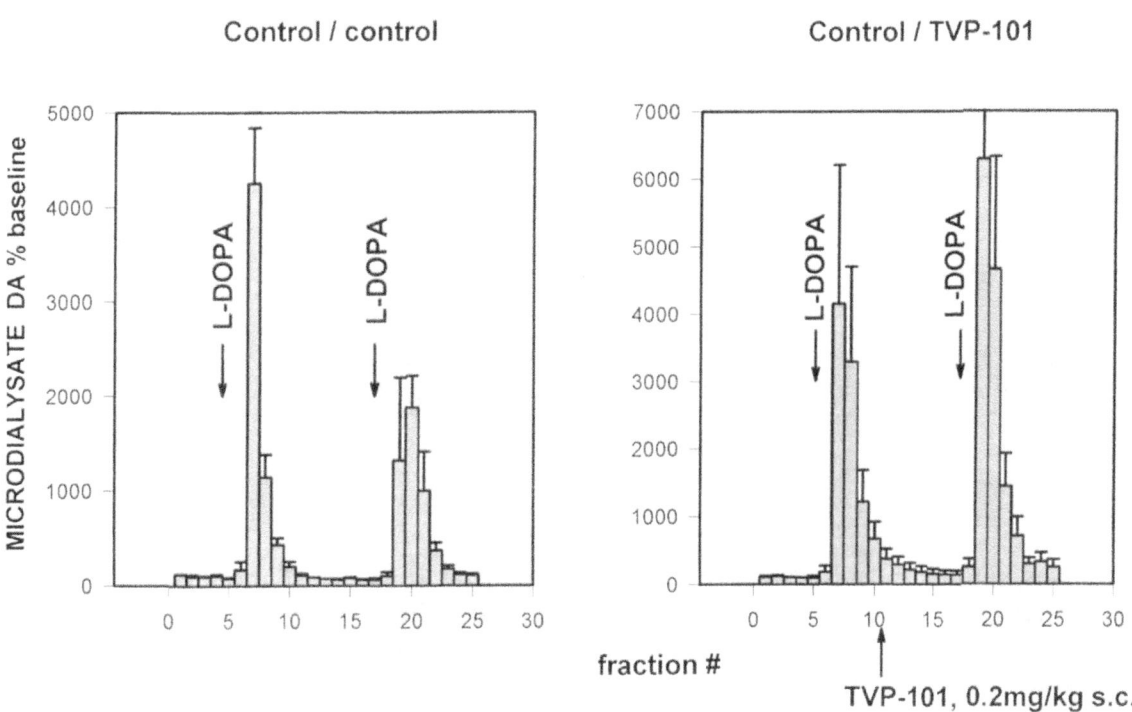

Fig. 1. Extracellular dopamine (DA) levels in right caudate nucleus following administration of L-DOPA bolus (0.5 mM for 15 min) through microdialysis probe in three rhesus monkeys treated with saline or rasagiline (TVP-101). Data expressed as percentage of basal level in first 4 collections ± SEM

Table 1. Basal microdialysate concentrations (pmol/20μl) of dopamine and metabolites
in caudate nucleus of the rhesus monkey

	DA	DOPAC	HVA
Run 1 (control/control)	0.37 ± 0.05	13.3 ± 1.8	90 ± 14
Run 2 (control/drug)	3.76 ± 0.91*	6.8 ± 0.62*	86 ± 5.5

Data are means ± SEM of the first four basal microdialysis collections in left and
right caudate samples in each of the three monkeys (i.e. a total of 24 samples). * = P <
0.01 for difference between run 1 and run 2

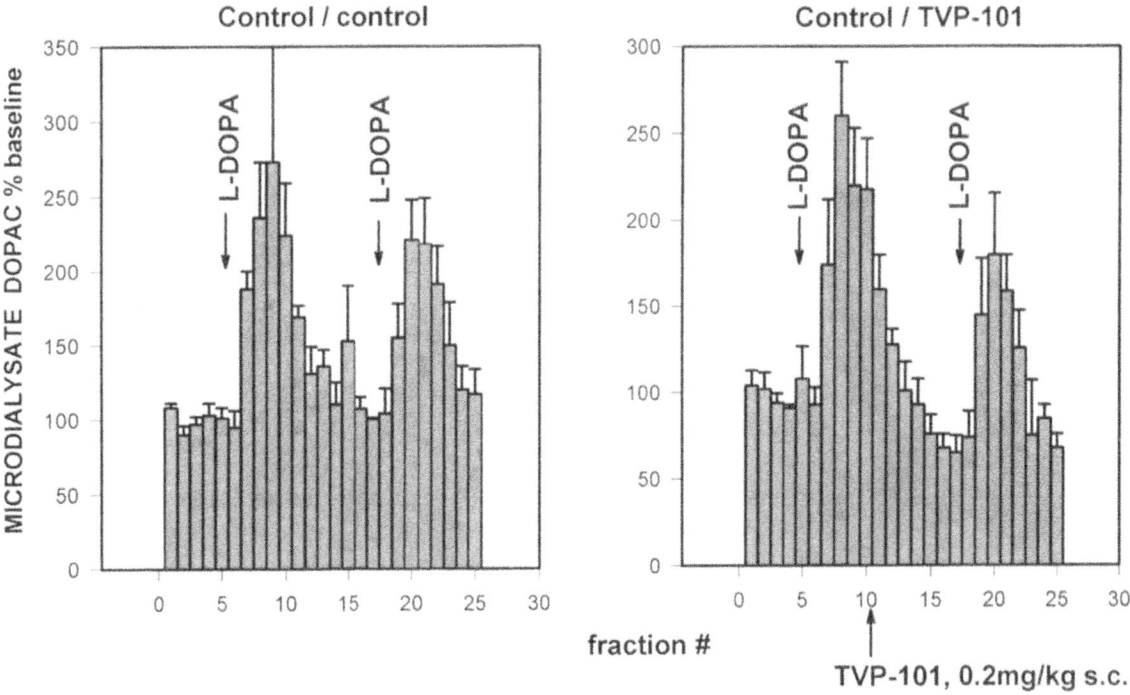

Fig. 2. Extracellular DOPAC levels in right caudate nucleus following administration of
L-DOPA bolus (0.5 mM for 15 min) through microdialysis probe in three rhesus monkeys
treated with saline or rasagiline (TVP-101). Data expressed as percentage of basal level
in first 4 collections ± SEM

was similar in control/control and control/drug runs. The amount of DA
recovered in perfusate, however, was markedly different in control/control
and control/drug runs. In the first run, substantially less DA entered the
microdialysate after the second L-DOPA bolus, while following rasagiline,
more DA was recovered (Fig. 1).

In the case of DA metabolites, whereas no significant change could be
detected in HVA levels following L-DOPA bolus, a marked efflux of
DOPAC was seen (Fig. 2). The post-DOPA DOPAC efflux was slightly lower
following the second bolus in both control and drug experiments, and no

marked difference could be seen in post-DOPA DOPAC efflux following rasagiline, however, baseline DOPAC showed a reduction in each animal following the MAO-inhibitor (Fig. 2). Neither 5-HIAA nor 3-methoxytyramine levels showed any significant alteration following the L-DOPA bolus, or treatment with rasagiline, and 3-O-methyldopa levels showed a very slight increase following L-DOPA, amounting to about 1% of the simultaneous L-DOPA level. The picture of amine and metabolite levels was similar in putamen and caudate nucleus samples.

Monoamine oxidase activity measured in cerebral cortex of the treated monkeys at the end of the second experiment, ie about 6h after injection of rasagiline, averaged 134 ± 4.8 and 7.8 ± 1.0% of the activity in control monkey cortex (MAO-A and MAO-B respectively).

Discussion

This limited study in 3 monkeys has obvious problems which limit the conclusions which can be drawn; however certain aspects of the results are clear-cut, and considering the difficulty of mounting such a study in primates, we felt it worthwhile to present the data as they stand. An initial aim of the study was to establish whether two striatal L-DOPA infusions result in a similar release of DA. In fact, we found a lower recovery of L-DOPA following the second bolus, together with a lower DA efflux. The reason for this effect is not clear, but we cannot rule out in vitro factors, such as some degradation of the L-DOPA solution between the two infusions, as well as in vivo factors such as reduction in activity of L-DOPA uptake and metabolising systems following exposure to the high concentration of substrate in the first bolus. Another complicating factor is that basal DA levels were substantially higher in the second experimental run, indicating either a local or systemic reaction to the surgical procedure. As a result, we cannot conclude for certain that the observed increase in DA efflux following rasagiline injection was in fact a drug effect, but the reduction in DOPAC level seen following the MAO inhibitor points in this direction.

In a previous study in rats, we used a similar technique of administration of L-DOPA directly to striatal tissue in microdialysate to estimate the effects of selective MAO inhibitors on DA produced from exogenous L-DOPA. We found a significant increase in striatal microdialysate DA following selective inhibition of MAO-A with clorgyline, but not after selective MAO-B inhibition with deprenyl or rasagiline (Finberg et al., 1995), in agreement with other workers who administered L-DOPA systemically (Wachtel and Abercrombie, 1994). Da Prada and coworkers (Colzi et al., 1990) showed clearly in the rat that striatal extracellular DA levels are increased following inhibition of MAO-A but not MAO-B. There are, however, very few studies in primate species on the effect of MAO-A or MAO-B inhibition on striatal metabolism and release of DA. An early study by Riederer and Youdim (1986) showed an increase in tissue levels of DA in Parkinsonian patients who had been treated with deprenyl and L-DOPA. Recently, a study in squirrel monkeys showed an effect of both deprenyl and clorgyline (in MAO-A and MAO-B inhibiting doses respectively) to enhance DA and decrease DOPAC and HVA levels in the nigro-striatal tract (Di Monte et al., 1996). Paterson et al. (1995) studied the effect of deprenyl on DA metabolism in the macaque, and found that the drug inhibited DA metabolism in caudate nucleus and frontal cortex. In the primate brain, therefore, by contrast with the rat, inhibition of MAO-B alone is apparently capable of enhancing DA produced from exogenous L-DOPA, as shown also by the clinical L-DOPA sparing effect of deprenyl.

An interesting question arising from our study is the lack of increase in HVA extracellular levels following the L-DOPA bolus. The pronounced rise in DOPAC levels indicates that most of the formed DA is being deaminated intraneuronally. If so, the effect of MAO-B inhibitors to enhance formed DA is surprising in view of the primary location of MAO-A in the dopaminergic nigral neuron (Westlund et al., 1985; Richards et al., this volume). This and other important questions will have to await more detailed experimentation.

References

Birkmayer W, Riederer P, Youdim MBH, Linauer W (1975) The potentiation of the antiakinetic effect after L-DOPA treatment by an inhibitor of MAO-B, deprenyl. J Neural Transm 36: 303–326

Colzi A, D'Agostini F, Kettler R, Borroni E, Da Prada M (1990) Effect of selective and reversible MAO inhibitors on dopamine outflow in rat striatum; a microdialysis study. J Neural Transm [Suppl] 32: 79–84

Di Monte DA, De Lanney LE, Irwin LE, Royland JE, Chan P, Jalowec MW, Langston JW (1996) Monoamine oxidase-dependent metabolism of dopamine in the striatum and substantia nigra of L-DOPA-treated monkeys. Brain Res 738: 53–59

Finberg JPM, Tenne M, Youdim MBH (1981) Tyramine antagonistic properties of AGN-1135, an irreversible inhibitor of monoamine oxidase type B. Br J Pharmacol 73: 65–74

Finberg JPM, Wang J, Goldstein DS, Kopin IJ, Bankiewicz KS (1995) Influence of selective inhibiton of monoamine oxidase A or B on striatal metabolism of L-DOPA in hemiparkinsonian rats. J Neurochem 65: 1213–1220

Kalir A, Sabbagh A, Youdim MBH (1981) Selective acetylenic "suicide" and reversible inhibitors of monoamine oxidase types A and B. Br J Pharmacol 73: 55–64

Knoll J (1978) The possible mechanisms of action of (−)-deprenyl in Parkinson's disease. J Neural Transm 43: 177–198

Parkinson's Study Group (1993) Effects of tocopherol and deprenyl on the progression of disability in early Parkinson's disease. N Engl J Med 321: 1364–1371

Paterson IA, Davis BA, Durden BA, Juorio AV, Yu PH, Ivy G, Milgram W, Mendonca A, Wu P, Boulton AA (1995) Inhibition of MAO-B by (−)-deprenyl alters dopamine metabolism in the macaque (macaca fascicularis) brain. Neurochem Res 20: 1503–1510

Richards JG, Saura J, Luque JM, Cesurs AM, Gottowik J, Malherbe P, Borroni E, Gray J (1997) Monoamine oxidases: from brainmaps to physiology and transgenics to pathophysiology (this volume)

Riederer P, Youdim MBH (1986) Monoamine oxidase activity and monoamine metabolism in brains of Parkinson patients treated with l-deprenyl. J Neurochem 46: 1359–1365

Riederer R, Reynolds JP, Youdim MBH (1981) Selectivity of MAO inhibitors in human brain and their clinical consequences. In: Youdim MBH, Paykel ES (eds) Monoamine oxidase inhibitors: the state of the art. Wiley, Chichester, pp 63–76

Tipton KF, Houslay MD, Mantle TJ (1976) The nature and locations of the multiple forms of monoamine oxidase. Elsevier, Amsterdam, pp 5–16 (Ciba Foundation Symposium 39)

Wachtel SR, Abercrombie E (1994) L-3,4-Dihydroxyphenylalanine-induced dopamine release in the striatum of intact and 6-hydroxydopamine-treated rats: differential effects of monoamine oxidase A and B inhibitors. J Neurochem 63: 108–117

Wang J, Skirboll S, Aigner TG, Saunders RC, Bankiewicz KS (1990) Methodology of microdialysis of neostriatum in hemiparkinsonian non-human primates. Exp Neurol 110: 181–186

Westlund KN, Denney RM, Kochersperger LM, Rose RM, Abell CW (1985) Distinct monoamine oxidase A and B populations in primate brain. Science 230: 181–183

Authors' address: Prof. J. P. M. Finberg, Pharmacology Unit, Faculty of Medicine, Technion, P.O. Box 9649, Haifa, Israel

Effects of N-propargyl-1-(R)aminoindan (rasagiline) in models of motor and cognition disorders

Z. Speiser[1], **R. Levy**[2], and **S. Cohen**[1]

[1]Department of Physiology and Pharmacology, Tel Aviv University, and
[2]Teva Pharmaceutical Industries, Ltd., Corporate R&D Division, Kiryat Nordau,
Netanya, Israel

Summary. N-propargyl-1-(R)aminoindan (rasagiline) is a new and selective irreversible MAO-B inhibitor, currently being considered as the mesylate salt for potential therapy in certain neurological disorders. It has been studied in animal models of cognition and motor dysfunction. Its ability to restore normal motor activity was determined in models of acute drug-induced dopaminergic dysfunction: Its effects in improving cognition and memory deficits was studied in adult and senescent rats that had been exposed to prolonged hypoxia, then subjected to the passive and active avoidance tests. In α-methyl-p-tyrosine (α-MpT)-induced hypokinesia (100–120 mg/kg, i.p.) pretreatment with rasagiline at 2.5 mg/kg i.p. restored motor activity to control level. But pretreatment with reserpine abolished the protective effect of rasagiline. Rasagiline at 0.5 mg/kg/day was protective against α-MpT also in hypoxia-lesioned rats. In haloperidol-induced catalepsy in rats (1.5 mg/kg, s.c.) or mice (4–6 mg/kg s.c.), rasagiline improved recovery of normal locomotion, gait and coordination at 0.4–2.4 mg/kg i.p. and 1.8–15 mg/kg i.p., respectively. In amphetamine-induced stereotypy (0.6 mg/kg s.c.), rasagiline potentiated this effect at 1.5 mg/kg i.p. In hypoxia-induced impairment of memory and learning, rasagiline at 0.32–0.5 mg/kg/day per os improved performance of adult rats in passive and active avoidance, and of senescent rats in active avoidance. Selegiline was either ineffective or less effective at equivalent doses. Racemic N-propargyl-1-aminoindan (AGN-1135), besides being of lower potency, had a different dose-dependency than rasagiline in antagonizing haloperidol-induced catalepsy or α-MpT-induced hypokinesia. 1-(R)aminoindan ((R)AI), a metabolite of rasagiline, in relatively high doses produced effects that were distinct in certain respects from those of rasagiline.

Introduction

Among the few selective inhibitors of monoamine oxidase type B is N-propargyl-1-aminoindan (PAI), appearing in literature reports under the code name AGN-1135 (Finberg et al., 1981; Kalir et al., 1981; Finberg and Youdim,

1985; Heikkila et al., 1985; Riederer and Youdim, 1986). The findings by these authors prompted the resolution of AGN-1135 into its two optical enantiomers, N-propargyl-1-(R)(+)aminoindan ((R)PAI), and N-propargyl-1-(S)(−)aminoindan ((S)PAI). (R)PAI being the more potent and also the more selective member, having an IC_{50} ratio MAO-A/MAO-B of about 24 in rat brain homogenates and a corresponding ex vivo ED_{50} ratio in the range of 20–70 depending on the experimental system used (Youdim et al., 1995). (R)PAI, henceforth referred to as rasagiline is currently being considered for potential therapy in certain neurological disorders, especially Parkinson's disease as TVP-1012 which is rasagiline mesylate. The present study is a report of the effects of rasagiline and related drugs in animal models of experimental dopaminergic dysfunction.

Methods

Study design

The study comprised two phases. *Phase I* consisted of a preliminary assessment of the effects of an acute parenteral administration of (R)PAI and related drugs in drug-induced motor disorders. Dopaminergic dysfunction was caused by acute administration of agents that affect endogenous dopamine function at presynaptic or postsynaptic sites. In *Phase II*, cognitive disorders were caused by exposure of adult or senescent rats to prolonged hypoxia, following which the rats show impaired performances in learning and memory tasks (Speiser et al., 1990). Chronic medication with rasagiline was used in Phase II.

Animals

Male Wistar rats (Olac, Jerusalem) were of three different age groups: In Phase I, adult young rats 250–300 g each, or male ICR mice weighing 20–25 g each were used; in Phase II, one group consisted of 10–15 month-old rats weighing 400–540 g each; another group consisted of senescent rats, 20–24 month-old weighing 450–560 g each. All rats were kept throughout this study at a room temperature of $24 \pm 2°C$, RH of 60% and a light/darkness cycle of 11/13 hours, with free access to standard food (Purina) and water. At the completion of each experiment, the animals were killed by exposure to an atmosphere of CO_2.

Hypoxia (Phase II)

Rats were exposed to a single hypoxic episode as follows (Speiser et al., 1990): Four to five rats were kept for six hours in a glass chamber equipped with an inlet and outlet tubes for the admission of an atmosphere of premixed nitrogen (92%) and oxygen (8%) at a flow rate of 3 L/min. Control rats received room air from a compressed tank under similar conditions.

Drugs and drug treatment

Rasagiline as the HCl or mesylate salt (TVP-1012), racemic PAI as the HCl salt, 1-(R)aminoindan ((R)AI) (HCl) and selegiline (as the HCl salt) were supplied by Teva

Pharmaceutical Industries, Ltd. All other drugs were obtained from commercial sources. In the acute medication series (Phase I), the drugs were administered i.p., unless otherwise stated. In the hypoxia-lesioned senescent rats (Phase II) rasagiline HCl was administered orally in drinking water immediately following the conclusion of the hypoxic episode at the standard dose of 0.5 mg (HCl salt)/kg/day for a minimum duration of 40 days.

Locomotor tests (Phase I)

Haloperidol-induced rigidity, and loss of motor coordination, mice. Male, ICR mice 25–30 g each, were pretreated with either 1 or 3 mg/kg of each of the test drugs. All drugs were administered i.p. in a volume of 0.2 mL. Two hours later, haloperidol was injected s.c. at a dose of 4–6 mg/kg in a volume of 0.1–0.2 mL. Motor coordination tests were made at 1 or 2 hours after haloperidol, that is 5 or 6 hours after the presumed protective drug administration. Motor coordination tests and rigidity were quantified according to three different parameters: (a) ability to walk the length of a horizontal rod, 80 cm-long; (b) ability to climb down, face down, a vertical rod, 80 cm-long; (c) duration of immobility in an unnatural sitting posture whereby the abdomen of the mouse is pressed against a "wall". Full performance as in haloperidol-untreated mice is given the score of 4 in each test, or a total of 12 in all tests. Poor performance is given a score from 1 to 3 in accordance with the severity of the motor dysfunction produced.

Haloperidol-induced bradykinesia, rats. Rats were injected i.p. with each of the test drugs or saline at doses in the range of 0.3–30 mg/kg. After two hours, the animals received s.c. haloperidol at a dose of 1.5 mg/kg. One hour later, motor activity was recorded for the next 10 hours. In this and following tests, locomotion scores were taken in seven fully-computerized cages (26 × 25 cm) having a grid of infra-red beams at 4 cm-intervals. Crossing of a beam initiated an electric signal which was fed into a computer. For long periods of recording (night) the cages were provided with Purina chow and water. The number of crossings over a given period gave a measure of locomotion.

α-MpT-induced hypokinesia. Rats were injected with the test drugs immediately before receiving α-methyl-p-tyrosine (α-MpT) 100–120 mg/kg i.p. Motor activity was scored for the duration of 10 hours in activity cages, as given under haloperidol. In a variant of this test (Skuza et al., 1994), rats were treated with reserpine (5 mg/kg s.c.) then a-MpT (250 mg/kg i.p.) 24 and 3 hours before scoring for motor activity. The test drugs were injected i.p. immediately before taking the motor activity test.

Behavioral tests (Phase II)

In the chronic medication regimen all behavioral tests were performed 90–120 min after administration of the last dose of the test compound.

Amphetamine-induced stereotypy. Rats received the test drugs, followed two hours later by D-amphetamine sulfate (0.5 mg/kg s.c.). Counts of lateral head movements per minute were taken over 45–60 minutes following amphetamine injection and averaged.

α-MpT-induced hypokinesia. This test was performed as described under Phase I.

Passive avoidance and active avoidance tests. These were performed as described in earlier reports (Speiser et al., 1989, 1990).

Results

Phase I

Prevention of haloperidol-induced catalepsy. In mice, full protection could be achieved with rasagiline at about 7.5–15 mg/kg. Racemic PAI which is an equimolar mixture of (R) and (S) enantiomers never matched the protective efficacy of rasagiline even at the highest dose used (Fig. 1), suggesting some negative contribution from the presumably inactive (S)enantiomer. In rats, either rasagiline and selegiline significantly improved motor activity when administered 2 hours before haloperidol (1.5 mg/kg s.c.). The effective range for either was 0.5–3 mg/kg (HCl salts), following which there seemed to be a decline in protective efficacy. With selegiline, a second protective phase appeared at 10 mg/kg, and is probably due to an amphetamine-like effect (Fig. 2).

Reversal of a-MpT-induced hypokinesia. At 2 mg/kg, rasagiline mesylate restored motor activity to control levels, the effective range being 2–5 mg/kg, followed by a steep decline. Again, there was a marked difference in dose-dependency between rasagiline and racemic PAI, suggesting

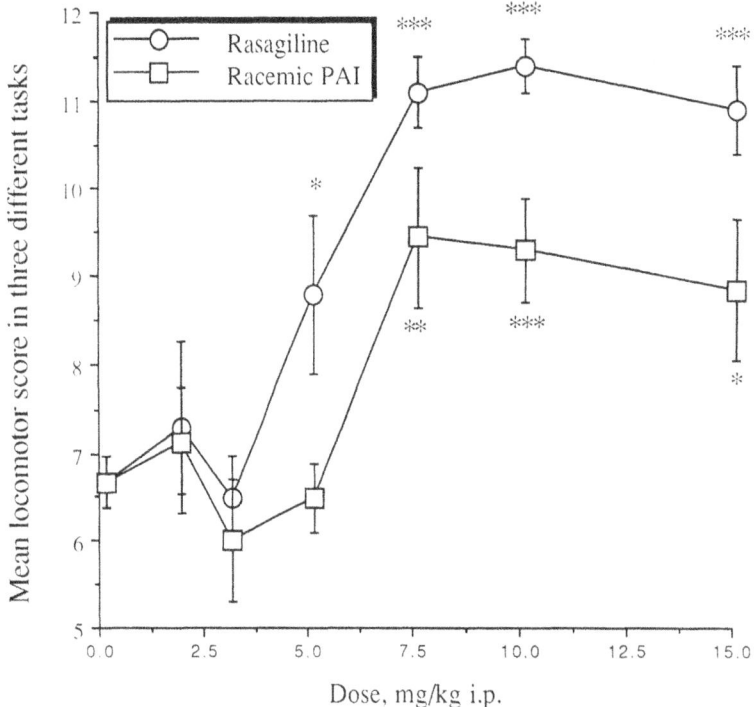

Fig. 1. Protection by rasagiline ((R)PAI) and racemic PAI against haloperidol-induced catalepsy in mice. The drugs were injected as mesylate salts i.p. 2 h before haloperidol 6 mg/kg s.c. Doses are free base equivalents. Locomotor scores were taken 3 h after haloperidol. Maximum score in each of three different motor activity tasks was 4, total 12. Results are mean ± SEM for n = 6 per dose. Statistical significance of difference between drug-treated groups and haloperidol control in this and following figures by Student's "t"-test. *P ≤ 0.05; **P ≤ 0.01; ***P ≤ 0.001

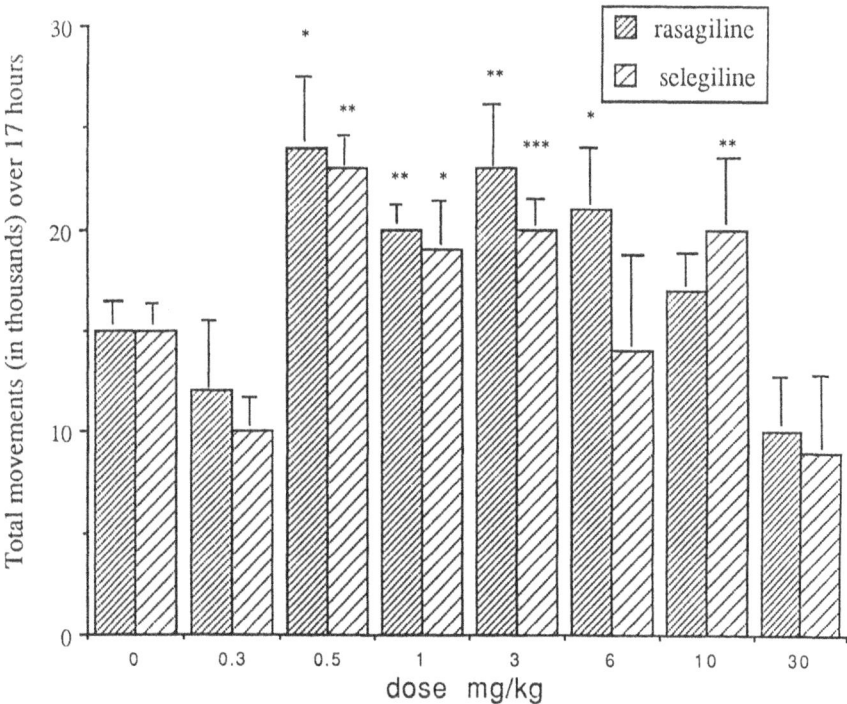

Fig. 2. Protection against haloperidol-induced catalepsy in rats by i.p. rasagiline and selegiline (HCl salts) given 2 h before haloperidol 1.5 mg/kg s.c. Results are mean scores ± SEM for 4–7 rats per dose. Mean score ± SEM for normal rats was 37.1 ± 3.2 ($\times 10^3$)/17 h. Significant difference from haloperidol alone is indicated by asterisks

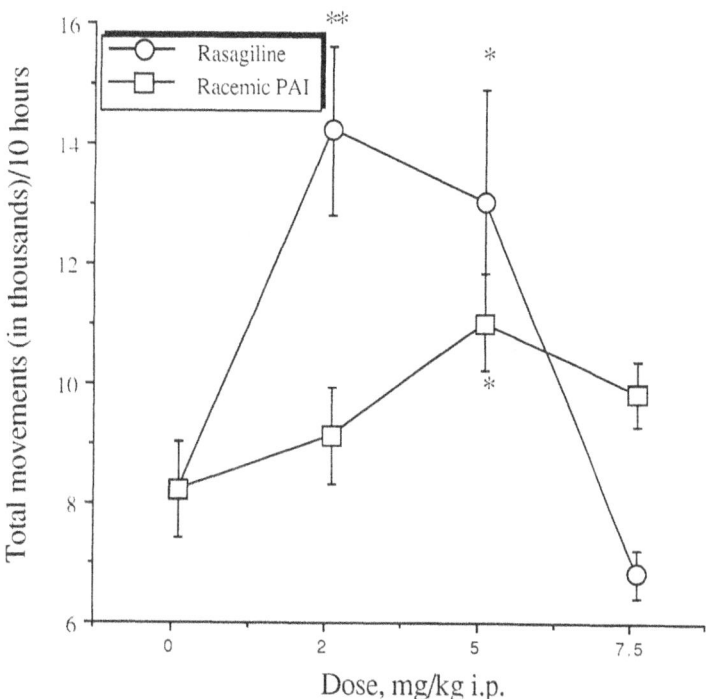

Fig. 3. Protection against α-MpT-induced hypokinesia in rats by rasagiline (R-PAI) and racemic PAI mesylates. Doses are the free base equivalents. Drugs were injected i.p. immediately before α-MpT 100 mg/kg i.p. Activity was scored for 10 h in a computerized activity cage. Normal rats scored 16 ± 1.4 ($\times 10^3$). α-MpT-treated rats scored 8 ± 0.8 ($\times 10^3$). Comparison is with respect to α-MpT alone for n = 4–8 per dose

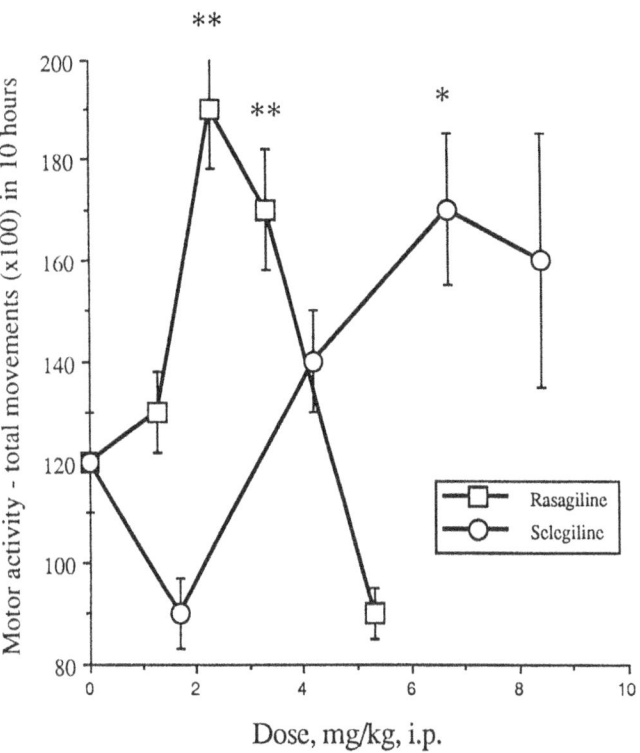

Fig. 4. Protection against α-MpT-induced hypokinesia in rats by rasagiline and selegiline (as free base equivalents) injected i.p. 2h before α-MpT 120 mg/kg i.p. Total movements were recorded in a computerized activity cage for the duration of 10h, starting immediately after α-MpT. Normal rats recorded 200 ± 20 ($\times 10^2$). Comparison is with respect to α-MpT alone, for n = 4–6 per dose

significant interference from the (S)enantiomer (Fig. 3). In a comparison between rasagiline and selegiline, rasagiline seemed to be more potent than selegiline on a weight to weight basis with respect to the corresponding base equivalents (Fig. 4). But in either case, pretreatment of rats with reserpine greatly reduced the protective effect with respect to total movements of control: untreated (31%), rasagiline, (39%), selegiline (44%) and (R)AI (46%). In contrast, apomorphine restored motor activity almost to control levels (87%).

Phase II, 10–15 month-old rats

α-MpT-induced hypokinesia. This test was carried on days 70–80 following hypoxia. Figure 5 shows the mean scores for total movements over a period of 10 hours, with respect to control (hypoxia + saline), for α-MpT alone and after pretreatment with each of rasagiline, (R)AI or selegiline. All three drugs appeared to prevent the hypokinesia elicited by α-MpT, but to different degrees. The least active was (R)PAI which nevertheless brought activity back to control level (hypoxia + saline). (R)AI and selegiline seemed to exceed the control level.

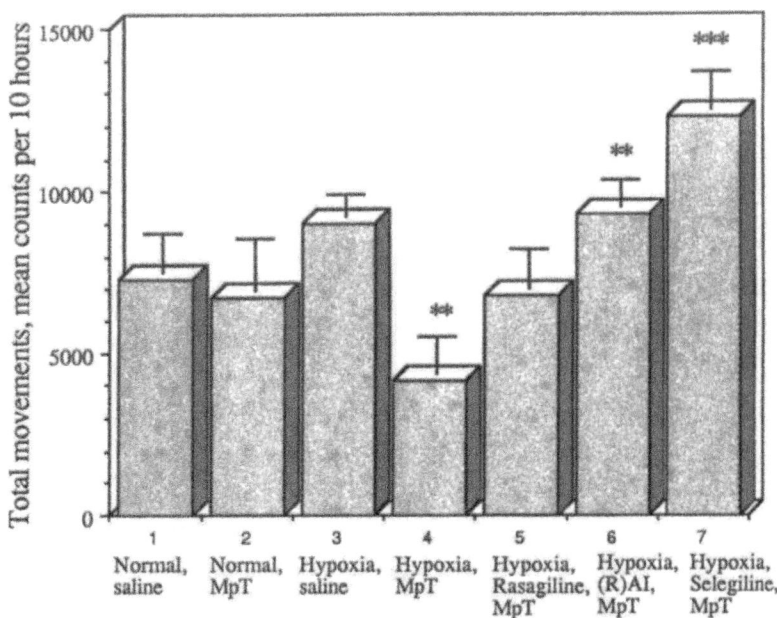

Fig. 5. Protection against α-MpT-induced hypokinesia in 10–15 month-old hypoxia-lesioned rats by rasagiline mesylate, selegiline HCl and (R)AI HCl, given orally 0.5 mg/kg/day for 70–80 days following hypoxia. α-MpT was given i.p. 120 mg/kg then activity was recorded in a computerized activity cage for 10 h. Statistical significance: hypoxia/MpT vs. hypoxia/saline group; hypoxia/drug/MpT vs. hypoxia/MpT group, for n = 7–9 per group

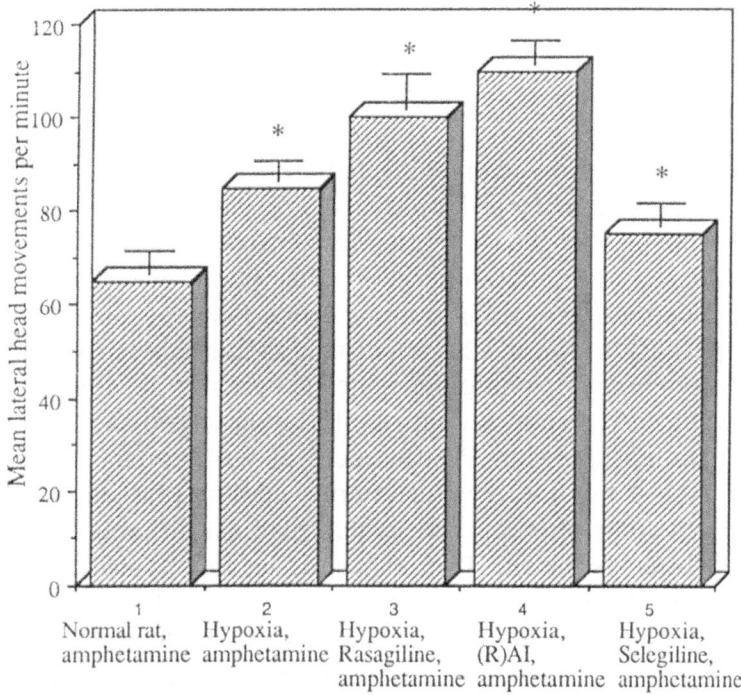

Fig. 6. Amphetamine-induced stereotype behavior in 10–15 month-old hypoxia-lesioned rats that received chronic oral doses of rasagiline mesylate, selegiline HCl or (R)AI HCl 0.5 mg/kg/day for 29 days. Statistical significance is with respect to amphetamine alone for n = 11–13 per group

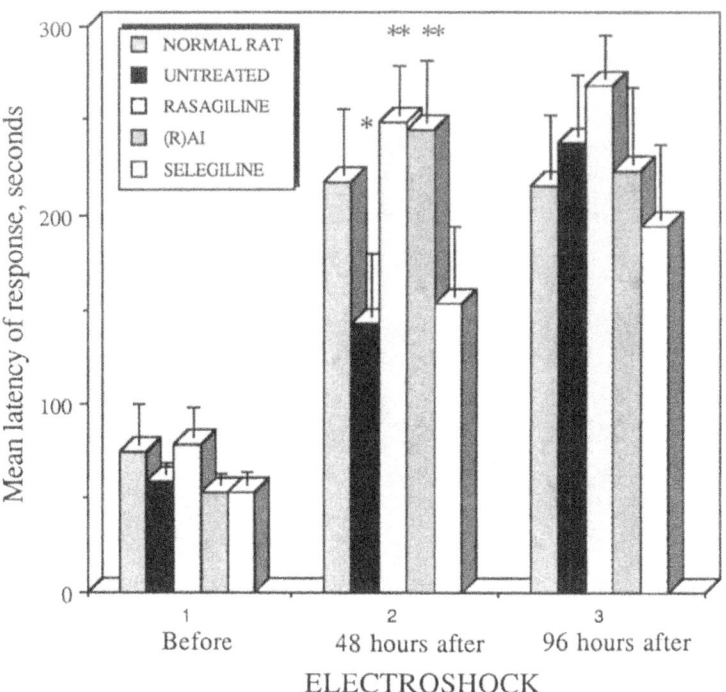

Fig. 7. Mean latency of response for entering dark compartment in the passive avoidance test of hypoxia-lesioned 10–15 month-old rats that received an oral dose of rasagiline mesylate, selegiline HCl or (R)AI HCl, 0.5 mg/kg/day for 13–17 days immediately after prolonged hypoxia. Statistical significance is with respect to hypoxic untreated rats for n = 11–13 per group

Amphetamine-induced stereotypy. This test was performed on day 29 following hypoxia. The scores are given in Fig. 6. In the hypoxia group, pretreatment with either rasagiline or (R)AI produced a significant potentiation of the stereotypic behavior induced by amphetamine with respect to their respective control (hypoxia + amphetamine). At this point, we recall that drug-untreated hypoxia rats are usually more active than drug-untreated control rats, owing perhaps to the development of dopamine hypersensitivity in response to presynaptic dopamine deficiency (Speiser et al., 1990).

Passive avoidance. At 48 hours, control rats, but not hypoxia rats, had already achieved maximum latencies. Pretreatment of the hypoxia rats with either rasagiline or (R)AI, but not selegiline, improved their performance to control level (Fig. 7). In contrast, selegiline-treated hypoxia rats showed no improvement. At 96 hours after the first exposure to electroshock, performance improved in the drug-untreated hypoxia group and the selegiline-treated hypoxia group, reaching the same level as with rasagiline or (R)AI.

Active avoidance, senescent rats. The mean avoidance scores in 60 trials is given in Fig. 8. Rasagiline caused a significant increase in the mean avoidance score and a corresponding decrease in the latency of response with respect to the untreated hypoxia-lesioned group. In fact, the rasagiline-treated group was not different from the normal rat group.

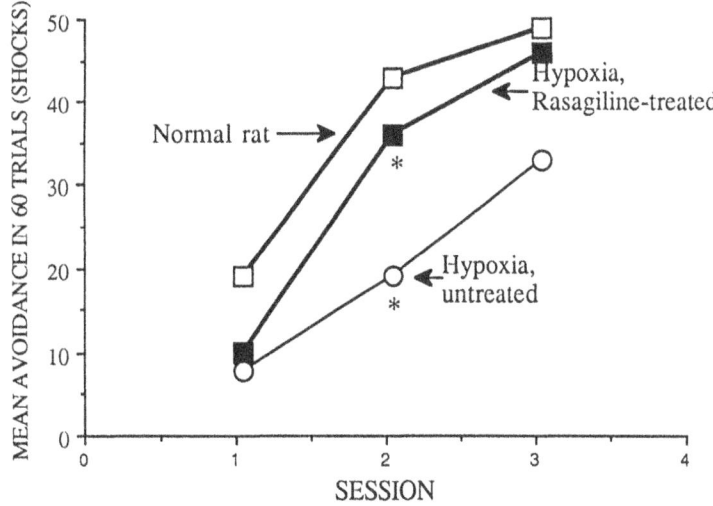

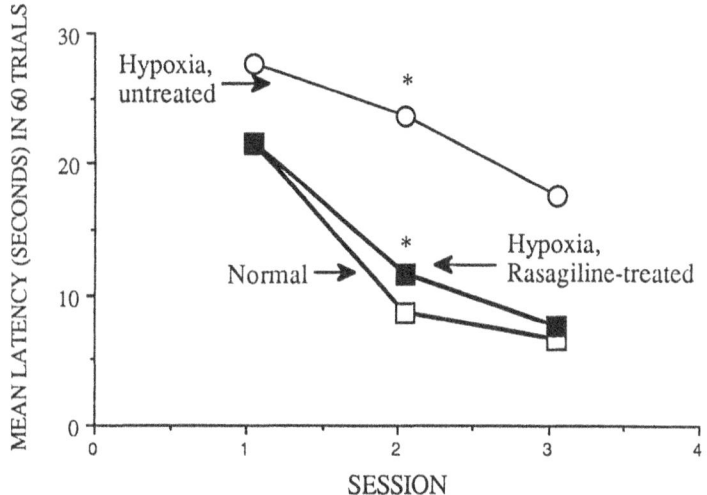

Fig. 8. Active avoidance performance of hypoxia-lesioned 2 year-old senescent rats, treated orally with rasagiline HCl, 0.5 mg/kg/day for about 60 days following hypoxia. Intervals between sessions was 14 days. Response on session 2 of normal and rasagiline-treated rats was significantly different from response of untreated hypoxia rats, for n = 7–8 per group

Discussion

Phase I

1. Prevention of haloperidol-induced catalepsy by rasagiline in either mice (Fig. 1) or rats (Fig. 2) can be reasonable ascribed to an enhancement of dopaminergic activity in the basal ganglia. It has also been shown that racemic PAI cannot be considered as a diluted form of rasagiline when both are tested

in the cataleptic intact mouse. The dose-dependencies of these two forms of PAI in reversing the effects of haloperidol are different. With each, a "saturation" dose level is reached, beyond which further improvement cannot be achieved. One can assume some negative contribution on behalf of (S)PAI found in the racemic form.

2. In otherwise normal rats, rasagiline and selegiline were almost equipotent in antagonizing haloperidol-induced catalepsy in the dose range of 0.5–3 mg/kg (Fig. 2), both showing decline in activity beyond this range. Selegiline showed a second phase of activity at 10 mg/kg which is believed to be due to its amphetamine-like metabolites.

3. The difference in dose-dependency between rasagiline and racemic PAI found in mice was further confirmed in the α-MpT-induced hypokinesia test in rats (Fig. 3). Here, rasagiline restored normal motor activity at doses in the same range as found effective in the haloperidol-induced catalepsy test in rats (Fig. 2). This was not possible with racemic PAI, even when used at the highest dose. In this test, selegiline was also effective, but also and consistently less potent than rasagiline: 2 mg/kg rasagiline base were equipotent with 6.6 mg/kg selegiline base (Fig. 4). Since the decrease in locomotion after α-MpT is ascribed to the inhibition of central dopamine synthesis, the restorative effects of either rasagiline or selegiline could possibly arise from a sparing action on residual dopamine stores after inhibiton of MAO or from a direct agonistic effect at the dopamine receptor site. That the latter possibility is unlikely was shown in reserpine-pretreated rats that received also α-MpT. In these, none of the two agents, nor 1-(R)AI was effective in restoring motor activity to control level, but apomorphine, a direct dopaminergic agonist, did.

Phase II

4. In the hypoxia-lesioned rats, the effects produced by the drugs under consideration seem to be of a different nature than in the otherwise normal rats. In the antagonism of α-MpT-induced hypokinesia in hypoxia-lesioned adult rats by the test drugs (R)AI seemed to be more effective than rasagiline, but almost equal to selegiline at the uniform dose used (Fig. 5). Due account must be taken of the fact that the animal model (hypoxia-lesioned) is different from the one used in the acute test, that exposure to the test drugs was chronic and not acute and that use of equal doses of the drugs studied on a weight basis causes a positive bias with respect to the lower MW aminoindan. Thus, a dose of 1 mg of rasagiline mesylate is equivalent to 3.74 mmol; 1 mg rasagiline HCl, to 4.80 mmol; 1 mg AI HCl to 6.82 mmol; and 1 mg selegiline HCl to 4.48 mmol. Still, it is obvious that the effect perceived cannot be rationalized entirely in terms of relative MAO-B inhibitory potencies of the three agents. Selegiline may increase synaptic dopamine perhaps also by causing its release or preventing its reuptake through one of its metabolites (Engberg et al., 1991). Rasagiline lacks this effect (M. Rehavi, personal communication). It is not unlikely that the hypoxia episode had altered in a

differential manner various functions associated with uptake and release and which is reflected in the difference between rasagiline and selegiline. It is also possible that the low activity of rasagiline in this test is in fact due to its partial and incomplete conversion into its metabolite 1-(R)AI which was highly effective in this test. The case of (R)AI, deserves some attention. Unlike 2-aminoindan, racemic 1-aminoindan was reported as being devoid of amphetamine-like activity. The reported order of the IC_{50} values in the inhibition of catecholamine uptake into the striatum was: ($\pm$)amphetamine, 0.4 µM; 2-aminoindan, 3.3 µM; ($\pm$) 1-aminoindan, 1 mM (Horn and Snyder, 1972). On the other hand, it is not unlikely that (R)AI which is structurally related to benzylamine but is otherwise a poor substrate for MAO-B, might also behave as a reversible inhibitor of this enzyme in analogy with other benzylamines (Fowler et al., 1980). The overall effect of such inhibition could be more pronounced after prolonged chronic therapy as in the present case. Indeed, a preliminary assessment of the relative percent inhibition of the two forms of MAO in brain homogenates of the hypoxic rats yielded the following approximate limits at 0.5 mg/kg/day for 80 days: (R)AI HCl, 6 (A), 43 (B); rasagiline mesylate, $\geq$50(A), 93(B); selegiline HCl, 23 (A), 80 (B). Similar levels of inhibition of MAO-A and B were reported by Gelowitz et al. (1994) after prolonged i.p. treatment of normal rats with selegiline at a dose of 0.25 mg/kg.

5. The effect of the test drugs on amphetamine-induced stereotypy may give further insight into the mechanism involved in the action of the individual drugs. The fact that (R)AI greatly enhances the effect of amphetamine speaks against any possible competition between the two agents in release and reuptake processes (Fig. 6). Rather, it would seem that supplies of dopamine or a dopaminergic agent became more available after pretreatment with (R)AI. (R)AI itself lacks an amphetamine-like effect and is not capable of eliciting stereotype behavior, a finding consistent with the much earlier report on racemic 1-aminoindan as a poor inhibitor of catecholamine uptake into hypothalamus and striatum as compared to 2-aminoindan (Horn and Snyder, 1972). The lack of effect of selegiline can be ascribed to its competition with amphetamine for the common carrier, thus denying it entry into the neurone. The moderate effect of rasagiline may be due either to the increase in striatal dopamine or to the in vivo generation of (R)AI, or to both effects.

6. In learning and memory tests, normal performance in the passive avoidance response (Fig. 7) is generally related to unimpaired cholinergic function (Speiser et al., 1988, 1989). But neither rasagiline nor (R)AI have a demonstrable direct cholinergic effect, at least in organ bath preparations. Yet, evidence from microdialysis studies in awake unrestrained rats showed a 170% increase in hippocampal ACh release after systemic apomorphine, and a 400% increase after amphetamine (Nilsson et al., 1992). An alternative explanation is that rasagiline improves memory consolidation in this test by an increase in central norepinephrine concentrations, following partial inhibition of MAO-A (Zornetzer, 1985). Indeed, moclobemide which is a reversible MAO-A inhibitor, corrected scopolamine-induced memory impairment in

humans and experimental animals (Wesnes et al., 1990). The proposed mechanism may account for the cognition-enhancing activity of rasagiline which is a partial MAO-A inhibitor at the prescribed non-selective dosage used; and also for the apparent lack of activity of selegiline at the same dosage, but does not account for the activity of (R)AI which is not a MAO-A inhibitor.

A different or additional mechanism may focus on the activation of site-specific D2 receptors which are involved in consolidation processes, with regional specialization of the caudate nucleus depending on the nature of the conditioning stimulus (Izquierdo, 1992). Rasagiline is expected to increase local dopamine concentration by inhibition of MAO-B. But the lack of activity of selegiline which is likewise a MAO-B inhibitor remains an open question. One might speculate that rasagiline is instrumental in the release of dopamine in the nucleus accumbens in a process that has been linked to active conditioned response (McCullough et al., 1993). The improvement in performance was even more pronounced in senescent rats in the active avoidance (Fig. 8). These performed almost as normal rats already in session 1 and improved even further in session 2. At the same time, prolonged hypoxia did not seem to impair to any significant extent their performance in the passive avoidance test. A rational explanation of these results can be approached from a consideration of the spontaneous decline of dopamine turnover with age on the one hand and the further impairment of dopamine synthesis after a single episode of hypoxia. The data in Table 1 show a sharp fall in caudate dopamine turnover rate constant in 20 month-old rats, with little or no decrease in steady-state dopamine concentration (unpublished). This fall corresponds to a turnover-time fall from about 2h to 4h. Hypoxia can further increase the turnover time by a factor of 2, as shown in the case of adult rats (Speiser et al., 1990). Dopamine turnover seems to be particularly susceptible to aging, also in other parts of the CNS. In aging rats it was found to decrease in the lateral geniculate nucleus and visual cortex, while noradrenaline turnover was unaltered (Herrera et al., 1993). Thus, senescent rats, especially after hypoxic lesion, must suffer from very limited supplies of dopamine when faced with a task that requires full mobilization of the striatal dopamine system. Rasagiline seems to afford a partial remedy to such shortage. All that can be concluded at this stage is that improved performance in the active avoidance test is not conditional on effective MAO-B inhibition, at least not entirely. Recently, Lamensdorf et al. (1995) suggested that enhanced dopamine release by chronic MAO-B inhibition with selegiline or rasagiline (reported as TVP-1012) may be caused by an increase in endogenous β-phenylethylamine or a slowly developing inhibition of dopamine uptake.

Acknowledgements

We thank Profs. J. Finberg and M. Youdim for stimulating discussion and for providing the MAO inhibition data.

References

Engberg G, Elebring Th, Nissbrandt H (1991) Deprenyl (selegiline), a selective MAO-B inhibitor with active metabolites; effects on locomotor activity, dopaminergic neurotransmission and firing rate of nigral dopamine neurons. J Pharmacol Exp Ther 259: 841–847

Finberg J, Youdim MBH (1985) Modification of blood pressure and nictitating membrane response to sympathetic amines by selective monoamine oxidase inhibitors, types A and B, in the cat. Br J Pharmacol 85: 541–546

Finberg JPM, Tenne M, Youdim MBH (1981) Tyramine antagonistic properties of AGN-1135, an irreversible inhibitor of monoamine oxidase type B. Br J Pharmacol 73: 65–74

Fowler CJ, Callingham BA, Mantle TJ, Tipton KF (1980) The effect of lipophilic compounds upon the activity of rat liver mitochondrial monoamine oxidase A and B. Biochem Pharmacol 29: 1177–1183

Gelowitz DL, Richardson JS, Wishart ThB, Yu PH, Lai C-T (1994) Chronic L-deprenyl or L-amphetamine: equal cognitive enhancement, unequal MAO inhibition. 47: 41–45

Heikkila RE, Davoisin JP, Finberg J, Youdim MBH (1985) Prevention of MPTP-induced neurotoxicity by AGN-1133 and AGN-1135, selective inhibitors of monoamine oxidase-B. Eur J Pharmacol 116: 313–317

Herrera AJ, Machado A, Cano J (1993) Ageing and monoamine turnover in the lateral geniculate and visual cortex of the rat. Neurochem Int 22: 531–539

Horn AS, Snyder SH (1972) Steric requirements for catecholamine uptake by rat brain synaptosomes: studies with rigid analogs of amphetamine. J Pharmacol Exp Ther 180: 523–530

Izquierdo I (1992) Dopamine receptors in the caudate nucleus and memory processes. Trends Pharmacol Sci 13: 7–8

Kalir A, Sabbagh A, Youdim MBH (1981) Selective acetylenic "suicide" and reversible inhibitors of monoamine oxidases types A and B. Br J Pharmacol 73: 55–64

Lamensdorf I, Youdim MBH, Finberg JPM (1995) Increase in striatal dopamine following chronic selective inhibition of MAO A or B. Isr J Med Sci 31: 739

McCullough LD, Sokolowsky JD, Solomone JD (1993) A neurochemical and behavioral investigation of the involvement of nucleus accumbens dopamine in instrumental avoidance. Neurosci 52: 919–925

Nilsson OG, Leanza G, Björklund A (1992) Acetylcholine release in the hippocampus: regulation by monoaminergic afferents as assessed by in vivo microdialysis. Brain Res 584: 132–140

Riederer P, Youdim MBH (1986) Monoamine oxidase activity and monoamine metabolism in brains of parkinsonian patients treated with l-deprenyl. J Neurochem 46: 1359–1365

Speiser Z, Amitzi-Sonder J, Gitter S, Cohen S (1988) Behavioral differences in the developing rat, following postnatal anoxia or postnatally-injected AF-64A, a cholinergic neurotoxin. 30: 89–94

Speiser Z, Reicher S, Gitter S, Cohen S (1989) Tacrine or arecoline mediates reversal of anoxia- or AF-64A-induced behavioural disorders in the developing rat. Neuropharmacol 28: 1325–1332

Speiser Z, Amitzi-Zonder J, Ashkenazi R, Gitter S, Cohen S (1990) Central catecholaminergic dysfunction and behavioral disorders following hypoxia in adult rats. Behav Brain Res 37: 19–27

Skuza G, Rogoz Z, Quack G, Danysz W (1994) Memantine, anmantadine and L-deprenyl potentiate the action of L-Dopa in monoamine-depleted rats. J Neural Transm 98: 57–67

Wesnes (1990) Potential of moclobemide to improve cerebral insufficiency identified using a scopolamine model of aging and dementia. Acta Psychiatr Scand [Suppl] 360: 71–72

Youdim MBH, Finberg JPM, Levy R, Sterling J, Lerner D, Berger-Paskin T, Yellin H (1995) R-Enantiomers of N-propargyl-aminoindan compounds. Their preparation and pharmaceutical compositions containing them. United States Patent 5,457,133

Zornetzer, ST (1985) Catecholamine system involvement in age related memory dysfunction. Ann NY Acad Sci 444: 242–254

Authors' address: Dr. Z. Speiser, Department of Physiology and Pharmacology, Tel Aviv University, Tel Aviv 69978, Israel

(R)(+)-N-Propargyl-1-aminoindan (rasagiline) and derivatives: highly selective and potent inhibitors of monoamine oxidase B

J. Sterling[1], A. Veinberg[1], D. Lerner[1], W. Goldenberg[1], R. Levy[1], M. Youdim[2], and J. Finberg[2]

[1] Teva Pharmaceutical Industries, Jerusalem, and [2] Department of Pharmacology, Rappaport Faculty of Medicine, Technion, Haifa, Israel

Summary. (+)-N-Propargyl-1-aminoindan (rasagiline) and a series of derivatives have been synthesized and screened for monoamine oxidase inhibitory activity. Rasagiline and several analogues were found be highly selective and potent inhibitors of the B form of the enzyme in contrast to the levorotatory enantiomer which was not active. The results indicate that rasagiline has potential for the treatment of Parkinson's Disease. This compound is currently under development for that indication.

Introduction

The theoretical justification for the use of type B-selective monoamine oxidase (MAO) inhibitors in the treatment of Parkinson's disease is well known (Birkmayer, 1975). This therapeutic potential is supported by the experimental and clinical data for selegiline (Parkinson Study Group, 1989, 1993), an irreversible inhibitor of MAO-B. Selegiline remains the only drug approved for Parkinson's therapy based on this mechanism of action. The principle metabolites of selegiline, amphetamine and methamphetamine, are biologically active molecules which could cause adverse side-effects in patients treated with selegiline (Engberg, 1991). MAO-B selectivity is an important property for an irreversible inhibitor of MAO to minimize risk of the "cheese effect" from the ingestion of tyramine-rich foods that are metabolized by MAO-A in the gastrointestinal tract (Youdim, 1995).

Rasagiline (compound 1, Table 1) shares with selegiline certain structural features (both are propargylamines) and also a similar activity profile in the inhibition of MAO. However, rasagiline cannot be metabolized to amphetamine derivatives.

This paper reports the MAO inhibitory activity profile of rasagiline and of some structural analogues.

Material and methods

The synthesis and characterization of the compounds have been reported elsewhere (Youdim, 1995; Sterling, 1996).

J. Sterling et al.

Table 1. Synthesized propargylamines ([a]Hazelhoff, 1985)

Compd. #	R	R'
1 (rasagiline)	H	H
2	H	CH_3
3	4-F	H
4	5-F	H
5	6-F	H
6	4-Cl	H
7	6-Cl	H

Compd. #	Position	R	R'
8	1	H	H
9[a]	2	H	CH_3

MAO activity was determined using a modification of the method of Otsaka and Kobayashi (1964), as described by O'Carrol et al. (1983). Substrate concentrations were $10\,\mu M$ ^{14}C-phenylethylamine and $100\,\mu M$ ^{14}C-5-HT. Deaminated products were extracted into toluene/ethyl acetate 1:1, to which was added 2% 2,5-diphenyloxazole for estimation by liquid scintillation counting.

Results and Discussion

Rasagiline (TVP-1012, the mesylate salt) is the R(+) enantiomer of N-propargyl-1-aminoindan. The racemic compound has been described in the literature as AGN-1135 (Finberg et al., 1981; Kalir et al., 1981). The absolute stereochemistry was assigned based on its synthesis from R-aminoindan (Ghislandi, 1976). The S(−)enantiomer is nearly four orders of magnitude less potent an inhibitor of MAO in vitro (see Table 2). The data in Table 2 shows that this high degree of stereospecificity is unusual in compounds with MAO inhibitory activity. The N-methyl derivatives (compounds 2R and 2S) differ only by a factor of 5 in MAO-B inhibitory activity and R- and S-deprenyl differ by a factor of only 2.6.

Rasagiline is an irreversible inhibitor of MAO with a proposed mechanism of action analogous to that of selegiline (Polymeropoulos, 1993). Thus, a single 1 mg/kg dose (i.p. rat) almost completely inhibits MAO-B activity measured ex vivo in the brain. The slow recovery of activity is determined by the

Table 2. In vitro MAO inhibitory activity (IC$_{50}$: nmol/L)

| Compd # | Inhibition (IC$_{50}$: nmol/L) | | Relative potency | | |
	A	B	A/B	R/S A	R/S B
1R	73	2.5	30	356	6,800
1S	26,000	17,000	1.5		
2R	3	10	0.3	28	5
2S	70	50	1.4		
3R	340	3	113		
4R	140	9	16		
5R	2,800	2	1,400	14	6,000
5S	40,000	12,000	3.3		
6 (rac)	90	4	22		
7 (rac)	6,900	9	767		
8R	900	90	10	4	0.5
8S	3,800	50	76		
9(−)[a]	46	88	0.5	3	0.2
9(+)[a]	140	16	8.8		
R-deprenyl[a]	450	6	75	80	2.6
S-deprenyl[a]	3,600	30	120		

[a] Hazelhoff (1985)

rate of de-novo synthesis of MAO-B enzyme. Full recovery of the activity of rat brain MAO-B activity takes about 3 weeks (after a single 1 mg/kg i.p. dose).

The structure activity relationship of this series of N-propargyl-1-aminoindans was explored by the synthesis and screening of selected analogs (see Tables 1 and 2). The addition of an N-methyl group (compound 2) greatly reduces the degree of MAO selectivity. Enlarging the five-membered ring to six as in compound 8 reduces the potency of MAO-B inhibition, as well as the selectivity of for MAO-B over A. In this case, the S isomer is the more potent and selective for MAO-B. Likewise, compound 9 has a similar profile, although the absolute stereochemistry was not assigned. For comparison, R-deprenyl (selegiline) is slightly less potent than compound 1 but somewhat more selective.

Of this series of analogues, only some of the halogenated derivatives offered comparable potency and selectivity to the parent compound. In fact, compound 5R seemed to have superior properties to compound 1.

We also examined the ex vivo MAO inhibitory activity of these compounds in rat brain 2 h after dosing with the various compounds i.p. In general, the results are analogous to those obtained in vitro. Also here, compound 5R seems to be a particularly promising compound.

Table 3. Ex vivo MAO inhibitory activity IC$_{50}$ (mg/kg) in rat brain after ip injection

Compd #	MAO-A Inhibition (IC$_{50}$: mg/kg)	MAO-B Inhibition (IC$_{50}$: mg/kg)	Relative potency A/B
1R	1.2	0.07	17
1S	>10	>10	
3R	>1	0.33	>3
4R	0.8	0.07	11
5R	21.3	0.14	152
5S	23	0.45	51
6 (rac)	22	0.13	169
7 (rac)	10	1.7	6
8R	5	<0.5	>10
8S	5–10	5–50	
deprenyl	11.5	1.12	10

Based on these data, it is difficult to generalize on the steric requirements for MAO inhibitory potency and selectivity. Furthermore, Yu et al. (1992) have shown that an aromatic moiety on the propargyl amine is not necessary for comparable activity. Thus, three rather different compounds have similar inhibitory properties (potency and selectivity) for MAO: rasagiline (a secondary cyclic benzylamine), selegiline (a tertiary aralkylamine) and N-2-hexyl-N-methylpropargylamine (a tertiary branched alkylamine). In contrast, other small changes (stereochemistry, alternative substitution or ring size) may have a great effect on those properties. The observed differences between the various substituted halogenated derivatives likely arises from electronic effects, perhaps changes in the pKa or lipophilicity (Martin, 1975).

We are continuing the study of the SAR of these compounds to better understand the factors that affect potency and selectivity and to more fully develop their potential for the treatment of Parkinson's and other CNS disorders.

Rasagiline is currently in phase 3 clinical trials for the treatment of Parkinson's disease.

References

Birkmayer W, Riederer P, Youdim MBH, Linauer W (1975) Potentiation of antiakinetic effect after L-dopa treatment by an inhibitor of MAO-B, l-deprenyl. J Neural Transm 36: 303–323

Engberg G, Elebring T, Nissbrandt H (1991) Deprenyl (selegiline), a selective MAO-B inhibitor with active metabolites. J Pharmacol Exp Ther 259: 841–847

Finberg JPM, Tenne M, Youdim MBH (1981) Tyramine antagonistic properties of AGN-1135, an irreversible inhibitor of monoamine oxidase type B. Br J Pharmacol 73: 65–74

Ghislandi V, Vercesi D (1976) Scissione ottica E configurazione dell'1-aminobenzociclobutaene e dell'1-aminoindano. Boll Chim Farm 115: 489–500

Hazelhoff B, De Vries J, Dijkstra D, de Jong W, Horn A (1985) The neuro-pharmacological profile of N-Methyl-N-propargyl-2-aminotetralin. Naunyn Schmied-ebergs Arch Pharmacol 330: 50–59

Kalir A, Sabbagh A, Youdim MBH (1981) Selective acetylenic "suicide" and reversible inhibitors of monoamine oxidase types A and B. Br J Pharmacol 73: 55–64

Martin Y, Martin W, Taylor J (1975) Regression analysis of the relationship between physical properties and the in vitro inhibition of monoamine oxidase by propynylamines. J Med Chem 18: 883–888

O'Carrol AM, Fowler CJ, Philips JP, Tobbia I, Tipton KF (1983) The deamination of dopamine by human brain monoamine oxidase. Specificity for the two enzyme forms in seven brain regions. Naunyn Schmiedebergs Arch Pharmacol 332: 198–202

Otsaka S, Kobayashi Y (1964) A radioisotopic assay for monoamine oxidase determinations in human plasma. Biochem Pharmacol 13: 995–1006

Parkinson Study Group. DATATOP (1989) A multicenter controlled clinical trial in early Parkinson's disease. Arch Neurol 46: 1052–1060

Parkinson Study Group. DATATOP (1993) Effects of tocopherol and deprenyl on the progression of disability in early Parkinson's disease. N Engl J Med 328: 176–183

Polymeropoulos E (1993) 1-Deprenyl: a unique MAO-B inhibitor. In: Szelenyi I (ed) Inhibitors of monoamine oxidase. Birkhäuser, Basel, pp 109–124

Sterling J, Levy R, Veinberg A, Goldenberg W, Finberg J, Youdim MBH, Gutman A (1996) Monofluorinated derivatives of N-propargyl-1-aminoindan and their use as inhibitors of monoamine oxidase. USP 4,486,541

Youdim MBH (1995) The advent of selective monoamine oxidase A inhibitor antidepressants devoid of the cheese reaction. Acta Psychiatr Scand 91 [Suppl 386]: 5–7

Youdim MBH, Finberg J, Levy R, Sterling J, Lerner D, Berger-Peskin T, Yellin H (1995) R-Enantiomers of N-propargyl-aminoindan compounds, their preparation and pharmaceutical compositions containing them. USP 5,457,133

Yu P, David B, Boulton A (1992) Aliphatic propargylamines: potent, selective, irreversible monoamine oxidase B inhibitors. J Med Chem 35: 3705–3713

Authors' address: Dr. J. Sterling, Teva Pharmaceutical Industries, P.O. Box 1142, Jerusalem, Israel

Function of endogenous monoamine oxidase inhibitors (tribulin)

V. Glover

Department of Paediatrics, Queen Charlotte's and Chelsea Hospital,
London, United Kingdom

Summary. Recent research on tribulin [low molecular weight endogenous inhibitory activity of monoamine oxidase (MAO)] has confirmed that its level is increased in both human urine and rat tissues by stress or anxiety, and by anxiogenic drugs. However tribulin is now known to contain several different molecules. The raised inhibitory activity in rat tissues is selective for MAO-A. There is a parallel decrease in MAO-A activity ex vivo, suggesting a possible functional effect. Increase in endogenous MAO I may competitively inhibit the binding of irreversible MAO I drugs, and may also help to mediate some mood altering effects of other drugs, or procedures such as ECT. In human urine both MAO-A I and MAO-B I have been found to be increased in mild stress. Similar findings have been made with human saliva. Selective inhibitors of MAO-A have been identified from human urine, and pig brain, but it is not yet clear to what extent they account for the MAO-A I activity increased in stress.

Isatin is an endogenous selective inhibitor of MAO-B ($K_i \sim 3\,\mu M$). It has a distinct distribution in rat brain, with highest concentration in the hippocampus of $0.1\,\mu g/g$. Its level is increased by pentylene tetrazole, and isatin is itself anxiogenic in rodent models. Its administration also increases monoamine levels in the brain. It is a potent antagonist of the ANP receptor, and it may act to link the control of monoamine function and natriuresis.

Introduction

There is clear evidence for the existence of a range of different molecules occurring in vivo, which can inhibit MAO in vitro. Several such molecules have now been described by different groups (e.g. Egashira et al., 1989). Our research has focused on tribulin. This is now best described as a family of low molecular weight molecules, extractable into ethyl acetate at acid pH, which are inhibitors of MAO in vitro. Inhibitors of benzodiazepine binding to central and peripheral receptors are also extracted under the same conditions. Many independent studies have shown that the level of tribulin in human urine, and in rat brain, is increased by various conditions of stress or anxiety, or by anxiogenic drugs (Glover, 1993). Earlier studies measured the effects on

MAO using the common substrate, tyramine. More recent studies have studied MAO-A inhibitory components (assayed with 5-HT), separately from those that inhibit MAO-B (assayed with PEA).

Tribulin: the effects of stress, anxiety and drugs

Armando and her co-workers have shown that the acute stress of 1.5 hours of cold restraint caused a significant increase in MAO-I in rat heart and kidney, a significant decrease in MAO in heart and kidney (in homogenates) and a significant increase in noradrenaline in the heart and dopamine in the kidney (Armando et al., 1988). No changes in any of these parameters were observed in other tissues. These results provide evidence for parallel functional changes in MAO, MAO-I and catecholamine substrates in specific tissues, and are compatible with a functional effect of endogenous MAO-I in rat tissues. In a subsequent study the same group looked at the stress of footshock, and examined its effects on MAO-A and B separately. Acute footshock reduced MAO-A activity in heart and brain, with a concomitant increase in tribulin activity (still measured non selectively). Clorgyline, the irreversible selective MAO-A inhibitor caused significantly less inhibition in tissues in which tribulin was increased, suggesting that the endogenous MAO-I may be competing with the irreversible drug. (Lemoine et al., 1990). This is a similar finding to that of a previous study which showed that cold restraint stress reduced the in vivo binding of phenelzine to rat brain MAO (Clow et al., 1989b). In a further study the Armando group found that footshock reduced the MAO-A activity in the heart by 23% in a final assay mixture diluted 25 fold from the tissue, suggesting substantially more inhibition in situ. The MAO-B levels were unchanged. This would fit with selective increase of only MAO-A I (Armando et al., 1993).

Table 1. Effects of stress, anxiety, or drugs on MAO I in rat brain

Treatment	Effect	Reference
Footshock	MAO A ↓ MAO B ↔ MAO I ↑	Lemoine et al. (1990)
Pentylene tetrazole and yohimbine	MAO I ↑	Bhattacharya et al. (1991b)
Audiogenic seizures	MAO A I ↑ MAO B I ↑	Medvedev et al. (1992)
Drug withdrawal (alcohol, morphine, nicotine)	MAO A I ↑ MAO B I ↑	Bhattacharya et al. (1995)
Rhazya Stricta (0.5 g/kg)	MAO A I ↑ MAO B I ↔	Ali, Sandler, Glover (in preparation)
Rhazya Stricta (2 g/kg)	MAO A I ↓ MAO B I ↔	Ali, Sandler, Glover (in preparation)

There is increasing evidence that both anxiogenic agents and electrical activity cause an increase in tribulin in selected rat tissues, with MAO-A I being increased more than MAO-B I. Both the anxiogenic agents pentylene tetrazole and yohimbine caused such an increase in brain, and the effects of both were counteracted by diazepam (Bhattacharya et al., 1991b). ECT also caused increased MAO-I in rat brain (Bhattacharya et al., 1991a). In experimental audiogenic seizures in rats, strong seizures caused increase in MAO-I, with a significantly greater increase in MAO-A I than MAO-B I (Medvedev et al., 1992). During withdrawal from various drugs such as morphine, ethanol and nicotine, rats showed increased anxiety in the elevated plus maze. This was paralleled by an increase in the brain MAO-A I, with a smaller increase in MAO-B I. Withdrawal from drugs such as cannabis and ondansetron, which showed no withdrawal anxiety, produced no increase in tribulin (Bhattacharya et al., 1995).

Rhazya stricta extracts are used in folk medicine in the Arab emirates for a range of maladies including diabetes. It is also antidepressant at certain doses in animal models. We have recently found that extracts from this plant, in doses similar to used by the local population, have potent effects on tribulin in rat brain. At 0.5 g/kg it caused a doubling of tribulin A activity; at 2 g/kg it caused its disappearance (unpublished observations). There were no comparable effects on MAO-B I. These results raise the possibility that the antidepressant, or tranquillising activities of certain herbal remedies or other drugs may be via an indirect effect on endogenous MAO-I.

Clow and her group have examined MAO-A I, MAO-B I and isatin output in human urine under mild conditions of everyday stress (Doyle et al., 1996a; Doyle et al., 1996c). Both mean MAO-A I and MAO-B I output were increased. However neither were correlated with isatin level, and this in turn was not correlated with stress (Pang et al., 1996). This group has also recently shown that human saliva contains both MAO-A I and MAO-B I the output of both of which is related to stress (Doyle et al., 1996b).

Thus there is considerable evidence for the existence of selective MAO-A I in rat brain and certain other tissues, increased by stress, that is quite distinct from isatin. In human studies, MAO-A I, MAO-B I and isatin have all been shown to be under separate control.

Structures of components of tribulin

We have now characterised several components of tribulin. Of these the best studied is isatin, which is a selective inhibitor of MAO-B (K_i ~3 μM); it inhibits MAO-A with a K_i of about 60–70 μM (Glover et al., 1988). We have also characterised several selective inhibitors of MAO-A extracted from human urine, and one from brain. The selective MAO-A I isolated from human urine, and identified by gas chromatography-mass spectrometry were all esters of tryptamine and tyramine metabolites. They were ethyl indole-3-acetate, methyl indole-3-propionate, methyl indole-3-acetate and ethyl 4-hydoxyphenylacetate. K_i values for MAO-A were 44, 105, 88, and 120 μM respectively, whilst those for MAO-B were all above 1 mM (Medvedev et al., 1995). The levels of these esters in urine or tissues has not yet been quantified, under either normal or stressful conditions, and it is too early to say whether these account for the MAO-A I activity which is increased in stress. Also, the esters found in human urine were not detectable in extracts from pig brain. However, Medvedev and co-workers have purified a different selective MAO-A inhibitor from pig brain, 4-hydroxyphenylethanol. This was also found in sheep brain. Its K_i for MAO-A was very high, about 1.4 mM. (Medvedev et al., 1995).

Thus although there is good evidence for endogenous inhibitors of MAO, particularly MAO-A, increased by stress or anxiety, their relationship with the selective MAO-A I so far identified remains uncertain. Also, it is not clear whether it is the same molecules that are being increased in animal brains and human urine. Although there is evidence that the MAO-A I in rat tissues is having some functional effects, it is possible that those in urine or saliva are reflecting metabolic activity in other pathways, possibly involving tryptamine or tyramine.

Isatin

It is likely that isatin does have direct physiological effects, though it is not yet clear what role its inhibition of MAO-B has in these. It is currently quantified by gas chromatography-mass spectrometry (Glover et al., 1988; Halket et al., 1991). It has a distinct distribution in rat tissues, and also within the brain (Watkins et al., 1990). In the brain highest levels have been found in the hippocampus and cerebellum (about 0.1 μg/g or 1 μM), and in tissues, highest levels are in the vas deferens and seminal vesicles (1 μg/g); levels in the heart are about 0.1 μg/g. These values are quite high and could well be physiologically active (see below). A recent immunohistochemical stain for isatin has also shown a highly specific localisation within particular cells in the brain, again particularly apparent in the hippocampus and cerebellum (Clow unpublished observations), suggesting that local concentrations could be much higher than in the tissue as a whole.

The pathway of synthesis of isatin is not yet known, but its level in rat urine is clearly largely dependent on the presence of the gut flora (Sandler et al., 1991). Thus germ free rats have much lower urinary levels; however the concentrations in brain and heart of germ free animals is the same as that of conventional ones. This suggests that urinary isatin is largely made by the gut flora, and is unlikely to be a valid reflection of output from brain or heart, which have an independent synthetic pathway.

The anxiogenic agent pentylene tetrazole caused an increase of isatin levels in rat (Bhattacharya, et al., 1991b) and rabbit brain but not liver (Clow et al., 1989a). Isatin is anxiogenic itself in animal models at doses of 10–20 mg/kg; at does of 50 mg/kg and above it becomes sedative (Bhattacharya et al., 1991c; Bhattacharya and Acharya, 1993). It is also antidiuretic (Hota and Acharya, 1994). Several studies have shown that peripheral administration of

Table 2. Properties of isatin

Property	Reference
Selective inhibitor of MAO B (IC50 ~ 3 µM)	Glover et al. (1988)
Potent antagonist of the ANP receptor (IC50 ~0.4 µM)	Glover et al. (1995)
Causes increase in brain monoamine level from 10 mg/kg	Bhattacharya et al. (1993)
Anxiogenic at low dose (10–20 mg/kg)	Bhattacharya et al. (1991)
Sedative at higher doses (50 and above mg/kg)	Bhattacharya et al. (1993)
Distinct distribution in rat tissues and brain Levels in hippocampus and cerebellum ~0.1 µg/g Levels in seminal vesicles and vas deferens >1 µg/g	Watkins et al. (1990)
Distinct distribution in cells. High in Purkinje cells in cerebellum and in layers of hippocampus	A. Clow (personal communication)
Can counteract anxiolytic effects of ANP in vivo	Bhattacharya et al. (1996)
Can counteract diuretic effects of ANP in vivo	S. K. Bhattacharya (personal communication)

isatin (from as low as 10 mg/kg) causes an increase of monoamines such as 5-HT and noradrenaline in the brain (Yuwiler, 1990; McIntyre and Norman, 1990; Bhattacharya and Acharya, 1993; Hamaue et al., 1994). The mechanism for this is unknown. It cannot to be due to selective inhibition of MAO-B, and at the lower doses it is unlikely that the concentrations reached cause a functional inhibition of MAO-A. Isatin passes into the brain from the periphery, but a peripheral dose of 50 or 100 mg/kg resulted in a brain concentration of about 9 mg/kg or 60 µM (Bhattacharya et al., 1991b).

The most potent action of isatin in vitro determined to date, is as an inhibitor of the antinatriuretic peptide (ANP) receptor (K_i ~0.4 µM) (Glover et al., 1995). It acts as an antagonist at this receptor, causing a decrease in ANP induced cyclic GMP. It has also been shown to counteract ANP in vivo, antagonising both its anxiolytic effects (Bhattacharya et al., 1996), and its antidiuretic effects (S. K. Bhattacharya, personal communication).

In cerebral-spinal fluid (CSF) isatin is significantly increased in patients with Bulimia nervosa (Brewerton et al., 1995). This is a condition in which both anxiety and water balance may well be disturbed. It will be of interest to study its CSF levels in other diseases.

Links between stress and natriuresis

There is considerable evidence for the close relationship between the control of stress and the control of fluid balance. Corticotrophin releasing factor (CRF) and AVP are often co-localised. Both cortisol and aldosterone are present in the adrenal cortex; there is an overlap in substrate specificity in the glucocorticoid receptors. In several conditions, including pre-menstrual tension, migraine, and the postpartum period, stress is associated with water retention, and its relief with diuresis. Hormones associated with water balance can themselves have psychological effects. For example ANP is anxiolytic (Bhattacharya et al., 1996), and AVP is anxiogenic in animal models (S. K. Bhattacharya, personal communication). Thus isatin may provide a link between the function of monoamines involved in stress, and the control of the natriuretic system by ANP (Medvedev et al., 1996). Whether its inhibition of MAO, particularly MAO-B, is also physiologically relevant, remains to be determined.

References

Armando I, Levin G, Barotini M (1988) Stress increases endogenous benzodiazepine receptor ligand-monoamine oxidase inhibitory activity. J Neural Transm 71: 29–37

Armando I, Lemoine AP, Segur ET, Barotini MB (1993) The stress-induced reduction in monoamine oxidase (MAO) A activity is reversed by benzodiazepines: role of peripheral benzodiazepine receptors. Cell Mol Neurobiol 13: 593–600

Bhattacharya SK (1995) Anxiogenic activity of centrally administered scorpion (mesobuthus tamulus) venom in rats. Toxicon 33: 1491–1499

Bhattacharya SK, Acharya SB (1993) Further investigations on the anxiogenic effects of isatin. Biogen Amines 9: 453–463

Bhattacharya SK, Banerjee PK, Glover V, Sandler M (1991a) Augmentation of rat brain endogenous monoamine oxidase inhibitory activity (tribulin) by electroconvulsive shock. Neurosci Lett 125: 65–68

Bhattacharya SK, Clow A, Przyborowska A, Halket J, Glover V, Sandler M (1991b) Effect of aromatic amino acids, pentylene tetrazole and yohimbine on isatin and tribulin activity in rat brain. Neurosci Lett 132: 44–46

Bhattacharya SK, Mitra SK, Acharya SB (1991c) Anxiogenic activity of isatin, a putative biological factor, in rodents. J Psychopharmacol 5: 202–206

Bhattacharya SK, Chakrabarti A, Sandler M, Glover V (1995) Rat brain monoamine oxidase A and B inhibitory (tribulin) activity during drug withdrawal anxiety. Neurosci Lett 199: 103–106

Bhattacharya SK, Chakrabarti A, Sandler M, Glover V (1996) Anxiolytic activity of intraventricularly administered atrial natriuretic peptide in the rat. Neuropsychopharmacology 15: 199–206

Breweron TD, Zealberg JZ, Lydiard RB, Glover V, Sandler M, Ballenger JC (1995) CSF isatin is elevated in bulimia nervosa. Biol Psychiat 37: 481–483

Clow A, Davidson J, Glover V, Halket JM, Milton AS, Sandler M, Watkins PJ (1989a) Isatin and tribulin concentrations are increased in rabbit brain but not liver following pentylenetetrazole administration. Neurosci Lett 107: 327–330

Clow A, Glover V, Oxenkrug GF, Sandler M (1989b) Stress reduces in vivo inhibition of monoamine oxidase by phenelzine in rat brain. Neurosci Lett 107: 331–334

Doyle A, Hucklebridge Evans P, Clow A (1996a) Urinary output of endogenous monoamine oxidase inhibitory activity is related to everyday stress. Life Sci 58: 1723–1730

Doyle A, Hucklebridge, Evans P, Clow A (1996b) Salivary monoamine oxidase A and B inhibitory activities correlate with stress. Life Sci 59: 1357–1362

Doyle A, Pang F-Y, Bristow M, Hucklebridge F, Evans P, Clow A (1996c) Urinary cortisol and endogenous monoamine oxidase inhibitor(s) but not isatin are raised in anticipation of stress and/or arousal in normal individuals. Stress Med 12: 43–50

Egashira T, Obata T, Nagai T, Kimba Y, Takano R, Yamanaka Y (1989) Endogenous
    monoamine oxidase inhibitor-like substances in monkey brain. Biochem Pharmacol
    38: 597–602
Glover V (1993) Trials and tribulations with tribulin. Biogen Amines 9: 443–452
Glover V, Halket JM, Watkins PJ, Clow A, Goodwin BL, Sandler M (1988) Isatin:
    identity with the purified monoamine oxidase inhibitor tribulin. J Neurochem 51:
    656–659
Glover V, Medvedev A, Sandler M (1995) Isatin is a potent endogenous antagonist of
    guanylate cyclase-coupled atrial natriuretic peptide receptors. Life Sci 57: 2073–2079
Halket JM, Watkins PJ, Przyborowska A, Goodwin BL, Clow A, Glover V, Sandler M
    (1991) Isatin (indole-2,3-dione) in urine and tissues. Detection by mass spectrometry.
    J Chromatogr Biomed Appl 562: 279–287
Hamaue N, Minami M, Kanamaru Y, Ishikura M, Yamazaki N, Saito H, Parvez SH
    (1994) Identification of isatin, an endogenous MAO inhibitor, in the brain of stroke
    prone SHR. Biogen Amines 10: 99–110
Hota D, Acharya SB (1994) Studies on peripheral actions of isatin. Ind J Exp Biol 32: 710–
    717
Lemoine AP, Armando I, Brun JC, Segura ET, Barotini M (1990) Footshock affects heart
    and brain MAO and MAO inhibitory activity (tribulin) in rat tissues. Pharmacol
    Biochem Behav 36: 85–88
McIntyre IM, Norman TR (1990) Serotonergic effects of isatin: an endogenous MAO
    inhibitor related to tribulin. J Neural Transm [Gen Sect] 79: 35–40
Medvedev AE, Gorkin VZ, Fedotova IB, Semiokhina AF, Glover V, Sandler M (1992)
    Increase of brain endogenous monoamine oxidase inhibitory activity (tribulin) in
    experimental audiogenic seizures in rats: evidence for a monoamine oxidase A in-
    hibitory component. Biochem Pharmacol 44: 1209–1210
Medvedev AE, Goodwin BL, Halket J, Sandler M, Glover V (1995a) Monoamine oxidase
    A-inhibiting components of urinary tribulin: purification and identification. J Neural
    Transm [PD-Sect] 9: 225–237
Medvedev AE, Kamyshanskaya NS, Halket J, Glover V, Sandler M (1995b) Endogenous
    monoamine oxidase inhibitor (tribulin A) from brain: purification and identification.
    Biochemistry (Moscow) 60: 489–495
Medvedev AE, Clow A, Sandler M, Glover V (1996) Isatin: a link between natriuretic
    peptides and monoamines? Biochem Pharmacol 52: 385–391
Pang F-Y, Hucklebridge FH, Forster G, Tan K, Clow A (1996) The relationship between
    isatin and monoamine oxidase-B inhibitory activity in urine. Stress Med 12: 35–42
Sandler M, Przyborowska A, Halket JM, Watkins P, Glover V, Coates ME (1991)
    Urinary but not brain isatin levels are reduced in germ-free rats. J Neurochem 57:
    1074–1075
Watkins P, Clow A, Glover V, Halket J, Przyborowska A, Sandler M (1990) Isatin,
    regional distribution in rat brain and tissues. Neurochem Int 17: 321–323
Yuwiler A (1990) The effect of isatin (tribulin) on metabolism of indoles in rat brain and
    pineal: in vitro and in vivo studies. Neurochem Res 15: 95–100

Author's address: Dr. V. Glover, Department of Paediatrics, Queen Charlotte's and
Chelsea Hospital, Goldhawk Road, London W6 OXG, United Kingdom

Long term administration of (−)-deprenyl increases mortality in male Wistar rats

I. M. Gallagher[1]**, A. Clow**[2]**, and V. Glover**[1]

[1] Department of Paediatrics, Queen Charlotte's & Chelsea Hospital, and
[2] School of Biological & Health Sciences, University of Westminster, London,
United Kingdom

Summary. Long term administration of the monoamine oxidase inhibitor (−)-deprenyl (0.5 mg/Kg) for up to 20 months significantly increased mortality in the male Wistar rat, whereas the dopamine agonist pergolide (0.4 mg/Kg) and the antioxidant diethyldithiocarbamate (400 mg/Kg) had no significant effect on mortality. The increased mortality was not related to dietary intake or body weight of the rats. This is of interest in the light of recent evidence that (−)-deprenyl increases mortality in humans.

Introduction

Several attempts have been made to intervene pharmacologically in the life span of animals. However, there has been no pharmaceutical agent which has been shown to be reproducibly effective. The only means for significantly prolonging the life span of rodents is dietary restriction (Weindruch and Walford, 1988). However, in 1988, Knoll reported a dramatic effect of the monoamine oxidase inhibitor (−)-deprenyl, on the life span of male Logan-Wistar rats. After 24 months of age, when the animals started to receive the drug, the remaining life span increased twofold in the (−)-deprenyl treated group as compared to the control. Following this claim, Milgram et al. (1990) also examined the effect of the drug on longevity in male F-344 rats, also starting treatment at 24 months of age. They too found that (−)-deprenyl treated rats showed a significant increase in life span, but the increase was only 16% in the rat strain they used. More recent studies by Kitani et al. (1992) and Knoll et al. (1994), who started deprenyl treatment earlier, at 18 months and 8 months respectively, have confirmed the increase in life span (34% and 22% respectively) after (−)-deprenyl treatment in rats. The extension of life span by 16%–35%, has also been shown in mice treated from 12 months with (−)-deprenyl (Yen and Knoll, 1992).

However, the effects of (−)-deprenyl on life span in Parkinsonian patients appears controversial. Birkmayer et al. (1985) reported that (−)-deprenyl in combination with levodopa showed a 30% reduction in mortality compared with a group of patients taking levodopa alone. This claim was flawed by a retrospective design that paid no heed to the need for randomization. However, more recently Lees (1995) has reported that mortality among patients receiving (−)-deprenyl in combination with levodopa was about 60% higher than in those receiving levodopa alone. This recent finding does conflict with the previous human and rodent studies.

The present work was part of a larger investigation into the effects of (−)-deprenyl, the dopamine agonist pergolide and the antioxidant diethyldithiocarbamate (DDC) on neuroprotection and ageing in the rat brain. Here, we report on the effects these drugs have on mortality after 20 months of drug treatment.

Materials and methods

Animals and treatment

Male Wistar rats purchased from B&K Universal, UK, were used in this study. The rats were housed in groups of five in a temperature controlled room. The study began when the rats were 3 months old.

Drug administration

(−)-Deprenyl (0.5 mg/Kg, dissolved in physiological saline) was administered by s.c. injection to rats (n = 40) three times a week as this was the route used by Carrillo et al. (1994) and has previously been found to be effective in enhancing activity of the nigrostriatal system. In control rats (n = 40) (−)-deprenyl was replaced with physiological saline solution. DDC (400 mg/Kg) was administered orally to rats (n = 30) in their drinking water because during a 3 week administration of DDC by i.p. injection the rats became irritated by the daily injection and were extremely difficult to handle. Pergolide (0.4 mg/Kg) was also administered to the rats (n = 42) in their drinking water as it needed to be administered daily to be effective over a long period for continued post synaptic stimulation. In control rats (n = 40) pergolide and DDC were replaced with fresh drinking water daily. All animals were weighed weekly and the animals receiving drugs in their drinking water were monitored daily for their water intake. The drug concentrations were adjusted weekly to ensure that the rats received the established doses.

The endpoint of the experiment was morbidity or death. In some instances animals were sacrificed because of persistent stress or discomfort due to the growth of tumours. These animals were assumed to have died naturally and were included in the analysis of the data.

After 20 months of drug administration all surviving animals were culled for histological studies of the brain (Results not reported here).

Food intake study

The rats were placed for one week in cages with a wire mesh on the base of the cage so that unconsumed food could be collected in a tray underneath the cage. Rats were given a known quantity of pelleted food daily and 24 hours later the unconsumed pellets were weighed along with crumbs which had fallen through the wire mesh in the animals cage. Daily food intake was then calculated.

Statistics

Body weights and food intake study. Deprenyl data was statistically analyzed by the two-tailed Student's t-test. Pergolide and DDC data was initially subjected to analysis of variance (ANOVA). When overal ANOVA was significant at the level of 95%, the data was further analysed by the two-tailed Student's t-test. Significant differences were accepted at $p < 0.05$.
Mortality rate. Data was analyzed using 2 sided Fisher's Exact Test. Significant differences were accepted at $p < 0.05$.

Results

After 20 months of drug administration the ($-$)-deprenyl treated rats showed a significant increase in mortality compared to the control group (20% and 2.5% mortality respectively), whereas there was no significant difference in mortality between the pergolide and DDC treated rats and their control group (4.8%, 0% and 7.5% mortality respectively) (Fig. 1).

Body weights were recorded weekly and Fig. 2 (A, B) summarizes monthly mean body weights for all the groups during the 20 month study. The ($-$)-deprenyl treated rats and their control group showed that there was very little difference in mean body weight (Fig. 2A). However, Fig. 2B shows that

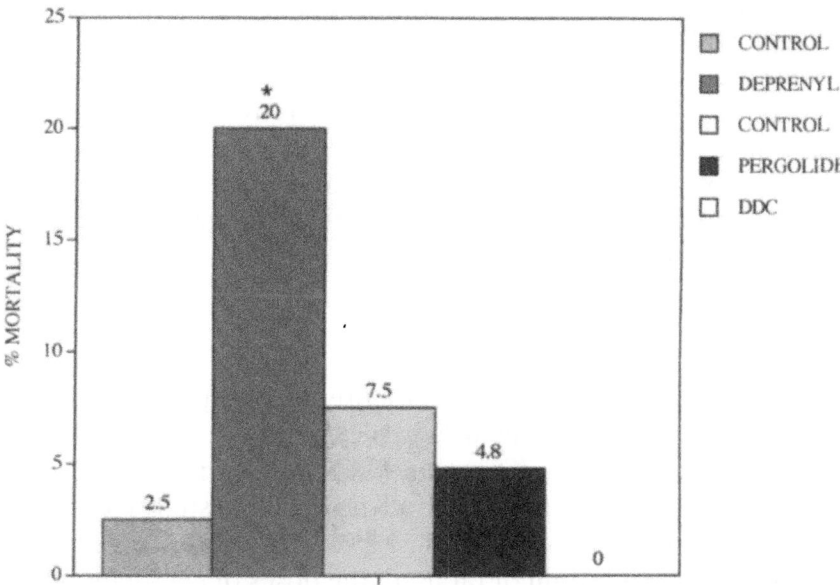

Fig. 1. % Mortality of ($-$)-deprenyl (0.5 mg/Kg, n = 40), pergolide (0.4 mg/Kg, n = 42) and DDC (400 mg/Kg, n = 30) treated rats and controls (n = 40) after 20 month drug administration. *Significantly different from corresponding control ($p = 0.03$)

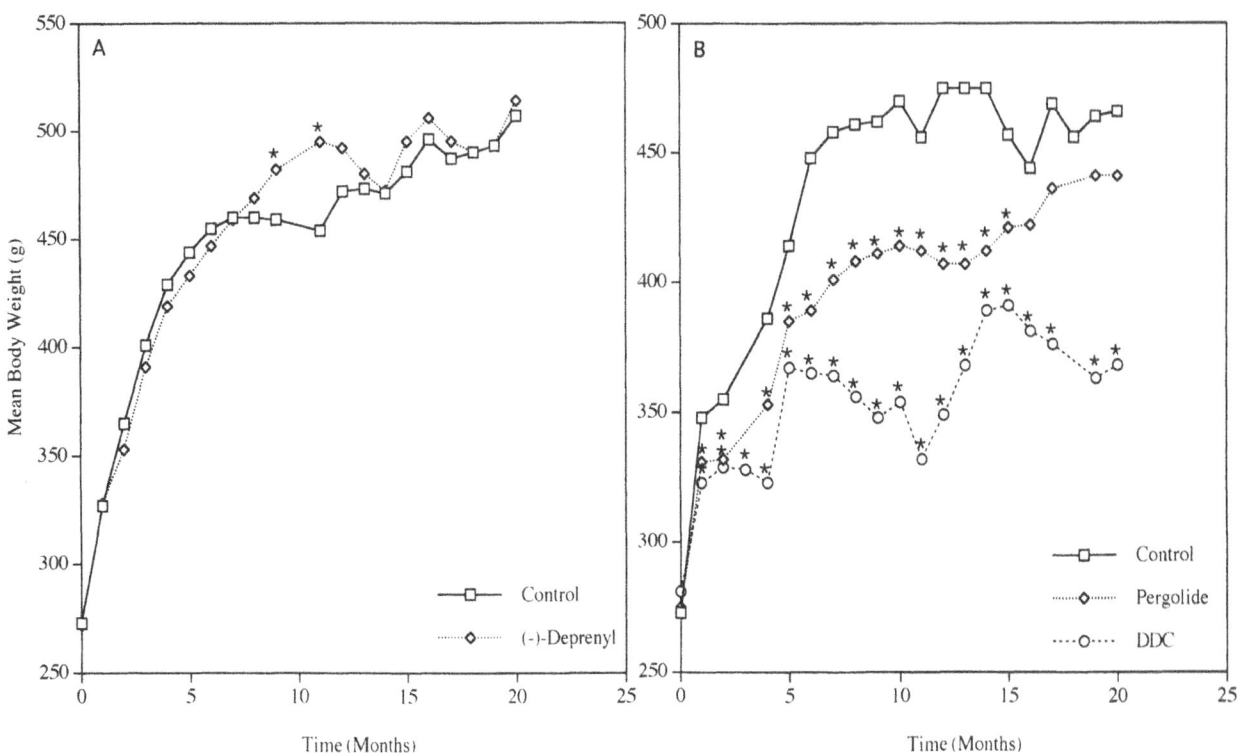

Fig. 2. Changes in mean body weight in (−)-deprenyl (0.5 mg/Kg, n = 40) treated rats and controls (n = 40) (**A**), and in pergolide (0.4 mg/Kg, n = 42) and DDC (400 mg/Kg, n = 30) treated rats and controls (n = 40) (**B**). *Significantly different from corresponding control (p < 0.05)

there were highly significant differences in mean body weight between the pergolide and DDC treated rats and their control group.

In order to examine if a reduced food intake was a possible cause for the difference in mean body weights a food intake study was carried out. There was no significant difference in food intake between the (−)-deprenyl treated rats and the control group, whereas there was a significant difference between the pergolide and DDC treated rats and their control group (Table 1).

Discussion

Previous studies have shown that (−)-deprenyl has a significant effect in prolonging the life span of ageing male rats (Knoll et al., 1988, 1994; Milgram et al., 1990; Kitani et al., 1992). These studies have shown differing degrees of increased life span, ranging from 210% in Knoll's initial study (Knoll, 1988) to 16% in Milgram's study (Milgram et al., 1990). Differences have been attributed to the differences in strain of rat used, the doses of (−)-deprenyl and the ages of the rats at the start of the studies. However, our long term study has shown a significant increase in mortality in (−)-deprenyl treated rats compared to the saline treated controls, whereas no significant difference in

Table 1. Effects of (−)-deprenyl (0.5 mg/Kg), pergolide (0.4 mg/Kg) and DDC (400 mg/Kg) on daily food intake

Group	Daily food intake (g/rat)
Control	25.69 ± 0.72
(−)-Deprenyl	24.42 ± 0.55
Control	24.04 ± 0.56
Pergolide	20.96 ± 0.45*
DDC	17.61 ± 1.49*

Data represent mean ± SD. * Significantly different from corresponding control (p < 0.05)

mortality was observed in the pergolide or DDC treated rats. In this study we used a dose of 0.5 mg/Kg (−)-deprenyl, which is comparible with that used by Kitani et al. (1992), and they observed a significant increase in life span. The major difference between our study and the previous studies is that we used a different strain of rat, the Wistar rat, and we started the administration of (−)-deprenyl when the rats were young, only 3 months old. In the other studies the youngest age at which (−)-deprenyl treatment started was 8 months old (Knoll, 1994). It is possible that the young rat is more sensitive to (−)-deprenyl and its metabolites and prolonged exposure from an early age may have deleterious effects on the long term health of the animal.

It is the general conclusion of experimental gerontology that a reproducible way to make animals live longer is by dietary restriction. Measurements of body weights provided no evidence that the (−)-deprenyl treated rats had an increased food intake or that it was suppressed in the control group. Analysis of daily food intake confirmed that the (−)-deprenyl treated rats were not significantly different from the control group. Both pergolide and DDC treated rats had significantly lower body weights than the control group. Analysis of daily food intake in these groups showed that reduced body weights were due to a reduced food intake. This was probably due to the taste of the drugs, which were delivered to the animals in their drinking water, which affected their appetite. However, the reduced food intake and body weights did not significantly affect mortality in these groups. If the experiment was allowed to have continued longer until natural death occured in all the animals, then we may possibly have seen some differences in mortality in these groups.

Our finding of increased mortality in the (−)-deprenyl treated rats is of interest in the light of recent evidence that (−)-deprenyl increases mortality in humans (Lees, 1995). The precise nature of death in humans remains to be established, but it has been suggested that (−)-deprenyl may have deleterious effects on the cardiovascular and cerebrovascular system. Unfortunately we do not know the exact nature of death of the rats in this study.

This study was not specifically designed to investigated the effects of these drugs on mortality, and it remains possible that our observations are a chance finding. However, given these results together with those of Lees (1995) further research on the effects of (−)-deprenyl on mortality in both rats and human is clearly warranted.

Acknowledgement

This study was supported by the Medical Research Council.

References

Birkmayer W, Knoll J, Rieder P, Youdim MBH, Hars V, Marton J (1995) Increased life expectancy resulting from L-deprenyl addition to Madopar treatment in Parkinson's disease: a long term study. J Neural Transm 64: 113–127

Carrillo M-C, Kitani K, Kanai S, Sato Y, Miyasaka K, Ivy GO (1994) The effect of a long term (6 months) treatment with (−)deprenyl on antioxidant enzyme activities in selective brain regions in old female Fischer 344 rats. Biochem Pharmacol 47(8): 1333–1338

Kitani K, Kanai S, Sato Y, Ohta, Ivy GO, Carrillo M-C (1992) Chronic treatment of (−)-deprenyl prolongs the life span of male Fischer 344 rats. Further evidence. Life Sci 52: 281–288

Knoll J (1988) The striatal dopamine dependency of life span in male rats. Longevity study with (−)deprenyl. Mech Ageing Dev 46: 237–262

Knoll J (1994) Angiohypotensin. In: Szekebres L, Papp JG (eds) Pharmacology of smooth muscle. Springer, Berlin Heidelberg New York Tokyo (Handbook Exp Pharmacol 111)

Lees AJ (1995) Comparison of therapeutic effects and mortality data of levodopa and levodopa combined with selegiline in patients with early, mild Parkinson's disease. BMJ 311: 1602–1607

Milgram NW, Racine RJ, Nellis P, Mendonca A, Ivy GO (1990) Maintenance on L-deprenyl prolongs life in aged rats. Life Sci 47: 415–420

Weindruch R, Walford RL (1988) The retardation of aging and disease by dietary restriction. Charles C. Thomas, Springfield

Yen TT, Knoll J (1992) Extension of life span in mice treated with Dinh lang (Policias fruticosum L) and (−)deprenyl. Acta Physiol Hungarica 79: 119–124

Authors' address: Dr. I. Gallagher, Department of Paediatrics, Queen Charlotte's and Chelsea Hospital, Goldhawk Road, London W6 OXG, United Kingdom

Selegiline as immunostimulant — a novel mechanism of action?

Th. Müller, W. Kuhn, R. Krüger, and **H. Przuntek**

Department of Neurology, St. Josef-Hospital, University of Bochum,
Federal Republic of Germany

Summary. In clinical studies the MAO-B inhibitor selegiline appears to slow the progression of neurological deficits in Parkinson's disease (PD) and the cognitive decline in Alzheimer's disease (AD). The mechanisms of action remain unclear. Several lines of evidence indicate an immune-mediated pathophysiology of PD and AD. According to animal trials, selegiline increases the survival rate of immune suppressed mice. Stimulation of the immune response to bacterial or viral infection or in chronic inflammatory processes is managed by an increased synthesis of the cytokines interleukin-1β (IL-1β) and subsequent interleukin-6 (IL-6). Outcome of viral or bacterial infections in the brain highly correlates with levels of the cytotoxic cytokine tumor-necrosis-factor-alpha (TNF). The aim of our study was to characterize the influence of selegiline on the biosynthesis of IL-1β, IL-6 and TNF in human peripheral blood mononuclear cells (PBMC) from healthy blood donors. After isolation and washing PBMC were cultured without and with selegiline in three different concentrations ($0.01\,\mu$mol/l, $0.001\,\mu$mol/l, $0.0001\,\mu$mol/l) in a humidified atmosphere (7% CO_2). Then cultures were centrifuged and supernatants were collected for IL-1β, IL-6 and TNF ELISA-assays. Treatment of cultured PBMC with various concentrations induced an increased synthesis of IL-1β (ANOVA $F = 9.703$, $p = 0.0007$), IL-6 (ANOVA $F = 20.648$, $p = 0.0001$) and a reduced production of TNF (ANOVA $F = 3.770$, $p = 0.040$). These results indicate, that the influence of selegiline on the cytokine biosynthesis may also contribute to its putative neuroprotective properties.

Introduction

It is still under discussion, whether the monoamine oxidase-B (MAO-B) inhibitor selegiline appears to slow the progression of neurological deficits in Parkinson's disease (PD) and the cognitive decline in Alzheimer's disease (AD) (Stoll et al., 1994; Knoll, 1995; Calne, 1995). Besides the central effect of selegiline to reduce metabolism of dopamine via MAO-B, additional mechanisms of action of selegiline independent of MAO-B inhibition have partially been demonstrated by in vitro and animal trials (for review: Knoll, 1995;

Tatton et al., 1995; Gerlach et al., 1995). It has been suggested, that those results may be due to an altered expression of mRNAs and proteins in nerve and glial cells by selegiline (Tatton et al., 1995). Recently it has been shown, that selegiline influences the endocrine system (Thyagarajan et al., 1995) and also the immune system. Both intraperitoneal and intracerebroventricular (i.c.v.) administration of selegiline markedly attenuated restraint stress-induced gastric ulcers in rats. L-Deprenyl given i.c.v. attenuated stress ulcers in microgram doses and virtually abolished ulcer formation at a dose of 2.0 micrograms (Glavin et al., 1986). Immunosuppressed NMRI-mice were raised and kept under germ-reduced conditions and fed with a germ-reduced diet (14 animals = controls). For a further group of 14 mice 4 mg of selegiline were added to 10 kg of the diet. The 50% survival time of the latter group was 160% that of the control group measured from birth or 220% measured from the beginning of the study. The survival time in weeks of the selegiline group finally reached 350%, and the area under the curve 250% that of the control group (Freisleben et al., 1994).

Former studies revealed, that organoselenium compounds, which show similarities in the chemical structure with selegiline, have been described as anti-inflammatory, antioxidant, glutathione peroxidase-like agents and inhibitors of prostaglandin synthesis. These compounds were also inducers of the cytokines interferon-gamma (IFN-gamma) and tumor-necrosis-factor-alpha (TNF) in human peripheral blood leukocytes (Inglot et al., 1990). Cytokines have been suggested as messengers in the communication between the immune system and the nervous system, in which signals travel only short distances (Bartfai and Schultzberg, 1993; Goujon et al., 1996). Neuroimmune interactions have been discussed in view of findings that nervous signals are important for the immune response (Bartfai and Schultzberg, 1993). The occurrence of neurotransmitter receptors on lymphocytes and cytokine receptors on nerve cells or glia has initiated further studies e.g. on the localization of different cytokines in the nervous system and on long and short term actions of cytokines in the nervous system (Bartfai and Schultzberg, 1993). Interleukin-1β (IL-1β) has been studied extensively along these lines, and found to occur in the nervous and endocrine system. Furthermore, several findings indicate their role as growth promoting factors, and for example the induction of NGF production by IL-1β suggests involvement of this cytokine in regeneration and development in the nervous system (Bartfai and Schultzberg, 1993; Woodroofe, 1995; Strijbos and Rothwell, 1995). In order to characterize putative effects of selegiline on the immune system we investigated the influence of selegiline on the biosynthesis of the cytokines TNF, IL-1β and Interleukin 6 (IL-6).

Material and methods

According to Imamura et al. (1993) peripheral blood mononuclear cells (PBMC) from 3 healthy blood donors (age: 26 years male, 23 male, 28 female) were isolated, suspended (10^6/ml) and cultured in buffered (25 mM Hepes) RPMI 1640 medium supplemented with 10% fetal calf serum and gentamycin 1.6 µg/ml at 37°C without and with selegiline in three different concentrations (10^{-8} mol/l, 10^{-9} mol/l, 10^{-10} mol/l) in a humidified atmosphere (7% CO_2). After 48 h of culturing, supernatants were collected, centrifuged and frozen in −80°C until testing. Each experiment was replicated three times.

IL-1β, IL-6 and TNF enzyme-linked immunosorbent assay (ELISA)

The levels of cytokines in the supernatants were measured using commercial available ELISA kits (Quantikine, R&D Systems, DPC Biermann GmbH, Bad Nauheim, Germany) for IL-1β, IL-6 and TNF. Collection and handling of specimens were performed and a standard curve was generated according to the manufacturers recommendations. All measurements were done in duplicate.

Statistical evaluation

ANOVA and subsequent Tukey-Kramer Multiple Comparisons Test were performed for comparison between standard (without selegiline) and selegiline treated cultures. Linear regression was used for evaluation of dose dependency of selegiline. Results are expressed as means ± SEM. Level of significance was $p < 0.05$.

Results

Treatment of cultured PBMC with selegiline induced a significant increased synthesis of IL-1β (ANOVA F = 9.609, p = 0.0016) in all three different concentrations of selegiline (Fig. 1). Regarding IL-6 a significant augmented biosynthesis (ANOVA F = 20.648, p = 0.0001) (Fig. 2) was also found, especially when treating the cultures with concentrations of 0.001 µmol/l and 0.01 µmol/l of selegiline.

In the case of TNF a significant reduced production (ANOVA F = 3.770, p = 0.040) (Fig. 3) appeared, the post test revealed, that level of significance was only reached in a concentration of 0.001 µmol/l.

No significant correlation between dose of selegiline and increased production of IL-1β (r^2 = 0.214, p = 0.537) or IL-6 (r^2 = 0.863, p = 0.071) or decreased synthesis of TNF (r^2 = 0.000289, p = 0.983) in these cultures was found.

Discussion

The results of this study indicate, that immunological properties of selegiline may cause its influence on the synthesis of cytokines IL-1β, IL-6 and TNF. These effects of selegiline were most prominent in the case of IL-1β. The elevated levels of IL-6 may be explained on the one hand by direct stimulation by selegiline. On the other hand an indirect effect caused

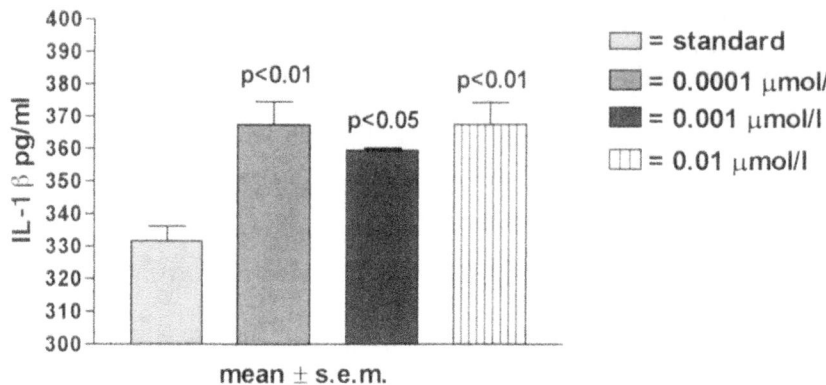

Fig. 1. IL-1β-levels in selegiline treated cultures

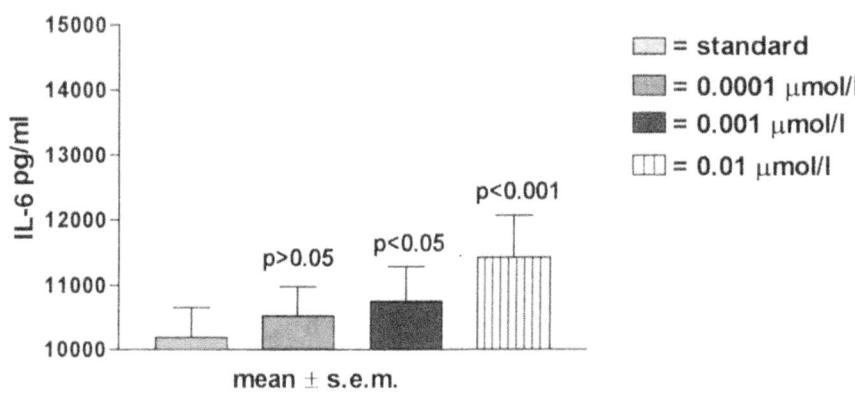

Fig. 2. IL-6 levels in selegiline treated cultures

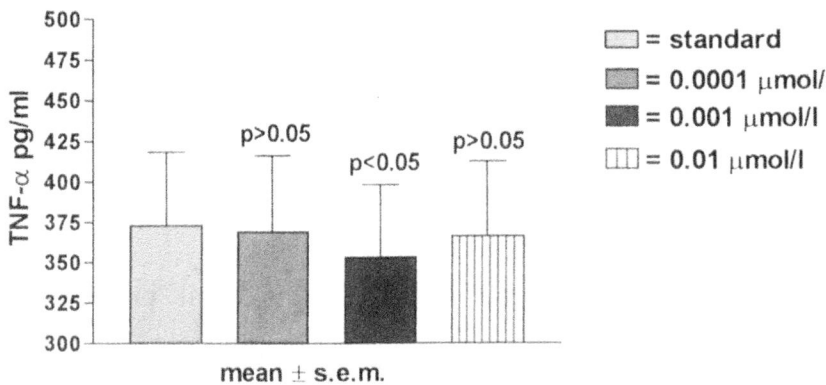

Fig. 3. TNF-levels in selegiline treated cultures

by IL-1β has to be considered, because IL-1β induces an increased release of IL-6 (Woodroofe, 1995) and subsequently gonadotropins FSH, LH (Yamaguchi et al., 1990). Our results indirectly confirm results of Thyagarajan et al. (1995), but we assume, that influence of selegiline on hormone secretion may be mediated both by its impact on cytokine synthesis and via suppression of monoamine metabolism.

There is growing evidence for a bidirectional interaction between the peripheral immune system and the brain (Gottschall et al., 1991; Goujon et al., 1996). Therefore it seems likely, that peripheral circulating cytokines may influence regenerative and degenerative processes in the brain (Wollman et al., 1992), e.g. via the organum vasculosum lamina terminalis, where the blood-brain-barrier is weak (Imura et al., 1991). In the context of AD (for review: McGeer and McGeer, 1995; Blum-Degen et al., 1995) and PD (for review: Kuhn and Müller, 1995) immune-related phenomena are also involved in the etiology of both diseases. Alterations in the content of several cytokines, e.g. IL-1β, IL-6 or TNF, and growth factors in the brain parenchyma, serum and/or cerebrospinal fluid or AD and PD patients were found (Griffin et al., 1989; Cacabelos et al., 1991; Mogi et al., 1994a,b, 1995; Blum-Degen et al., 1995). IL-1β and IL-6 are released from microglial cells and astrocytes (Lee et al., 1993). IL-1β stimulates the proliferation of astrocytes, and regulates the synthesis of nerve growth factor (Woodroofe, 1995). In the brain IL-6 induces acute phase protein synthesis, differentiation of neuronal cells and improves catecholaminergic and cholinergic cell survival (Hama et al., 1991; Bauer et al., 1991; Bartfai and Schultzberg, 1993; Blum-Degen et al., 1995). A recent study showed elevated IL-1β and IL-6 levels in the cerebrospinal fluid of de-novo-Parkinson's disease patients (Blum-Degen et al., 1995), which may reflect the original condition at the probable beginning of disease. Moreover due to this study and the influence of IL-1β and IL-6 on regenerative processes in the brain (Woodroofe, 1995) one may also speculate, that this cytokine status represents the regenerative potential of the brain in the fight against the neurodegenerative process. In contrast, increased levels of TNF in the striatum, substantia nigra and CSF of parkinsonian patients (Brosnan et al., 1988; Boka et al., 1994; Mogi et al., 1994b; Blum-Degen et al., 1995) may represent the cytotoxic and neurodegenerative disease inducing action of TNF, because in vivo intracerebral TNF levels highly correlate with the outcome in the case of viral and bacterial infections in the brain (Hofman et al., 1989; Poli et al., 1990; Gatti and Bartfai, 1993), and in vitro TNF may destroy myelin and oligodendrocytes (Selmaj and Raine, 1988a,b; Selmaj et al., 1988a,b, 1995).

On the background of these studies on cytokines and their possible impact on degenerative and regenerative mechanisms in the course of chronic neuronal death the results of our study fit to the postulated neuroprotective effects of selegiline, indicating that these mechanisms are possibly mediated by cytokines. But further in vitro and in vivo trials are necessary to provide evidence for the demonstrated influence of selegiline on cytokine biosynthesis.

References

Bartfai T, Schultzberg M (1993) Cytokines in neuronal cell types. Neurochem Int 22: 435–444

Bauer J, Strauss S, Schreiter Gasser U, Ganter U, Schlegel P, Witt I, Yolk B, Berger M (1991) Interleukin-6 and alpha-2-macroglobulin indicate an acute-phase state in Alzheimer's disease cortices. FEBS Lett 285: 111–114

Blum-Degen D, Müller Th, Kuhn W, Gerlach M, Przuntek H, Riederer P (1995) Interleukin-1-beta and interleukin 6 are elevated in the cerebrospinal fluid of Alzheimer's and de novo Parkinson's disease patients. Neurosci Lett 202: 17–20

Boka G, Anglade P, Wallach D, Javoy Agid F, Agid Y, Hirsch EC (1994) Immunocytochemical analysis of tumor necrosis factor and its receptors in Parkinson's disease. Neurosci Lett 172: 151–154

Brosnan CF, Selmaj K, Raine CS (1988) Hypothesis: a role for tumor necrosis factor in immune-mediated demyelination and its relevance to multiple sclerosis. J Neuroimmunol 18: 87–94

Cacabelos R, Barquero M, Garcia P, Alvarez XA, Varela de Seijas E (1991) Cerebrospinal fluid interleukin-1β (IL-1β) in Alzheimer's disease and neurological disorders. Meth Find Exp Clin Pharmacol 13: 455–458

Calne DB (1995) Selegiline in Parkinson's disease. BMJ 311: 1583–1584

Freisleben HJ, Lehr F, Fuchs J (1994) Lifespan of immunosuppressed NMRI-mice is increased by deprenyl. J Neural Transm [Suppl] 41: 231–236

Gatti S, Bartfai T (1993) Induction of tumor necrosis factor-alpha mRNA in the brain after peripheral endotoxin treatment: comparison with interleukin-1 family and interleukin-6. Brain Res 624: 291–294

Gabellec MM, Griffais R, Fillion G, Haour F (1995) Expression of interleukin-1α, interleukin-1β and interleukin 1 receptor antagonist mRNA in mouse brain: regulation by bacterial lipopolysaccharide (LPS) treatment. Mol Brain Res 31: 122–130

Gerlach M, Riederer P, Youdim MB (1995) Neuroprotective therapeutic strategies. Comparison of experimental and clinical results. Biochem Pharmacol 50: 1–16

Glavin GB, Dugani AM, Pinsky C (1986) L-deprenyl attenuates stress ulcer formation in rats. Neurosci Lett 70: 379–381

Gottschall PE, Komaki G, Arimura A (1991) Interleukin-1β activation of the central nervous system. In: Rothwell N, Dantzer R (eds) Interleukin in the brain. Pergamon Press, Oxford, pp 27–49

Goujon E, Parnet P, Laye S, Combe C, Dantzer R (1996) Adrenalectomy enhances proinflammatory cytokines gene expression, in the spleen, pituitary and brain of mice in response to lipopolysaccharide. Mol Brain Res 36: 53–62

Griffin WS, Stanley LC, Ling C, White L, MacLeod V, Perrot LJ, White CL, Araoz C (1989) Brain interleukin 1 and S-100 immunoreactivity are elevated in Down syndrome and Alzheimer disease. Proc Natl Acad Sci USA 86: 7611–7615

Hama T, Kushima Y, Miyamoto M, Kubota M, Takei N, Hatanaka H (1991) Interleukin-6 improves the survival of mesencephalic catecholaminergic and septal cholinergic neurons from postnatal, two-week-old rats in cultures. Neuroscience 40: 445–452

Hofman FM, Hinton DR, Johnson K, Merrill JE (1989) Tumor necrosis factor identified in multiple sclerosis brain. J Exp Med 170: 607–612

Imamura K, Suzumura A, Hayashi F, Marunouchi T (1993) Cytokine production by peripheral blood monocytes/macrophages in multiple sclerosis patients. Acta Neurol Scand 87: 281–285

Imura H, Fukata J, Mori T (1991) Cytokines and endocrine function: an interaction between the immune and neuroendocrine systems. Clin Endocrinol 35: 107–115

Inglot AD, Zielinska Jenczylik J, Piasecki E, Syper L, Mlochowski J (1990) Organoselenides as potential immunostimulants and inducers of interferon gamma and other cytokines in human peripheral blood leukocytes. Experientia 46: 308–311

Knoll J (1995) Rationale for (–)deprenyl (selegiline) medication in Parkinson's disease and in prevention of age-related nigral changes. Biomed Pharmacother 49: 187–195

Kuhn W, Müller Th (1995) Neuroimmune mechanisms in Parkinson's disease. J Neural Transm [Suppl] 46: 217–228

Lee SC, Liu W, Dickson DW, Brosnan CF, Berman JW (1993) Cytokine production by human fetal microglia and astrocytes. Differential induction by lipopolysaccharide and IL-1 beta. J Immunol 150: 2659–2667

McGeer PL, McGeer EG (1995) The inflammatory response system of brain: implications for therapy of Alzheimer and other neurodegenerative diseases. Brain Res Rev 21: 195–218

Mogi M, Harada M, Kondo T, Riederer P, Inagaki H, Minami M, Nagatsu T (1994a) Interleukin-1 beta, interleukin-6, epidermal growth factor and transforming growth factor-alpha are elevated in the brain from parkinsonian patients. Neurosci Lett 180: 147–150

Mogi M, Harada M, Riederer P, Narabayashi H, Fujita K, Nagatsu T (1994b) Tumor necrosis factor-alpha (TNF-alpha) increases both in the brain and in the cerebrospinal fluid from parkinsonian patients. Neurosci Lett 165: 208–210

Mogi M, Harada M, Kondo T, Riederer P, Nagatsu T (1995) Brain beta 2-microglobulin levels are elevated in the striatum in Parkinson's disease. J Neural Transm [Park Dis Dement Sect] 9: 87–92

Poli G, Kinter A, Justement JS, Kehrl JH, Bressler P, Stanley S, Fauci AS (1990) Tumor necrosis factor alpha functions in an autocrine manner in the induction of human immunodeficiency virus expression. Proc Natl Acad Sci USA 87: 782–785

Selmaj K, Raine CS (1988a) Tumor necrosis factor mediates myelin damage in organotypic cultures of nervous tissue. Ann NY Acad Sci 540: 568–570

Selmaj K, Raine CS (1988b) Tumor necrosis factor mediates myelin and oligodendrocyte damage in vitro. Ann Neurol 23: 339–346

Selmaj K, Bradbury K, Chapman J (1988a) Multiple sclerosis: effects of activated T-lymphocyte-derived products on organ cultures of nervous tissue. J Neuroimmunol 18: 255–268

Selmaj K, Nowak Z, Tchorzewski H (1988b) Interleukin-1 and interleukin-2 production by peripheral blood mononuclear cells in multiple sclerosis patients. J Neurol Sci 85: 67–76

Selmaj K, Papierz W, Glabinski A, Kohno T (1995) Prevention of chronic relapsing experimental autoimmune encephalomyelitis by soluble tumor necrosis factor receptor I. J Neuroimmunol 56: 135–141

Stoll S, Hafner U, Pohl O, Müller WE (1994) Age-related memory decline and longevity under treatment with selegiline. Life Sci 55: 2155–2163

Strijbos PJ, Rothwell NJ (1995) Interleukin-1 beta attenuates excitatory amino acid-induced neurodegeneration in vitro: involvement of nerve growth factor. J Neurosci 15: 3468–3474

Tatton WG, Ansari K, Ju W, Salo PT, Yu PH (1995) Selegiline induces "trophic-like" rescue of dying neurons without MAO inhibition. Adv Exp Med Biol 363: 15–16

Thyagarajan S, Meites J, Quadri SK (1995) Deprenyl reinitiates estrous cycles, reduces serum prolactin, and decreases the incidence of mammary and pituitary tumors in old acyclic rats. Endocrinology 136: 1103–1110

Wollmann EE, Kopmels B, Bakalian A, Delhaye-Bouchard N, Fradelizi D, Mariani J (1992) Cytokines and neuronal degeneration. In: Rothwell N, Dantzer R (eds) Interleukin in the brain. Pergamon Press, Oxford, pp 187–203

Woodroofe MN (1995) Cytokine production in the central nervous system. Neurology 45: S6–S10

Yamaguchi M, Matsuzaki N, Hirota K, Miyake A, Tanizawa O (1990) Interleukin 6 possibly induced by interleukin 1 beta in the pituitary gland stimulates the release of gonadotropins and prolactin. Acta Endocrinol Copenh 122: 201–205

Authors' address: Th. Müller, M.D., Department of Neurology, St. Josef-Hospital, University of Bochum, Gudrunstrasse 56, D-44791 Bochum, Federal Republic of Germany

Clorgyline effect on pineal melatonin biosynthesis in rats with lesioned suprachiasmatic nuclei

G. F. Oxenkrug[1] and P. J. Requintina[2]

[1] Department of Psychiatry and Pineal Research Laboratory, St. Elizabeth's Medical Center of Boston/Tufts University, Boston, MA, and [2] Providence VA Medical Center, Providence, R.I., USA

Summary. We have reported that clorgyline-induced stimulation of pineal melatonin biosynthesis could be augmented by the exposure to 24h of constant light in young but not in aged rats. Aging is associated with the declined integrity of the suprachiasmatic nuclei (SCN), the major station of the light signal passage from the retina to the pineal gland. The present study aimed to investigate whether SCN integrity is essential for clorgyline effects on pineal melatonin biosynthesis in light-primed rats. Clorgyline (2.5 mg/kg, s.c.) was administered to sham-operated and SCN-lesioned Sprague-Dawley rats kept under regular light/dark cycle or exposed to 24h of constant light. Pineal melatonin and related indoles were evaluated by HPLC-fluorimetric procedure.

Clorgyline stimulated pineal melatonin biosynthesis in both SCN-lesioned and sham-operated rats kept under regular light/dark cycle. Exposure to constant light (for 24h) augmented clorgyline-induced stimulation of melatonin biosynthesis in sham-operated rats, but not in SCN-lesioned animals. The obtained results suggest that decline in SCN activity (e.g., age-associated) might contribute to previously reported attenuation of clorgyline-induced stimulation of melatonin biosynthesis in light-primed aged rats.

Introduction

Clorgyline, like other monoamine oxidase (MAO-)A inhibitors, stimulates pineal melatonin biosynthesis (see for rev., Oxenkrug, 1991). This effect could be augmented in young rats by their exposure to 24h of light, most probably, due to the light-induced up regulation of the pineal β_1-adrenoceptors involved in the serotonin-N-acetylation, the rate limiting step of melatonin biosynthesis (Oxenkrug and McIntyre, 1991). It was reported that β_1-adrenoceptors of aged rats have decreased ability to respond by upregulation to light exposure (Greenberg and Weiss, 1989). In line with this suggestion we have shown that priming with light did not potentiate the clorgyline-induced stimulation of the melatonin biosynthesis in aged rats (Oxenkrug and McIntyre, 1991).

The suprachiasmatic nuclei (SCN) of the hypothalamus is the major station of the light signal passage from the retina to the pineal gland and the site of a circadian pacemaker in mammals (Rietveld, 1992). It is responsible for the generation and synchronization of the various mammalian endocrine, physiological and behavioral rhythms, including the pineal melatonin production. Ablation of the SCN results in the disruption of these rhythms.

In order to further examine the mechanism(s) of clorgyline-induced stimulation of the pineal melatonin biosynthesis we compared clorgyline effect in light-primed SCN-lesioned and sham-operated rats.

Methods

SCN-lesioned and sham-operated male Sprague-Dawley rats (3 months old, 250 g body weight) were obtained from Zivic Miller (Zelienople, PA). They were kept in a 12 h light: 12 h dark (L/D) schedule with free access to food and water for at least two weeks before the experiments.

Half of the animals were exposed to 24 h of light (L/L) after which they were treated with clorgyline (2.5 mg/kg s.c.) or saline. Another half were maintained in the L/D schedule. Clorgyline (2.5 mg/kg s.c.) or saline was administered 3 h after light onset. Rats were decapitated 90 min after injection; pineals were removed, immediately frozen in dry ice and stored at -70 C until assayed. Pineal melatonin, serotonin, N-acetyl serotonin (NAS) and 5-hydroxyindoleacetic acid (5-HIAA) were determined by HPLC-fluorimetric procedure. Each experimental group consisted of five rats. Data was statistically treated according to ANOVA and Student t-test. Results were expressed as mean ± S.D. (ng of the indole/pineal).

Results

Saline treated rats

Pineal NAS and melatonin levels were similar in SCN-lesioned and sham-operated rats kept under the L/D cycle. Light priming increased NAS and melatonin levels in SCN-lesioned rats but not in sham-operated rats (Table 1).

Clorgyline treated rats

Clorgyline administration increased pineal NAS and melatonin content in both SCN-lesioned and sham-operated rats kept under L/D cycle. Clorgyline further increased NAS and melatonin content in light-primed sham-operated animals as we have reported earlier (Oxenkrug and McIntyre, 1991). However, light priming did not augment clorgyline-induced stimulation of melatonin biosynthesis in SCN-lesioned rats (Table 1).

Discussion

The present study provides two interesting observations of the light priming effect on pineal melatonin biosynthesis in SCN-lesioned rats: increase of

Table 1. Clorgyline and pineal melatonin and related indoles of SCN-lesioned rats kept on light: dark and light:light cycle

	Melatonin	NAS	5-HT	5-HIAA
L:D cycle				
Sham-operated saline	$0.15 \pm 0.03^{\#}$	$0.17 \pm 0.04^{\#}$	$12.80 \pm 7.04^{*}$	$17.98 \pm 7.26^{\#}$
Sham-operated clorgyline	0.90 ± 0.42	0.60 ± 0.36	149.85 ± 41.96	8.10 ± 0.88
SCN lesioned saline	$0.18 \pm 0.08^{\#}$	$0.16 \pm 0.04^{\#}$	$14.12 \pm 8.77^{*}$	$16.64 \pm 3.63^{*}$
SCN lesioned clorgyline	0.72 ± 0.45	0.30 ± 0.12	54.58 ± 12.58	4.93 ± 1.22
L:L cycle				
Sham-operated saline	$0.16 \pm 0.03^{*}$	0.15 ± 0.05	$21.09 \pm 15.15^{*}$	$24.15 \pm 9.47^{\#}$
Sham-operated clorgyline	$2.56 \pm 0.57^{\dagger}$	$2.72 \pm 2.16^{\dagger}$	147.88 ± 46.19	8.98 ± 3.15
SCN lesioned saline	$1.58 \pm 0.12^{\triangle}$	$1.12 \pm 0.11^{\triangle}$	$15.13 \pm 4.72^{*}$	$24.13 \pm 4.64^{*\triangle}$
SCN lesioned clorgyline	1.28 ± 0.50	1.20 ± 0.68	110.52 ± 27.86	10.65 ± 2.33

$^{*}p < 0.005$; $^{\#}p < 0.05$ vs. clorgyline. $^{\dagger}p < 0.001$; $^{\triangle}p < 0.05$ vs. corresponding L:D group. Mean values (ng of the indole/pineal) shown $\pm$ SD for 5 rats per group

pineal NAS and melatonin levels in saline-treated rats, and lack of augmentation of clorgyline-induced elevation of NAS and melatonin levels. Both findings resembled the previously observed effects of light priming on aged rats, i.e., elevation of pineal melatonin levels in saline treated rats (McIntyre and Oxenkrug, 1991) and lack of augmentation of clorgyline-induced elevation of pineal NAS and melatonin levels (Oxenkrug and McIntyre, 1991). In effect, SCN lesioning of the young rats made them respond to clorgyline and light-priming similar to the aged non-operated rats. This is in line with the other reports indicating that partial lesions of the SCN produce changes in activity rhythms similar to that of aged animals (Pickard and Turek, 1982) and that specific cell populations undergo selective deterioration within the SCN of the aging rats suggesting the physical deterioration of the SCN during aging (Satinoff et al., 1993).

The lack of augmentation of clorgyline-induced stimulation of melatonin biosynthesis in light-primed aged rats has been explained by the decreased ability of pineal β_1-adrenoceptors of aged rats to be upregulated by light (Greenberg and Weiss, 1989). The results of the present study suggest that SCN integrity is necessary for clorgyline effect in light-primed rats and that failure of pineal β_1-adrenoceptors of aged rats to respond by upregulation to light exposure might depend upon the age-associated decline of the SCN activity. This hypothesis raises the possibility that the age-associated decline of pineal melatonin biosynthesis might depend upon changes at the extrapineal sites (e.g., SCN), and not that at the pineal gland itself (e.g., down regulation of β_1-receptors).

References

McIntyre IM, Oxenkrug GF (1991) Effect of aging on melatonin biosynthesis induced by 5-hydroxytrytophan and constant light in rats. Prog Neuropsychopharmacol Biol Psychiatry 15: 561–566

Oxenkrug GF (1991) The acute effect of monoamine oxidase inhibitors on serotonin conversion to melatonin. In: Cooper A, Sandler M, Harnett S (eds): 5-Hydroxytryptamine and mental illness. Oxford University Press, Oxford New York Tokyo, pp 99–108

Oxenkrug GF, McIntyre IM, Requintina PJ, Duffy JD (1991) The response of the pineal melatonin biosynthesis to the selective MAO-A inhibitor, clorgyline, in young and middle-aged rats. Prog Neuropsychopharmacol Biol Psychiatry 15: 895–902

Pickard GE, Turek FW (1982) Splitting of the circadian rhythm of activity is abolished by unilateral lesions of the suprachiasmatic nuclei. Science 215: 1119–1121

Rietveld WJ (1992) The suprachiasmatic nucleus and other pacemakers. In: Touitou Y, Haus E (eds) Biologic rhythms in clinical and laboratory medicine. Springer, New York, pp 55–64

Satinoff E, Li H, Tcheng TC, Liu C, McArthur AJ, Medanic M, Gillette MU (1993) Do the suprachiasmatic nuclei oscillate in old rats as they do in young ones? Am J Physiol 265: R1216–R1222

Authors' address: Prof. G. F. Oxenkrug, M.D. Ph.D. Department of Psychiatry, St. Elizabeth's Medical Center, QN-3P, 736 Cambridge st., Boston, MA, 02135, USA

The effect of MAO-A inhibition and cold-immobilization stress on N-acetylserotonin and melatonin in SHR and WKY rats

G. F. Oxenkrug[1] and P. J. Requintina[2]

[1] Department of Psychiatry and Pineal Research Laboratory, St. Elizabeth's Medical Center of Boston/Tufts University, Boston, MA., and [2] Providence VA Medical Center, Providence, R.I., USA

Summary. Selective monoamine oxidase (MAO) A inhibitors and cold-immobilization stress (which increases the production of the endogenous MAO-A inhibitor, tribulin) stimulate rat pineal melatonin biosynthesis in Sprague-Dawley and Fisher 344N rats. Considering the hyperactive sympathetic response of the hypertensive rats, it was interesting to compare the effect of clorgyline and cold-immobilization stress on pineal melatonin and related indoles levels in SHR and WKY rats (HPLC-fluorimetric method).

Clorgyline (0.5 mg/kg and 1.5 mg/kg, sc) induced a higher elevation of pineal melatonin and N-acetylserotonin (NAS) in SHR than in WKY rats. Cold immobilization stress resulted in lower serotonin, and higher NAS levels in SHR than in WKY rats with similar elevations in melatonin levels.

Our results suggest increased serotonin conversion into NAS and decreased NAS conversion into melatonin with decreased production of tribulin in SHR in comparison with WKY rats.

Introduction

Selective MAO-A inhibitors stimulate pineal melatonin biosynthesis in Sprague-Dawley and Fischer rats in vivo and in vitro, and increase human blood melatonin levels (Oxenkrug et al., 1984; for rev., see Oxenkrug, 1991). Increased production of the endogenous MAO inhibitor, tribulin, induced by cold-immobilization stress (Glover et al., 1991), might contribute to the stimulation of the pineal melatonin biosynthesis in stressed Sprague-Dawley rats (Oxenkrug and McIntyre, 1985). There are significant differences in reaction of spontaneously hypertensive (SHR) and normotensive, Wistar-Kyoto (WKY) rats to cold-immobilization stress (Pare and Schimmel, 1986). In this study, we compared the effect of clorgyline and cold-immobilization stress on the pineal melatonin biosynthesis of SHR and WKY rats.

Table 1. The effect of 2 doses of clorgyline on the pineal melatonin and related indoles of SHR and WKY rats

Treatment	NAS	Melatonin	Serotonin	5-HIAA
SHR rats				
Saline	ND	0.08 ± 0.02	57.51 ± 7.53	8.01 ± 1.73
0.5 mg/kg	0.47 ± 0.42	1.16 ± 0.34#	76.80 ± 17.68	1.07 ± 0.32
1.5 mg/kg	1.79 ± 0.30*	1.92 ± 0.36*	72.96 ± 4.86◇	0.80 ± 0.26
WKY rats				
Saline	ND	0.07 ± 0.00	69.57 ± 7.18	12.40 ± 4.33
0.5 mg/kg	0.18 ± 0.00	0.31 ± 0.13	77.57 ± 15.44	1.33 ± 0.42
1.5 mg/kg	0.47 ± 0.14	0.59 ± 0.17	89.86 ± 11.55	1.17 ± 0.55

*$p < 0.001$; #$p < 0.005$ vs. WKY; ◇$p < 0.02$ vs. corresponding WKY group. *ND* not detectable. Mean values (ng indole/pineal) shown ± S.D. of 5 rats per group

Table 2. The effects of stress on the pineal melatonin and related indoles of SHR and WKY rats

Treatment	NAS	Melatonin	Serotonin	5-HIAA
SHR rats				
non-stressed	ND	0.37 ± 0.11	71.24 ± 13.58	4.54 ± 0.83
stressed	4.94 ± 1.37*	0.81 ± 0.19	96.24 ± 26.97•	7.11 ± 1.93
WKY rats				
non-stressed	ND	0.31 ± 0.15	84.65 ± 11.29	5.02 ± 0.99
stressed	1.13 ± 0.68	0.75 ± 0.15	124.26 ± 14.13	6.98 ± 2.13

*$p < 0.001$, •$p < 0.01$ vs. WKY. *ND* not detectable. Mean values (ng indole/pineal) shown ± S.D. of 5 rats per group

Methods

Male SHR and WKY (200 g. b. w.) were kept at 12 hours light: 12 hours dark schedule with free access to food and water for at least two weeks before the experiments. The experiments were performed three hours after the light onset. Clorgyline (0.5 and 1.5 mg/kg, s.c.) was injected 90 minutes before decapitation. Cold-immobilization stress was performed by immobilizing the animals and exposing them to the cold (4°C) for two hours before decapitation. Pineals were immediately removed and frozen in dry ice and stored at −70°C until analyses. Melatonin, serotonin, N-acetyl serotonin (NAS) and 5-hydroxyindoleacetic acid (5-HIAA) were determined by HPLC-fluorometric procedure. Each experimental group consisted of five rats. Data was statistically treated according to ANOVA and Student t-test. Results were expressed as mean ± S.D. (ng of the indole/pineal).

Results

There was no difference in the daytime pineal melatonin and related indoles in SHR and WKY rats. Clorgyline, at both doses, induced higher increase of pineal melatonin and N-acetyl-serotonin (NAS) in both SHR and WKY rats and equal decrease of 5-HIAA levels (Table 1). Pineal NAS levels increased in SHR and WKY rats after cold-immobilization stress, with greater effect in the SHR rats. However, cold-immobilization stress induced an equal increase of pineal melatonin levels in SHR and WKY rats. The pineal serotonin content was higher in the WKY than in the SHR after cold-immobilization stress (Table 2). The ratio of 5-HIAA/serotonin, an index of MAO activity, was higher in stressed SHR (0.073) than in WKY rats (0.056).

Discussion

The NAS formation from serotonin (N-acetylation of serotonin) is the first step of the pineal melatonin biosynthesis which is catalyzed by N-acetyltransferase (NAT), and is adrenergically regulated. The observed higher NAS increase in SHR than in WKY after clorgyline administration and cold-immobilization stress is in line with the previously reported enhanced sympathetic activity and higher elevation of catecholamines in SHR than in WKY after exposure to various types of stresses (Kvetnansky et al., 1979; Nomura and Okamura, 1989). In the same vein, higher NAT activity in SHR than in WKY rats after challenge by isoproterenol suggested the supersensitivity of the pineal adrenoceptors (Illnerova and Albrecht, 1975). This might explain the higher rate of NAS formation in SHR than in WKY rats observed after clorgyline administration and exposure to cold-immobilization stress. However, the higher NAS levels in SHR were transformed into higher melatonin levels only in clorgyline treated but not in stressed SHR. The NAS conversion to melatonin represents the second step of melatonin biosynthesis and is catalyzed by hydroxy-indole-O-methyl transferase (HIOMT). We have previously shown that, even at daytime, NAS (administered in rather high doses) was converted into melatonin (Oxenkrug and Requintina, 1994). However, despite the almost three-fold higher level of NAS in stressed SHR than in stressed WKY rats, the melatonin levels were similar in both strains. This might suggest the deficiency of HIOMT in stressed SHR rats.

The other finding is the higher serotonin levels in stressed WKY than in stressed SHR rats. Pineal serotonin is metabolized either via MAO or N-acetylation pathways. Since NAS levels in WKY rats were, in fact, lower than in stressed SHR rats, it is likely that higher serotonin levels in WKY than in SHR rats might depend upon the lower rate of serotonin-N-acetylation in stressed WKY than stressed SHR rats. The other possible explanation of the higher serotonin levels in stressed WKY rats is the lower MAO activity due to higher tribulin production in WKY than SHR rats. In line with this suggestion is the lower 5-HIAA/serotonin ratio (i.e., the lower MAO activity) in stressed WKY than SHR rats.

The difference in the effect of the endogenous (tribulin) and exogenous (clorgyline) MAO-A inhibitors on pineal melatonin biosynthesis of SHR and WKY rats might depend on other stress-induced changes and not just on tribulin increase. This is in line with what we have previously speculated that clorgyline (and other selective MAO-A inhibitors) activates the physiological pathways of melatonin biosynthesis while stress induces non-physiological ways of pineal melatonin formation (see for review Oxenkrug, 1991).

In conclusion, the responses of pineal melatonin biosynthesis to clorgyline (exogenous MAO inihibitor) and tribulin (endogenous MAO inhibitor) are both similar (NAS increase) and different (melatonin elevation in clorgyline-treated SHR and WKY but only in stressed WKY). Our results suggest increased serotonin conversion into NAS and decreased NAS conversion into melatonin with decreased production of tribulin in SHR in comparison with WKY rats.

References

Glover V, Bhattacharya SK, Sandler M, File SE (1981) Benzodiazepines reduces a stress-augmented increase in rat urine monoamine oxidase inhibitor. Nature (London) 292: 347–349

Illnerova H, Albrecht I (1975) Isoproterenol induction of pineal serotonin N-acetyl-transferase in normotensive and spontaneously hypertensive rats. Experientia 31: 95–96

Kvetnansky R, McCarthy R, Thoa NB, Lake CR, Kopin IJ (1979) Sympatho-adrenal responses of spontaneously hypertensive rats to immobilization stress. Am J Physiol 236(3): H457–H462

Nomura M, Okamura K (1989) Catecholamine content changes in brain regions of spontaneously hypertensive rats under immobilization stress. J Neurochem 52: 933–937

Oxenkrug GF (1991) The acute effect of monoamine oxidase inhibitors on serotonin conversion to melatonin. In: Coppen A, Sandler M, Harnett S (eds) 5-Hydroxytryptamine and mental illness. Oxford University Press, Oxford New York Tokyo, pp 99–108

Oxenkrug GF, McIntryre IM (1985) Stress-induced synthesis of melatonin: possible involvement of the endogenous monoamine oxidase inhibitor (tribulin). Life Sci 37: 1743–1746

Oxenkrug GF, Requintina PJ (1994) Stimulation of rat pineal melatonin biosynthesis by N-acetyl serotonin. Int J Neurosci 77: 237–241

Pare WP, Schimmel GT (1986) Stress ulcer in normotensive and spontaneously hypertensive rats. Physiol Behav 36: 699–705

Authors' address: Prof. G. F. Oxenkrug, M. D., Ph. D., Department of Psychiatry, St. Elizabeth's Medical Center, QN-3P, 736 Cambridge st., Boston, MA, 02135, USA

The influence of the antidepressant pirlindole and its dehydro-derivative on the activity of monoamine oxidase A and GABA$_A$ receptor binding

A. E. Medvedev[1], **V. I. Shvedov**[2], **T. M. Chulkova**[1], **O. A. Fedotova**[2], **E. Saederup**[3], and **R. F. Squires**[3]

[1] Institute of Biomedical Chemistry, Russian Academy of Medical Sciences, and
[2] Centre of Drug Chemistry — Institute of Pharmaceutical Chemistry, Moscow, Russia
[3] Nathan Kline Institute for Psychiatric Research, Orangeburg, NY, USA

Summary. The influence of pirlindole and dehydro-pirlindole on GABA$_A$ receptor binding and MAO-A activity was investigated in vitro. Inhibition of rat brain and human placenta MAO-A by both compounds was much more potent (with IC$_{50}$ range 0.3–0.005 µM) than that of GABA$_A$ receptors. Pirlindole was inactive as a GABA antagonist. Dehydro-pirlindole exhibited selective blockade of GABA-A receptors with EC$_{50}$ 12 µM. Effects of both compounds on MAO-A activity were partially reversible. Data obtained suggest that in contrast to pirlindole dehydro-pirlindole may act not only as a MAO-A inhibitor but also as a potent GABA$_A$ receptor blocker.

Introduction

Pirlindole (2,3,3a,4,5,6-hexahydro-8-methyl-1-H-pyrazino [3,2,1-j,k]carbazole hydrochloride) (Fig. 1) is effective short-acting antidepressant employed in medical practice (Mashkovsky, 1993). The mechanism of its pharmacological activity includes selective inhibition of monoamine oxidase (MAO) type A (Gorkin, 1985). Pirlindole also exhibits some pharmacological effects that are partially independent of its MAO inhibitory activity (Simon et al., 1985). After administration in vivo pirlindole may cause convulsions in mice (Martorana et al., 1979) and rats (Verstakova et al., 1985) that can be blocked by diazepam. However in vitro pirlindole was a poor inhibitor of GABA$_A$ receptors. It did not reverse the inhibitory effect of 1 µM GABA on [^{35}-S] t-butylbicyclophosphorothionate (TBPS) binding even at concentration 100 µM (Squires and Saederup, 1988). The apparent discrepancy between in vivo and in vitro activity might result from biotransformation of pirlindole in vivo to metabolite with GABA$_A$ receptor blocking activity.

Some evidence suggests that in vitro MAO-A catalyses oxidative conversion of pirlindole into its dehydro-derivative (Fig. 1) (Baumanis et al., 1987), which may be a more potent, slowly reversible inhibitor of MAO-A. This

1

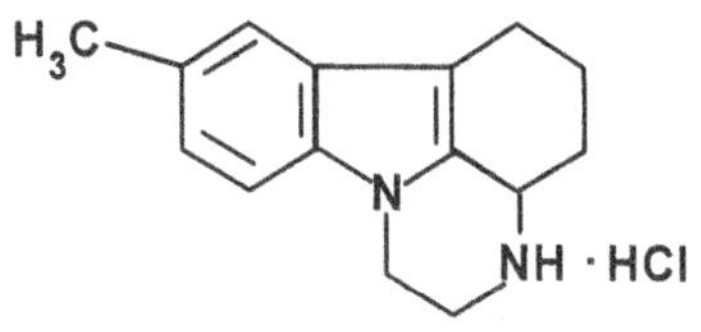

2

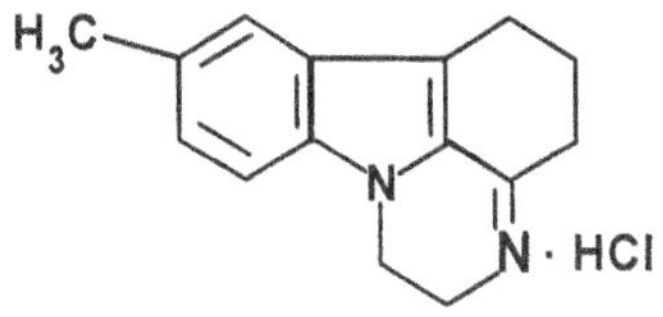

Fig. 1. Structural formulae of pirlindole (1) and dehydro-pirlindole (2)

might explain the persistence of MAO-A inhibition observed after pirlindole injection to rats, not only in brain and liver homogenate, but also in mitochondria isolated after administration of the drug (Mashkovsky et al., 1981; Medvedev et al., 1992).

In the present study we have compared the effects of pirlindole and dehydro-pirlindole on $GABA_A$ receptor binding and on the activity of MAO-A. We have also investigated the reversibility of MAO-A inhibition by these compounds.

Materials and methods

Pirlindole and dehydro-pirlindole were originally synthesised at the Centre for Drug Chemistry (Moscow, Russia) (Dvoryantseva et al., 1985). ^{14}C-Labelled 5-hydroxyl[side chain 2-^{14}C]tryptamine (5-HT) creatinine sulphate, 2-phenyl [1-^{14}C]ethylamine (PEA) HCl were obtained from Amersham Radiochemical Centre (Amersham, UK). [^{3}H]Pargyline and ^{35}S-TBPS were obtained from Du Pont NEN Products (Wilmington, Delaware, U.S.A.). Non-radioactive 5-HT and PEA were purchased from Sigma Chemical Co. (St Louis, MO, U.S.A.). Other chemicals of the highest grade available were produced by Reakhim (Moscow, Russia) or Fisher Scientific (U.S.A.).

Rat brain, liver and human placenta mitochondrial fractions were isolated by the conventional method of differential centrifugation and the activity of MAO-A was assayed radiometrically as described by Medvedev et al. (1994). The recovery of MAO after the inhibition by pirlindole and dehydro-pirlindole was examined by mitochondrial wash and by investigating [^{3}H]pargyline binding. In the first case the percent inhibition values were calculated with respect to the corresponding control values after washing. Specific binding of [^{3}H]pargyline to human placenta mitochondria was performed as described by Anderson and Tipton (1994). At saturating [^{3}H]pargyline concentration non-specific binding determined in the presence of 2 mM cold pargyline did not exceed 10%.

The binding of ^{35}S-TBPS to EDTA/water dialysed rat forebrain membranes was performed as described previously (Squires et al., 1983; Squires and Seaderup, 1987, 1988). Non-specific ^{35}S-TBPS binding defined as the binding obtained in the presence of 100 μM picrotoxin did not exceed 10–15% of total control binding. One μM GABA was added to suppress specific ^{35}S-TBPS binding to about 35% of control. Concentration ranges of pirlindole or dehydro-pirlindole were chosen so that, about half of the concentrations reversed the inhibitory effect of 1 μM GABA by less than 50% of maximal (optimum) and the other half gave greater than 50% of maximum reversal. Plateau values, or double-reciprocal plots were used to estimate ΔB_{opt} as defined previously (Squires and Saederup, 1987, 1988). Dehydro-pirlindole was also tested together with DMCM, Ro 5-4864 and clozapine, to determine the pattern of additivities with these compounds.

Results

In accordance with earlier results (Squires and Saederup, 1988) pirlindole was a relatively weak inhibitor of [^{35}S]TBPS binding with IC_{50} of 265 ± 21 μM. In contrast to pirlindole, dehydro-pirlindole exhibited more potent selective blockade of $GABA_A$ receptors. The concentration of the dehydro-pirlindole required to reverse 50% of the inhibition by 1 μM GABA (EC_{50}) was 12 ± 4.8 μM with $\Delta B_{opt} = 42 \pm 5.5\%$. Thus dehydro-pirlindole is slightly less potent but more selective, than clozapine ($EC_{50} = 7.5 \pm 0.75$ μM, $\Delta B_{opt} = 50 \pm 1.7\%$) as a $GABA_A$ receptor blocker (Squires and Saederup, 1991). When tested pairwise with DMCM (100 nM), Ro-5-4864 (5 μM) and clozapine (30 μM), dehydro-pirlindole was found to be non-additive with clozapine but partially additive with DMCM (22%) and Ro 5-4864 (33%) (Table 1).

Inhibition of rat brain and human placenta MAO-A by pirlindole was more potent than that of $GABA_A$ receptor binding with IC_{50} values of $0.32 \pm$

Table 1. Additivities of dehydro-pirlindole with DMCM and Ro 5-4864 as GABA antagonists

	Dehydro-pirlindole (50 μM)	Clozapine (30 μM)
Effect alone (%)	37 ± 8.6 (n = 8)	51 ± 6.8 (n = 37)
+100 nM DMCM (40 ± 5.7%)	59 ± 2.9 (n = 3)	75 ± 5.1 (n = 3)
+30 μM Clozapine (51 ± 6.8%)	33 ± 5.0 (n = 3)	—
+5 μM Ro 5-4864 (45 ± 5.9%)	70 ± 12 (n = 3)	76 ± 10 (n = 3)

GABA (1 μM) was present throughout. All values are given as % of ΔB control (control ^{35}S-TBPS binding minus binding in the presence of 1 μM $GABA_A$, mean ± SD). Dehydro-pirlindole (50 μM), clozapine (30 μM), DMCM (100 nM), and Ro 5-4864 (5 μM) all produced nearly maximal reversal of binding. DMCM (100 nM) together with Ro 5-4864 (5 μM) reversed 1 μM GABA binding by 75 ± 4.8%

Table 2. Effect of mitochondrial wash on the effect of MAO inhibitors

Inhibitor	Concentration (μM)	Inhibition (%)	Inhibition after wash (%)	Reactivation (%)
Pirlindole	1	69 ± 4	45 ± 3	27 ± 4*
Dehydro-pirlindole	1	92 ± 3	64 ± 3	28 ± 2**
Pargyline	10	95 ± 2	100	0
Clorgyline[a]	1	99	100	0

Rat liver mitochondria were preincubated with or without the compounds 1 h at 37°C in 0.1 M phosphate buffer, pH 7.5. MAO A activity was determined after resedimentation of mitochondrial fraction. *$P < 0.05$; **$P < 0.02$. Data represent mean of 3–4 separate experiments except a — where result of a single experiment is given

0.07 and $0.09 \pm 0.01\,\mu$M, respectively. Dehydro-pirlindole exhibited much more potent inhibition of MAO-A from these sources with IC_{50} values 6.4 ± 0.9 and 4.5 ± 0.5 nM, respectively.

The reversibility of the MAO-A inhibition was investigated by two independent methods. Table 2 shows that 1 μM pirlindole and dehydro-pirlindole incubated with rat liver mitochondria for 1 h at 37°C inhibited MAO-A activity by 73 and 92%, respectively. Subsequent wash of the mitochondrial fraction from these inhibitors resulted in partial reactivation of MAO-A by 28–29%. Under these conditions classical mechanism-based MAO inhibitors pargyline and clorgyline form covalent adducts with flavin component of MAO (Singer, 1985) irreversibly inhibiting enzyme activity, which did not recover after washing the mitochondrial preparation. The reversibility of MAO-A inhibition was also studied by investigating [3H]pargyline specific binding to human placenta mitochondria which contain only MAO-A. It was found that the specific binding of [3H]pargyline was 20–40% higher with dehydro-pirlindole and pirlindole compared with the irreversible mechanism-based MAO-A inhibitor clorgyline (data not shown). This finding also supports the conclusion that both pirlindole and dehydro-pirlindole interact reversibly with MAO-A.

Discussion

The results presented here suggest that, in contrast to pirlindole, dehydro-pirlindole may act as a potent partial GABA-A receptor blocker in vitro. It was previously shown that the incubation of pirlindole with mouse brain mitochondria containing catalytically active MAO-A caused a long-wave shift of its absorbance spectrum, which became similar to that of the standard dehydro-pirlindole. Although the product of MAO-A-dependent oxidation of pirlindole was not identified, some indirect data suggested the identity of the dehydro-pirlindole with this product (Baumanis et al., 1987).

Dehydro-pirlindole is a more potent inhibitor of MAO-A than pirlindole, and as a possible mechanism-based inhibitor it might explain the persistence of MAO-A inhibition throughout isolation of mitochondria after pirlindole administration in vivo (Mashkovsky et al., 1981; Medvedev et al., 1992). However, by analogy with such MAO inhibitors as moclobemide and brofaromine (Anderson et al., 1991; Keller et al., 1987), the latter phenomenon may also be due to the tight binding of pirlindole to MAO A.

If the MAO-dependent oxidative conversion of pirlindole into dehydro-pirlindole actually occurs in vivo, the possibility of its interaction with GABA(A) receptors would depend on its dissociation from the MAO-A-dehydro-pirlindole complex. Using two independent approaches we have demonstrated partial reversibility of MAO-A inhibition by pirlindole and dehydro-pirlindole. This suggests that dehydro-pirlindole may interact in the brain with other possible biological targets including GABA-A receptors. This proposed mechanism might (partly) explain the potentiation of clonidine-induced aggression by pirlindole (Kuksgaus et al., 1991).

GABA receptor inhibition by dehydro-pirlindole may represent a mechanism of the antidepressant action that is additional to MAO-A inhibition. Recently it was reported that the administration of supramaximal doses of the MAO-A inhibitors moclobemide and Ro 41-1049 (but not the MAO-B inhibitor lazabemide) markedly reduced (by 77 and 82% of controls, respectively) the in vitro binding of [3-H]SR 95531, a GABA-A receptor antagonist (Luque et al., 1994).

Acknowledgements

This work was supported by the Russian State Program "Design of New Drugs" (Grant No. 04.01.03), and New York State Office of Mental Health.

References

Anderson M, Tipton KF (1994) Estimation of monoamine oxidase concentration in soluble and membrane-bound preparations by inhibitor binding. J Neural Transm [Suppl] 41: 47–53

Anderson M, Waldmeier P, Tipton KF (1991) The inhibition of monoamine oxidase by brofaromine. Biochem Pharmacol 41: 1871–1877

Baumanis EA, Birska IA, Reikhman GO, Shvedov VI, Gorkin VZ (1987) On the mechanism of modification by pyrazidol of the catalytic activity of mitochondrial monoamine oxidase. Probl Med Chem N6: 90–96

Dvoryantseva GG, Pol'shakov VI, Sheikner YuN, Shvedov VI, Grynev AN, Mashkovsky MD, Karapetyan HA, Strukov YT (1985) Structure-activity relationship in pyrazino[3,2,1-j,k] carbazole derivatives. Eur J Med Chem-Chem Ther 230: 414–418

Gorkin VZ (1985) Studies on the nature and specific inhibition of monoamine oxidases. In: Kelemen K, Magyar K, Vizi ES (eds) Neuropharmacology-85. Akademiai Kiado, Budapest, pp 9–14

Keller HH, Kettler R, Keller G, Da Prada M (1987) Short-acting novel MAO inhibitors: In vitro evidence for the reversibility of MAO inhibition by moclobemide and Ro 1606491. Naunyn Schmiedebergs Arch Pharmacol 335: 12–20

Kuksgaus NE, Golovina SM, Adnreeva NI (1991) The influence of the Soviet antidepressants pyrazidol and incazan on the effect of clonidine. Farmacol Toksikol 54(1): 11–14

Luque JM, Erat R, Kettler R, Cesura A, Da Prada M, Richards JG (1994) Radioautographic evidence that the GABA A receptor antagonist SR 95531 is a substrate inhibitor of MAO A in the rat and human locus coerulens. Eur J Neurosci 6: 1038–1049

Martorana PA, Heucke O, Nitz R-E (1979) The new antidepressant pirlindole. Antagonism of acute overdose in the mouse. Arzneimittelforschung/Drug Res 29: 950–952

Mashkovsky MD (1993) Drugs: reference book for clinicians, vol 1. Meditzina, Moscow, pp 110–116

Mashkovsky MD, Gorkin VZ, Veryovkina IV, Asnina VV, Tupikina SM (1981) Selective inhibition of type A monoamine oxidase by pyrazidol. Buyl Eksper Biol Med N2: 169–171

Medvedev A, Gorkin V, Shvedov V, Fedotova O, Fedotova I, Semiokhina A (1992) Efficacy of pirlindole, a highly selective reversible inhibitor of monoamine oxidase type A, in the prevention of experimentally induced epileptic seizures. Drug Invest 4: 501–507

Medvedev AE, Kirkel AZ, Kamyhanskaya NS, Moskvitina TA, Axenova LN, Gorkin VZ, Andreeva NI, Golovina SM, Mashkovsky MD (1994) Monoamine oxidase inhibition by novel antidepressant tetrindole. Biochem Pharmacol 47: 303–308

Simon P, Poncelet M, Chermat R (1985) Persistence of central effects of pirindole, a short-acting monoamine oxidase inhibitor, in the presence of monoamine oxidase inhibitor. Drug Dev Res 5: 323–326

Singer TP (1985) Inhibitors of FAD-containing monoamine oxidases. In: Mondovi B (ed) Structure and funcitons of amine oxidases. CRC Press, Boca Raton, pp 219–229

Squires RF, Saederup E (1987) $GABA_A$ receptor blockers reverse the inhibitory effect of GABA on brain specific [^{35}S]TBPS binding. Brain Res 414: 357–364

Squires RE, Saederup E (1988) Antidepressant and metabolites that block GABA-A receptors coupled to ^{35}-S-t-butylbiscyclophosphothionate binding sites in rat brain. Brain Res 441: 15–22

Squires RF, Casida JE, Richardson M, Saederup E (1983) [^{35}S]t-Butylbicyclophosphorothionate binds with high affinity to brain-specific coupled γ-aminobutyric acid-A and ion recognition sites. Mol Pharmacol 23: 326–336

Verstakova OL, Gerchikov LN, Chicherina LA, Sharova SA, Liberman SS (1985) Toxicological characteristics of the Soviet antidepressant pyrazidol. Farmacol Toksikol 48: 57–60

Authors' address: Dr. A. E. Medvedev, Institute of Biomedical Chemistry, Academy of Medical Sciences, Pogodinskaya Str., 10, Moscow, 119832 Russia

"In vitro" effect of some 5-hydroxy-indolalkylamine derivatives on monoamine uptake system

J. A. Morón[1], V. Perez[1], E. Fernández-Alvarez[2], J. L. Marco[2], and M. Unzeta[1]

[1] Departament de Bioquímica i Biologia Molecular, Universitat Autonoma de Barcelona, Bellatera (Barcelona), and [2] Departament de Química Orgánica, C.S.I.C., Madrid, Spain

Summary. Three different indolalkylamine derivatives (FA 102, FA 69, FA 70) having in common an -OH group at 5 position of the indole ring and differing in the presence of a methyl group at the N or the acetylenic group of the side chain, have been synthesized and assayed as monoamine oxidase-A (MAO-A) [E.C.1.4.3.4] inhibitors. They were effective inhibitors with, in some cases, similar potencies to clorgyline. "In vitro" experiments were performed on rat brain synaptosomes to investigate whether these MAO-A inhibitors had any effect on noradrenaline (NA), dopamine (DA) and 5-hydroxytryptamine (5-HT) transport systems in different rat brain regions. The effect of these drugs were compared with those of clorgyline and l-deprenyl. FA 102, FA 69, FA 70 behaved as inhibitors of ^{3}H-monoamine uptake with similar rank of order of potency for amine uptake inhibition: 5-HT > DA > NA. The IC_{50} values for FA 102, FA 69, FA 70, respectively, were: 17 μM, 60 μM, 18 μM for 5HT uptake in cortex and 37 μM, 55 μM and 20 μM in hippocampus; 70 μM, 385 μM, 695 μM for NA uptake in cortex and 315 μM, 255 μM and 600 μM in hypothalamus; 270 μM, 160 μM, 40 μM for DA uptake in striatum. l-Deprenyl was a very poor inhibitor of monoamine uptake, whereas clorgyline behaved similarly to these indolalkylamine derivatives. Comparing these results with the IC_{50} values of citalopram, nisoxetine and GBR12909, specific and selective inhibitors of 5-HT, NA and DA transport systems respectively, indicated that these indolalkylamine derivatives interact more strongly with the 5HT uptake system.

Introduction

Inhibitors of MAO have been widely used in the therapy of neurological and psychiatric disorders such as Parkinson disease (The Parkinson Study Group, 1989) and depression (see Murphy et al., 1987). In the context of the monoaminergic hypothesis of depression (Shildkraut, 1965), the design of a new drug that combines a potent MAO-A inhibitory behaviour with a high affinity

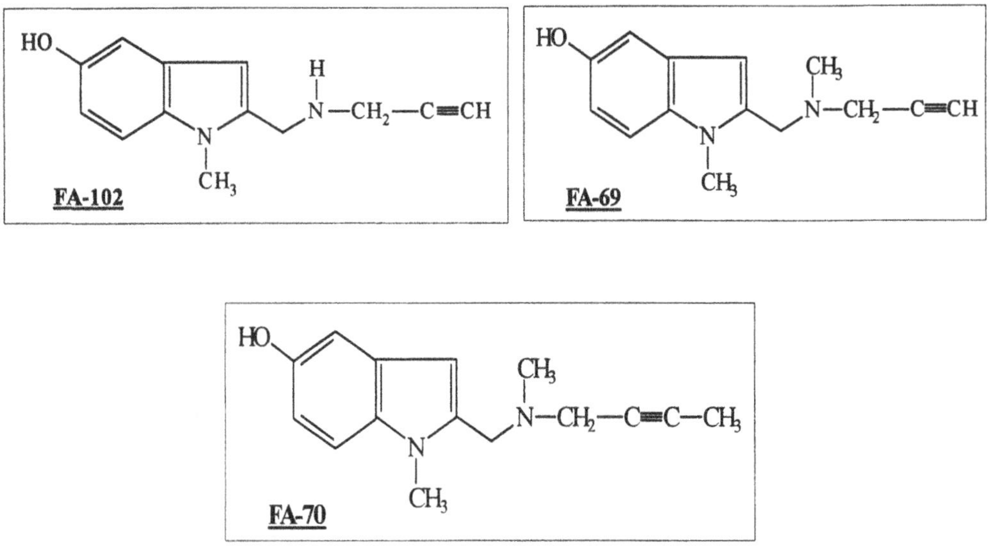

Fig. 1. Structures of the indolalkylamine derivatives used in this study

towards monoamine uptake systems should be useful in the antidepressive therapy.

Based on the structure of tryptamine, Cruces et al. (1988) synthetized several acetylenic derivatives as potential MAO inhibitors. In this context, we have selected the indolalkylamine derivatives FA 102, FA 69, FA 70 (Fig. 1) all having in common an-OH group at 5 position of the indole ring, but differing in the absence (FA 102) or the presence of a methyl group at the N of the side chain (FA 69, FA 70), and of a methyl group on the acetylenic group in the side chain (FA 70). All these compounds behaved as potent mechanism-based inhibitors of rat brain mitochondrial MAO-A (unpublished data).

In this work we have investigated whether these compounds, in addition to their MAO-inhibitory properties, also affect the monoamine uptake systems in brain tissue. This involved determination of the IC_{50} values (concentration that results in 50% inhibition) and the type of inhibition given by these compounds towards the different monoamine uptake systems. The effect of the selective amine uptake inhibitors citalopram (5-HT), nisoxetine (NA) and GBR12909 (DA), was also determined and used for references purposes.

Materials and methods

Synaptosomes preparation

All experiments were performed using male Sprague-Dawley rats weighing 250–300 g. Purified synaptosomes were obtained from pooled cortex, hippocampus (5-HT); hypothalamus, cortex (NA); striatum (DA) as described by Dodd et al. (1981). Tissues were homogenized in 40 volumes of Krebs buffer, pH 7.4, containing 0.32 M sucrose using a Teflon-glass homogenizer. The crude homogenate was centrifuged at 1,000 g for 10 min. The resulting supernatant was layered onto half of its volume of 1.2 M sucrose solution

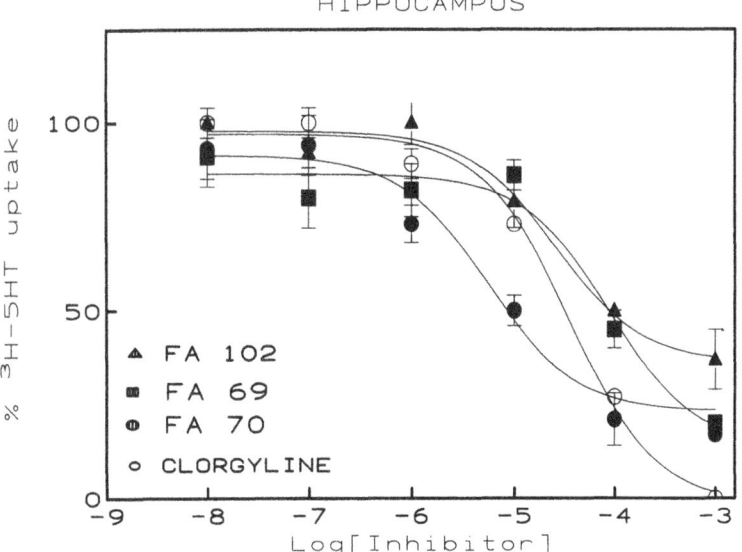

Fig. 2. Effect of the MAO inhibitors FA 102, FA 69, FA 70 and clorgyline on uptake of ^{3}H-5HT in hippocampus determined "in vitro". Synaptosomes obtained from these brain regions were incubated for 4 min in Krebs buffer, pH 7.4, in presence of 0.8 mM pargyline, 0.64 mM ascorbic acid, 0.06 μM ^{3}H-5HT and in presence or absence (control values) of these drugs at 37°C (or 4°C to assess non-specific uptake). Data are expressed as the percent inhibition of uptake with respect to control values. Each value is the mean ± SEM from 3 different experiments

Table 1. Kinetic constants for ^{3}H-monoamine uptake in different rat brain regions

	K_m (μM)	Vmax (pmol/min·mgr)
5-HT		
Hippocampus	0.07 ± 0.0064	2.031 ± 0.077
Cortex	0.120 ± 0.013	2.210 ± 0.0122
NA		
Hypothalamus	0.260 ± 0.250	0.50 ± 0.03
Cortex	0.180 ± 0.019	1.42 ± 0.074
DA		
Striatum	0.223 ± 0.016	23.16 ± 4.8

Values shown are the mean values ± SEM from 3 different experiments

and centrifuged at 100,000 g for 25 min to yield, at the sucrose interface, a crude fraction that contained synaptosomes, myelin and microsomes. This fraction was carefully collected, diluted in ice-cold 0.32 M sucrose solution, layered onto half of its volume of 0.8 M sucrose solution and centrifuged again under the same conditions to generate a pellet enriched in synaptosomes. The pellet was then resuspended in 1 ml of 0.32 M sucrose. Protein concentration was determined according to the Bradford method (Bradford, 1976).

Uptake determinations

Synaptosomal uptake was determined radiochemically following the method described by Valtier et al. (1992) with some modifications. Optimum assay conditions for monoamine uptake were determined by studying the time-courses of ^{3}H-monoamine acumulation at different protein concentrations. The kinetic parameters of the transporter systems were determined for each experiment. Aliquots of synaptosomal preparations were incubated at 37°C (or 4°C to assess passive diffusion) in Krebs buffer with 5-hydroxy-[G-^{3}H]-tryptamine creatinine sulphate [12.5 Ci/mmol] (^{3}H-5-HT), 1-[7,8-^{3}H]-noradrenaline [49 Ci/mmol] (^{3}H-NA) or [7,8-^{3}H]-dopamine [50 Ci/mmol] (^{3}H-DA) for 4 minutes in the presence of 0.8 mM pargyline (an inhibitor of MAO) and 0.64 mM ascorbic acid. The reaction was stopped by adding ice-cold Krebs buffer. Samples were then filtered through Whatman (GF/C) filters, washed twice with Krebs buffer and counted for radioactivity in a LKB Wallac 1409 liquid scintillator counter.

Results

The kinetic parameters for each monoamine uptake system were determined in different brain regions, in order to standardize the experimental conditions. The uptake of ^{3}H-5-HT, ^{3}H-NA and ^{3}H-DA into rat brain synaptosomes was time and temperature dependent and obeyed saturation kinetics with Km (the apparent affinity constant of the transporter) and Vmax (the maximum velocity of the transporter system) values in agreement with those published previously (Table 1). In all cases, concentrations of ^{3}H-monoamines that were close to their Km values were used to study the inhibitory effects of the indolalkylamine derivatives.

FA 102, FA 69 and FA 70 reduced the ^{3}H-5-HT, ^{3}H-NA, ^{3}H-DA uptake by rat brain synaptosomes, in a dose-dependent manner (Fig. 2). They were most potent as inhibitors of ^{3}H-5-HT transport (Table 2). FA 102 and FA 70 showed similar inhibitory effects to that of clorgyline towards the ^{3}H-5-HT uptake system. In contrast, l-deprenyl exhibited a relatively weak inhibitory

Table 2. IC$_{50}$ values (μM) of monoamine uptake inhibition by some indolalkylamine derivatives

	5HT		NA		DA
	Cortex	Hippocampus	Cortex	Hypothalamus	Striatum
FA 102	17 ± 2	37 ± 6	70 ± 6	315 ± 14	270 ± 12
FA 69	60 ± 3	55 ± 5	385 ± 15	255 ± 30	168 ± 10
FA 70	18 ± 2	20 ± 5	695 ± 20	600 ± 40	40 ± 5
Clorgyline	20 ± 4	23 ± 5	200 ± 20	118 ± 15	16 ± 2
Deprenyl	>1,000	>1,000	>1,000	>1,000	100 ± 15
Citalopram	0.028 ± 0.03	0.007 ± 0.0005			
Nisoxetine			0.035 ± 0.001	0.058 ± 0.002	
GBR12909					0.006 ± 0.001

Values shown are the mean values ± SEM from 3 different experiments

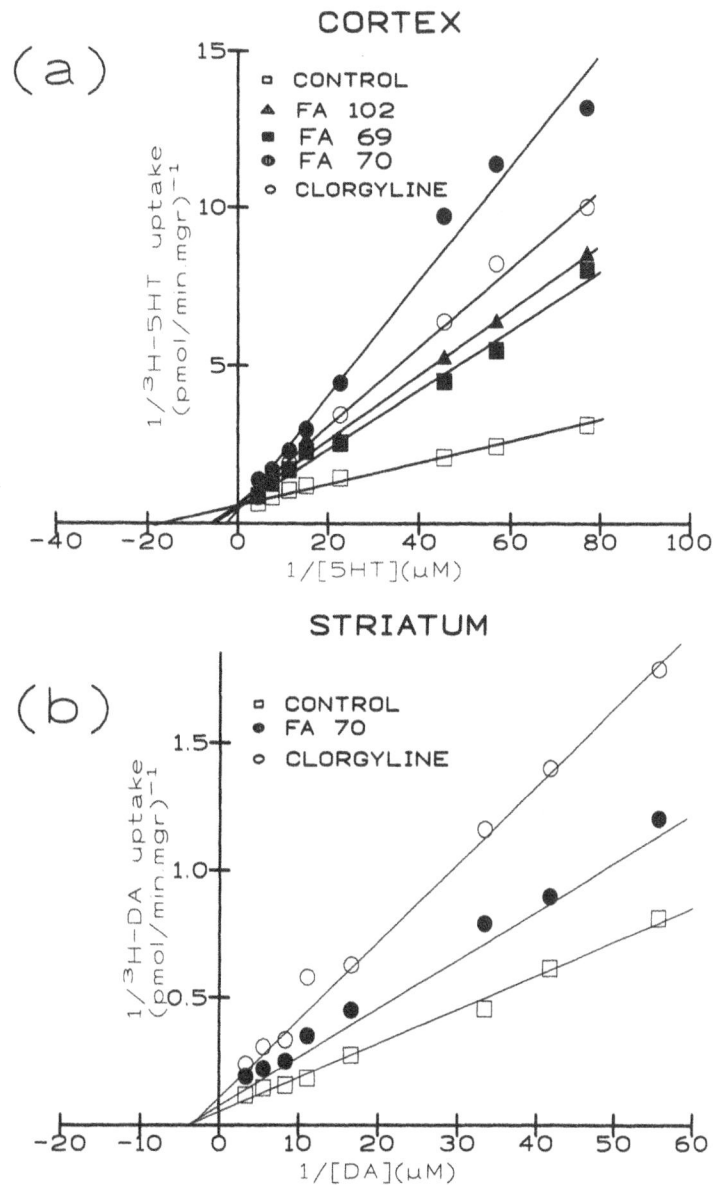

Fig. 3. Double-reciprocal plots of ^{3}H-monoamine uptake in presence or absence of some indolalkylamine derivatives. Synaptosomes isolated from pooled cortex (5HT)(a) or striatum (DA)(b) were incubated for 4 min in Krebs buffer pH 7.4 containing 0.8 mM pargyline, 0.64 mM ascorbic acid and different concentrations of ^{3}H-5HT (0.013–0.22 μM) or ^{3}H-DA (0.018–0.30 μM) in presence or absence (control values) of a tryptamine analogue (FA 102, FA69, FA70) or clorgyline at 37°C (or 4°C to assess non-specific uptake). Each value is the mean ± SEM from 3 different experiments. The type of inhibition that these drugs exert on ^{3}H-5HT (competitive) (**a**) or ^{3}H-DA (non competitive) (**b**) uptake was determined by linear regression fitting

effect, as previously reported by Fang and Yu (1994). The IC$_{50}$ values for ^{3}H-5-HT uptake in cortex and hippocampus showed no significant differences, suggesting that these derivatives present the same affinity for the transporter in both brain regions. All these indolalkylamine derivatives inhibited also ^{3}H-

DA uptake less potently than that of ³H-5-HT and were weaker inhibitors of ³H-NA transport. The apparent affinity of FA 70 for the ³H-DA uptake system was about half that of clorgyline. The introduction of a methyl group at the N of the side chain (FA 69) and also at the acetylenic group (FA 70) resulted in a 7 fold increase of the affinities of these compounds towards ³H-DA transport system in rat striatum, compared to FA 102 where these methyl groups are absent. However, the dimethyl-substituted derivative (FA 70) was a less potent inhibitor of ³H-NA transport in both cortex and hypothalamus and FA 69 was also decreased affinity in cortex. There was a significant difference between the synaptsomes from cortex and hypothalamus in their sensitivities to inhibition by FA 102. Double-reciprocal plots of ³H-5-HT uptake in cortex (Fig. 3a) showed these indolalkylamine derivatives and clorgyline to inhibit the uptake of this amine in a competitive manner. Whereas the double-reciprocal plots of ³H-DA uptake in striatum (Fig. 3b) showed that FA 70, as well as clorgyline, inhibited the transport of ³H-DA in a non-competitive manner.

Discussion

The present work shows that these indolalkylamine derivatives which are potent mechanism-based inhibitors of rat brain MAO-A, "in vitro" also inhibit the nerve terminal monoamine uptake systems. In some cases their inhibitory potencies were similar to that shown by the MAO-A inhibitor clorgyline.

FA 102, FA 69 and FA 70 as well as clorgyline were, however, weaker inhibitors of ³H-monoamine uptake by synaptosomes than the specific uptake inhibitors: citalopram (5-HT), nisoxetine (NA) and GBR12909 (DA). The indolalkylamine derivatives inhibited monoamine uptake in a dose-dependent manner. The rank order of potency for amine uptake inhibition, 5-HT > DA > NA, was similar for all these compounds. In contrast to the behaviour of these compounds and clorgyline, l-deprenyl was a very weak inhibitor of these amine uptake systems. The inhibitory effect towards ³H-5-HT uptake was similar in cortex and hippocampus. However there was a large difference between the sensitivities of the ³H-NA uptake systems in cortex and hypothalamus to inhibition by FA 102. All these indolalkylamine derivatives and clorgyline inhibited the ³H-5-HT uptake in synaptosomes obtained from cortex in a competitive manner, indicating that these drugs competed with 5-HT for transport sites. In contrast, FA 70 and clorgyline inhibited ³H-DA uptake in synaptosomes from striatum in a non-competitive manner, suggesting that these compounds bound to the transporter at a different site from that of the substrate.

This work shows that in addition to their MAO inhibitory properties these indolalkylamine derivatives also affect the monoamine uptake systems from different rat brain regions. Further research will be necessary in order to clarify the relative importance of these two factors in altering monoaminergic function "in vivo".

Acknowledgement

Support from the Programa de Quimica Fina(QFN93-4428) is gratefully acknowledged.

References

Bradford HJ (1976) A rapid and sensitive method for the quantitation of microgram quantities of protein utilizing the principle of protein-dye binding. Anal Biochem 72: 248–254

Cruces MA, Elorriaga D, Fernández-Alvarez E, Prieto O (1988) Synthesis and biochemical properties of N-acetylenic and N-allenic derivatives of 2-aminemethylindoles as selective inhibitors of monoamine oxidase. Pharmacol Res Commun 20: 102–107

Dodd PR, Hardy JA, Dakley AE, Edwarson JA, Perry EK, Dalaunoy JP (1981) A rapid method for preparing synaptosomes: comparison with alternative procedures. Brain Res 226: 107–118

Fang J, Yu PH (1994) Effect of l-deprenyl, its structural analogues and some monoamine oxidase inhibitors on monoamine uptake. Neuropharmacol 33: 763–768

Murphy DL, Aulakh CS, Garrick NA, Sunderland T (1987) Monoamine oxidase inhibitors as antidepressants; implications for the mechanism of action of antidepressants and psychobiology of the affective disorders and some related disorders. In: Meltzer H (ed) Psychopharmacology: the third generation of progress. Raven Press, New York, pp 545–552.

The Parkinson Study Group (1989) Effect of deprenyl on the progression of disability in early Parkinson's disease. N Engl J Med 321: 1364

Schildkraut JJ (1965) The catecholamine hypothesis of affective disorders: a review of supporting evidence. Am J Psychiatry 122: 509–522

Valtier D, Dement WC, Mignot E (1992) Monoaminergic uptake in synaptosomes prepared from frozen brain tissue samples of normal and narcoleptic canines. Brain Res 588: 115–119

Authors' adress: Dr. M. Unzeta, Departament de Bioquímica i Biologia Molecular, Facultat de Medicina, Universitat Autònoma de Barcelona, E-08193-Bellaterra (Barcelona), Spain

N-Alkyloxycarbonyl derivatives of ethylene diamine as monoamine oxidase inhibitors

V. F. Pozdnev, A. Z. Kirkel, N. S. Kamyshanskaya, L. N. Axenova,
and **A. E. Medvedev**

Institute of Biomedical Chemistry, Academy of Medical Sciences, Moscow, Russia

Summary.   A series of urethane type derivatives of ethylene diamine (EDA) was synthesised and tested as inhibitors of monoamine oxidase (MAO) A and B. Nature of aromatic ring and a position of substituents in it were important for the inhibitory activity. Chlorobenzyloxycarbonyl-EDA derivatives exhibited selective inhibition of MAO-A with $3,4\text{-}Cl_2\text{-}C_6H_4CH_2OCO\text{-}EDA$ being a most potent and selective MAO-A inhibitor (IC_{50} 4 μM). Within the compounds studied, 3,4- dichloro-benzyloxycarbonyl-EDA exhibited most potent inhibition of MAO-A. This compound inhibited the activity of rat liver MAO-A non-competitively with K_i (slope) value of 3.6 μM, whereas the inhibition of rat liver MAO-B was competitive with K_i (slope) value of 56 μM (not shown). 2,4-Dichlorobenzyloxycarbonyl-EDA also inhibited rat liver MAO-A in a non-competitive manner with K_i of 14.6 μM.

Introduction

Monoamine oxidase (E.C.1.4.3.4; MAO) plays an important role in the metabolism of monoamine neurotransmitters in the central nervous system. The enzyme exists as two isoenzyme forms, MAO-A and MAO-B, distinguished by substrate specificity, inhibitor sensitivity and different primary structures (Gorkin, 1983; Shih, 1993). Selective MAO-A inhibitors are antidepressants (Burrows and Da Prada, 1989; Medvedev et al., 1994) whereas selective MAO-B inhibitor deprenyl is used for the treatment of Parkinson's disease (Tetrud et al., 1989).

The development of reversible selective MAO-A inhibitor moclobemide (p-chloro-N-[2-morpholinoethyl]benzamide) approved for use as an antidepressant drug, stimulated interest to an investigation of MAO inhibition by various analogues and derivatives (Annan and Silverman, 1993). Moclobemide strongly inhibits MAO-A in vivo (Da Prada et al., 1989). In vitro, the inhibition of rat brain or human placenta MAO-A by moclobemide showed an initial competitive phase with very low affinity ($K_i = 0.2\text{--}0.4$ mM) (Cesura et al., 1992). However, the potency of the inhibitor increased with incubation time and its IC_{50} (582 μM) decreased about 20-fold after 90 min of exposure of the enzyme to the inhibitor (IC_{50} = 32 μM) (Cesura et al., 1992). This suggested that either a metabolite of moclobemide is responsible for the inhibitory effect or moclobemide has the characteristics of a slow-binding inhibitor (Cesura et al., 1992). None of the isolated metabolites was found to be as potent as moclobemide (Cesura and Pletscher, 1992). N-(2-aminoethyl)-4-chlorobenzamide hydrochloride (Ro-16-6491) exhibited very potent and selective time dependent, reversible MAO-B inhibition (Cesura et al., 1989). A compound structurally closely related to Ro-16-6491, N-(2-aminoethyl)-5-chloro-2-pyridine carboxamide hydrochloride (Ro-19-6327), shows more potent and selective MAO-B inhibition (Cesura et al., 1989). Some of halo- and nitro-substituted analogues of N-(2-aminoehyl)benzamide were shown to be potent, competitive, time-dependent inhibitors of MAO-B (Anan and Silverman, 1993).

In the present report we have examined the MAO inhibitory effect of some N-aryloxycarbonyl- and N-benzyloxycarbonyl-derivatives of ethylenediamine (EDA). They differed from N-(2-aminoethyl)benzamide inhibitors of MAO not only by the length of the molecule but also by structure of the amide bond. The position of O-atom or -CH_2O- group between the phenyl ring and carbonyl group in N-(2-aminoethyl)-benzamide leads to formation of carbonic acid derivatives.

Materials and methods

Synthesis of compounds

Diphenyl carbonate, di-p-cresyl carbonate, dipentafluorophenyl carbonate were purchased from "Reakhim" (Russia), di-tert-butyl pyrocarbonate was obtained from Fluka (Switzerland), p-methoxybenzyloxy-carbonyl azide was a product of Aldrich (U.K.). Dibenzyl carbonate was prepared as described previously (Pozdnev and Nuzhnova, 1976). Di-adamant-1-yl pyrocarbonate was synthesyzed from sodium adamant-1-yl carbonate and trichloroacetyl chloride (Pozdnev at al., 1979). Bis-2,4-dinitro-phenyl carbonate was prepared by nitration of diphenyl carbonate. 2-Chlorbenzyl, 4-chlorbenzyl, 2,4-dichlorbenzyl, 3,4-di-chlorbenzyl and 4-nitrobenzyl carbonates were synthesized from di-pentafluorophenyl carbonate and appropriate benzyl alcohol by routine method. 3-Pyridinemethyl-2,4-dinitrophenyl carbonate was obtained as viscous oil from bis-2,4-dinitrophenyl carbonate and 3-pyridinemethanol.

N-formyl-EDA, a key intermediate in the synthesis of most compounds in this work, was obtained from EDA and methylformate. N-phenyloxycarbonyl-EDA (Table 1, Compound 1) and N-4-methylphenyloxycarbonyl-EDA (Compound 2) were obtained by acylation of EDA with di-phenyl- or di-4-methylphenyl carbonates respectively. Acylation of EDA by di-benzyl carbonate yielded N-benzyloxycarbonyl-EDA (Compound 3). Compounds 4–8 were synthesized by acylation of N-formyl-EDA with substituted benzylpentafluorophenyl carbonates with subsequent deformylation of N,N'-diacyl-EDA by treatment with hydrochloric acid in methanol. Acylation of free EDA with 4-methoxybenzyloxycarbonyl azide in aqueous dioxane yielded the compound 9. N-1-Adamantyloxycarbonyl-EDA (10) was synthesized from N-benzyloxycarbonyl-EDA (3) by acylation with di-adamantyl-1 pyrocarbonate with subsequent debenzyloxycarbonylation of N,N'-diacyl-EDA intermediate by catalytic hydrogenation using Pd/C catalyst in the presence of hydrochloric acid. Acylation of EDA with N-3-picolyl-2,4-dinitrophenyl carbonate gave 3-picolyloxycarbonyl-EDA (11).

Table 1. Physico-chemical characteristics of EDA derivatives RO(C=O)-NHCH$_2$CH$_2$NH$_2$ HCl and inhibition of rat brain MAO A and B

	R	m.p. °C	R$_f$[b]	IR[c] (c=o) cm^{-1}	IC$_{50}$ (μM)[d]	
					MAO-A	MAO-B
1	C$_6$H$_5$	166–168	0.44	1,720	241	139
2	4-CH$_3$-C$_6$H$_4$	193–194	0.43	1,720	25	400
3	C$_6$H$_5$CH$_2$	167–168	0.44	1,690	200	500
4	4-Cl-C$_6$H$_4$CH$_2$[a]	198–200	0.47	1,690	180	350
5	2-Cl-C$_6$H$_4$CH$_2$	123–124	0.47	1,690	16	400
6	2,4-Cl$_2$-C$_6$H$_3$CH$_2$[a]	163–164	0.54	1,690	25	160
7	3,4-Cl$_2$-C$_6$H$_3$CH$_2$[a]	133–134	0.54	1,690	4	71
8	4-NO$_2$-C$_6$H$_4$CH$_2$	154–155	0.40	1,680	400	126
9	4-CH$_3$O-C$_6$H$_4$CH$_2$	180–181	0.45	1,685	100	450
10	Adamantyl-1	191–192	0.41	1,700	630	1,000
11	Pycolyl-3	183–184	0.15	1,715	>1,000	280
12	moclobemide	—	—	—	45	>1,000

[a] Satisfactory ^{1}H-NMR data were obtained for these compounds; [b] Thin-layer chromatography was run on silica gel plates (Merck) in chloroform — methanol — acetic acid 40:10:1 and developed with ninhydrin; [c] in KCl pellets. [d] Data represent mean of 3–4 determinations. SD in all cases never exceeded 10%

All the compounds 1–11 (Table 1) used as hydrochlorides were chromatographically homogenous.

Determination of MAO activity and its inhibition

Rat brain and liver mitochondria were used as a source of MAO-A and MAO-B (Medvedev et al., 1994). The activity of MAO was assayed radiometrically (Tipton and Youdim, 1976) with minor modifications (Medvedev et al., 1994) using 0.1 mM [^{14}C]-5-hydroxytryptamine (serotonin) and 2 μM [^{14}C]-benzylamine as substrates for MAO-A and -B respectively. These concentrations which are either close to or below K$_m$ value allow competitive inhibition to be detected. To obtain reliable comparative IC$_{50}$ values in our laboratory, rat brain mitochondria were incubated at 37°C for 60 min with all suspected MAO inhibitors and the enzymatic activity was measured thereafter by adding substrates of MAO-A and -B. Such an approach is widely used for pilot studies of potential MAO inhibitors (Da Prada et al., 1989). The incubation itself (without mitochondria) did not influence the inhibitory activity of all the compounds tested.

Results and discussion

Data on MAO inhibition are given in the Table 1. Most of all N-alkyloxycarbonyl-EDA derivatives were relatively selective MAO-A inhibitors. Phenyloxycarbonyl-EDA (1) was a weak and nonselective inhibitor of either form of MAO. Elongation of the compound 1 by inserting methylene group between the phenyl ring and carboxamide residue (3) had insignificant infuence on the manifestation of MAO-A inhibition and reduced inhibitory potency with respect to MAO-B. The presence of methyl, methoxy or chlorine atom(s) in the phenyl ring of compounds increased potency of MAO-A inhibition. When the compounds were tested for MAO-A inhibition the following rank of inhibitory activity was obtained: $1 < 4 < 9 < 2 = 6 < 5 < 7$. Except compound 7 all other compounds caused even weaker inhibition of brain MAO-B than N-phenyloxycarbonyl-EDA (1) itself. Apparently the position of chlorine atom in the phenyl ring is crucial for the inhibitory action with ortho-chloro derivative (5) being more potent MAO-A inhibitor than 4-chlorobenzyloxycarbonyl-EDA (4). The insertion of nitro group in the para-position (8) decreased inhibitory activity with respect to MAO-A. The presence of nitro group in the phenyl ring slightly increased inhibition of MAO-B (IC_{50} = 126 μM). This is consistent with IC_{50} value of 48.2 μM for MAO-B inhibition by N-(2-aminoethyl)-4-nitrobenzamide hydrochloride (Annan and Silverman, 1993). However, meta- and especially ortho-isomer were more potent MAO-B inhibitors with IC_{50} of 2.17 and 0.06 μM, respectively (Annan and Silverman, 1993). Substitution of the aromatic phenyl ring for highly hydrophobic non-planar adamantyl group decreased inhibitory potency of the compound on both MAO-A and B. Picolyloxycarbonyl-EDA (11) possessing aromatic pyridine ring also exhibited some selectivity towards MAO-B, however, IC_{50} values were much higher than those reported for MAO-B inhibition by Ro-19-6327 (Cesura and Pletscher, 1989).

Within the compounds studied, 3,4-dichloro-benzyloxycarbonyl-EDA (7) exhibited most potent inhibition of MAO-A. This compound inhibited the activity of rat liver MAO-A non-competitively with K_i (slope) value of 3.6 μM, whereas the inhibition of rat liver MAO-B was competitive with K_i (slope) value of 56 μM (not shown). 2,4-Dichlorobenzyloxycarbonyl-EDA (6) also inhibited rat liver MAO-A in a non-competitive manner with K_i of 14.6 μM.

Thus our results suggest that some of chloro-benzyloxycarbonyl derivatives of ethylene diamine are selective inhibitors of MAO-A.

References

Annan N, Silverman RB (1993) New analogues of N-(2-Aminoethyl)-4- chlorobenzamide (Ro 16-6491). Some of the most potent monoamine oxidase-B inactivators. J Med Chem 36: 3968–3970

Burrows CD, Da Prada M (eds) (1989) Reversible MAO A inhibitors as antidepressants. J Neural Transm 28 [Suppl]: 1–106

Cesura AM, Pletscher A (1992) The new generation of monoamine oxidase inhibitors. Prog Drug Res 38: 171–297

Cesura AM, Galva MD, Imhof R, Kubez E, Picotti GB, Da Prada M (1989) [³H]Ro 19-6327: a reversible ligand and affinity labelling probe for monoamine oxidase-B. Eur J Pharmacol 162: 457–465

Cesura AM, Kettler R, Imhof R, Da Prada M (1992) Mode of action and characteristics of monoamine oxidase-A inhibition by moclobemide. Psychopharmacology 106: S15–S16

Da Prada M, Kettler R, Keller HH, Burkland WP, Haefely WE (1989) Preclinical profiles of the novel reversible MAO-A inhibitors, moclobemide and brofaromine, in comparison with irreversible MAO inhibitors. J Neural Transm 28 [Suppl]: 5–20

Gorkin VZ (1983) Amine oxidases in clinical research. Pergamon Press, Oxford

Medvedev AE, Kirkel AZ, Kamyshanskaya NS, Axenova LN, Moskvitina TA, Gorkin VZ, Andreeva NI, Mashkovsky MD (1994) Monoamine oxidase inhibition by novel antidepressant tetrindole. Biochem Pharmacol 47: 303–308

Pozdnev VF, Nuzhnova IA (1976) Preparation of carbonic acid derivatives for peptide synthesis from alkyloxycarboxyl salts. Zh Organ Khim (Russ) 12: 1407–1410

Pozdnev VF, Smirnova EA, Podgornova NN, Zencova NK, Kalejs UO (1979) Synthesis of di-tert-butyl pyrocarbonate from carboxylic acid chloroanhydrides. Zh Organ Khim 15: 106–109

Shih J (1993) cDNA cloning of human liver MAO-A and MAO-B. In: Yasuhara H, Parvez H, Oguchi K, Sandler M, Nagatsu T (eds) Monoamine oxidase: basic and clinical aspects. VSP, Utrecht, pp 15–22

Tetrud VW, Lanston JW (1989) The effect of deprenyl (selegiline) on the natural history of Parkinson's disease. Science (Washington DC) 245: 519–522

Tipton KF, Youdim MBH (1976) Assay of monoamine oxidase. In: Wolstenholm GEW, Knight J (eds) Monoamine oxidase and its inhibition. Elsevier, Amsterdam, pp 393–403

Authors' address: A. E. Medvedev, Institute of Biomedical Chemistry, Russian Academy of Medical Sciences, 10 Pogodinskaya str., Moscow, 119832 Russia

S. E. Daniel, F. F. Cruz-Sánchez, A. J. Lees (eds.)

Dementia in Parkinsonism

1997. 29 figures. VII, 204 pages.

Cloth DM 185,–, öS 1295,–

ISBN 3-211-82960-1

Special edition of "Journal of Neural Transmission, Supplement 51, 1997"

(Soft cover edition of Supplement 51 only available for subscribers to "Journal of Neural Transmission")

This supplement compiles the views of international experts dealing with dementia in parkinsonism. The subject is covered from a clinical, morphological and biochemical point of view and the diversity of terminology for dementia with Parkinson's Disease is emphasized. Subjects covered include normal brain function and the influence of age, investigation and treatment of dementia, pathological nomenclature and criteria for diagnosis, the relationship between dementia in Parkinson's Disease and Alzheimer's Disease, and biochemical and morphological substrates of dementia.

P. Riederer, D. B. Calne, R. Horowski,

Y. Mizuno, W. Poewe, M. B. H. Youdim (eds.)

Advances in Research on Neurodegeneration
Volume 5

1997. 45 figures. VIII, 215 pages.

Cloth DM 198,–, öS 1386,–

ISBN 3-211-82933-4

Special edition of "Journal of Neural Transmission, Supplement 50, 1997"

(Soft cover edition of Supplement 50 only available for subscribers to "Journal of Neural Transmission")

The "International Winter Conferences on Neurodegeneration" have become an established forum to discuss various aspects of basic and clinical topics related to the underlying mechanisms of neurodegenerative disorders. This volume focuses on brain imaging, endogenous and exogenous neurotoxins, programmed cell death, apoptosis and necrosis, and immunoinflammatory mechanisms, infective diseases causing neurological disorders. These topics have been reviewed by invited experts and the articles give an up-to-date reflection of the state of the art in these research fields.

 SpringerWienNewYork

Sachsenplatz 4–6, P.O.Box 89, A-1201 Wien, Fax +43-1-330 24 26, e-mail: order@springer.at, Internet: http://www.springer.at
New York, NY 10010, 175 Fifth Avenue • D-14197 Berlin, Heidelberger Platz 3 • Tokyo 113, 3-13, Hongo 3-chome, Bunkyo-ku

Y. Mizuno, M. B. H. Youdim, D. B. Calne,
R. Horowski, W. Poewe, P. Riederer (eds.)

Advances in Research on Neurodegeneration
Volume 3 & 4

1997. 46 figures. VIII, 280 pages.
Cloth DM 215,–, öS 1505,–
ISBN 3-211-82935-0
Special edition of "Journal of Neural Transmission, Supplement 49, 1997"
(Soft cover edition of Supplement 49 only available for subscribers to "Journal of Neural Transmission")

The first part of the book focuses on disease models and mechanisms. The areas discussed include Alzheimer's disease, Parkinson's disease, glial and neuronal death, and demyelination/remyelination. The second part concentrates on the molecular biology of neurodegeneration. The topics include molecular genetics of neurological disorders, molecular biology of recognition sites, apoptosis, and neuroimmunology and multiple sclerosis. Leading experts have been invited to give state of the art presentations including their own recent data.

W. Kuhn, P. Kraus, H. Przuntek (eds.)

Deprenyl – Past and Future

1996. 16 figures. IX, 112 pages.
Cloth DM 130,–, öS 910,–
ISBN 3-211-82948-2
Special edition of "Journal of Neural Transmission, Supplement 48, 1996"
(Soft cover edition of Supplement 48 only available for subscribers to "Journal of Neural Transmission")

The clinical effect of L-Deprenyl was originally explained on the basis of irreversible and selective MAO-B inhibition and subsequent enhancement of dopaminergic neurotransmission. In recent years new experimental data have challenged this concept. In vitro and in vivo studies are suggesting that L-Deprenyl may have neuroprotective and/or neuroregenerative properties, too. Furthermore, controversial data of recently finished long-term clinical studies have brought forward an new discussion both on the clinical impact and the possible mode of action of L-Deprenyl in Parkinson's Disease and various other neurological and psychiatric disorders. This volume provides a forum for intensive discussions on the biochemical, pharmacological and clinical aspects of Parkinson's Disease.

 SpringerWienNewYork

Sachsenplatz 4-6, P.O.Box 89, A-1201 Wien, Fax +43-1-330 24 26, e-mail: order@springer.at, Internet: http://www.springer.at
New York, NY 10010, 175 Fifth Avenue • D-14197 Berlin, Heidelberger Platz 3 • Tokyo 113, 3-13, Hongo 3-chome, Bunkyo-ku

K. A. Jellinger, M. Windisch (eds.)

New Trends in the Diagnosis and Therapy
of Non-Alzheimer's Dementia

1996. 61 partly coloured figures. VIII, 288 pages.
Soft cover DM 190,–, öS 1330,–
Reduced price for subscribers to "Journal of Neural Transmission":
Soft cover DM 171,–, öS 1197,–. ISBN 3-211-82823-0
Journal of Neural Transmission, Supplement 47

This volume gives an overview of the present state of art on the classification, neuropathology, clinical presentation, neuropsychology, diagnosis, neuroimaging and therapeutic possibilities in non-Alzheimer's dementias, an increasingly important group of CNS diseases, which account for 7 to 30% of dementing disorders in adults and aged subjects, and thus, represent the second most frequent cause of dementia after Alzheimer's disease. The monograph provides the newest information for neurologists, psychiatrists, dementia research workers, dementia clinicians, neuropathologists, neurobiologists, and practicing physicians.

P. Riederer, W. Wesemann (eds.)

Parkinson's Disease: Experimental Models and Therapy

1995. 121 figures. XI, 466 pages.
Soft cover DM 240,–, öS 1680,–
Reduced price for subscribers to "Journal of Neural Transmission":
Soft cover DM 216,–, öS 1512,–. ISBN 3-211-82749-8
Journal of Neural Transmission, Supplement 46

Current research on Parkinson's disease is aimed at the goal of determining the underlying cause of this terrible disease and of developing adequate treatment strategies to deal with it. This volume focuses on models that mirror the progression of the symptoms of Parkinson's disease (iron, MPTP, 6-hydroxydopamine, "TaClo", etc.) while other topics are the evaluation of oxidative stress, calcium, excitotoxicity, nitric oxide, or nerve growth factors as possible pathophysiological candidates or causal parameters. Further topics are the interplay between exogenous and endogenous toxins, the potential of brain imaging by PET, MRI and SPECT, as well as promising therapeutic drug strategies. This volume represents a comprehensive survey of the state of the art for neurologists, biochemists, neuropharmacologists and toxicologists.

SpringerWienNewYork

Sachsenplatz 4-6, P.O.Box 89, A-1201 Wien, Fax +43-1-330 24 26, e-mail: order@springer.at, Internet: http://www.springer.at
New York, NY 10010, 175 Fifth Avenue • D-14197 Berlin, Heidelberger Platz 3 • Tokyo 113, 3-13, Hongo 3-chome, Bunkyo-ku

U. Bonuccelli, J. M. Rabey (eds.)

Old and New Dopamine Agonists in Parkinson's Disease

1995. 73 figures. VIII, 321 pages.
Soft cover DM 196.–, öS 1386.–
Reduced price for subscribers to "Journal of Neural Transmission":
Soft cover DM 176.40, öS 1247.40. ISBN 3-211-82717-X
Journal of Neural Transmission, Supplement 45

This book provides a comprehensive overview of the basic and clinical neuropharmacology of dopamine agonists and the rationale for their employment in PD. The authors have compiled an up-to-date guide, covering such topics as the pathophysiology of dopaminergic systems and the neuro-biochemistry of dopaminergic receptors, the clinical use of old and new dopamine agonists, both in the first-time treatment of PD patients and for reducing motor fluctuations in levodopa-treated ones, and the possible role of dopamine agonists as neuroprotective agents. Particular emphasis has been placed on apomorphine, an old dopamine agonist that has recently recaptured neurologists' interest for its use in both diagnostic use and therapeutic management of advanced parkinsonian patients. Articles discussing the results of ongoing clinical studies of newly developed dopamine agonists and the potential use of dopamine agonists, both new and old, as neuroprotectors should be of particular interest to the reader. The work is an exhaustive up-to-date compendium that assembles the entire spectrum of current basic and clinical research on dopaminergic systems and dopamine agonists in Parkinson's disease into a single authoritative source.

S. Hoyer, D. Müller, K. Plaschke (eds.)

Cell and Animal Models in Aging and Dementia Research

1994. 63 figures. VIII, 272 pages.
Soft cover DM 187,–, öS 1309,–
Reduced price for subscribers to "Journal of Neural Transmission":
Soft cover DM 168,30, öS 1178,10. ISBN 3-211-82549-5
Journal of Neural Transmission, Supplement 44

This supplement deals with data from cell cultures of different kinds, brain tissue slices and (brain) animal models related to aging and Alzheimer's disease. Results presented concern structural, metabolic and behavioral aspects with the aim to provide approaches which mimic biological variations in the aging and Alzheimer brain to better understand the pathogenesis of the latter disorder.

SpringerWienNewYork

Sachsenplatz 4-6, P.O.Box 89, A-1201 Wien, Fax +43-1-330 24 26, e-mail: order@springer.at, Internet: http://www.springer.at
New York, NY 10010, 175 Fifth Avenue • D-14197 Berlin, Heidelberger Platz 3 • Tokyo 113, 3-13, Hongo 3-chome, Bunkyo-ku

The manufacturer's authorised representative in the EU is Springer
Nature Customer Service Centre GmbH, Europaplatz 3, 69115 Heidelberg,
Germany. If you have any concerns regarding our products, please
contact ProductSafety@springernature.com

Printed and bound by CPI Group (UK) Ltd, Croydon, CR0 4YY

28/04/2026

02098462-0011